Yamaha XJ650 & 750 Owners Workshop Manual

by Pete Shoemark
with an additional Chapter on the 1983 US models
by Jeremy Churchill

Models covered

XJ650. 653cc. UK 1980 to 1984
XJ650 G Maxim. 653cc. US 1980
XJ650 H Maxim. 653cc. US 1981
XJ650 J Maxim. 653cc. US 1982
XJ650 K Maxim. 653cc. US 1983
XJ650 LH Midnight Maxim. 653cc. US 1981
XJ650 RJ Seca. 653cc. US 1982

XJ750. 748cc. UK 1981 to 1984
XJ750 J Maxim. 748cc. US 1982
XJ750 K Maxim. 748cc. US 1983
XJ750 MK Midnight Maxim. 748cc. US 1983
XJ750 RH Seca. 748cc. US 1981
XJ750 RJ Seca. 748cc. US 1982
XJ750 RK Seca. 748cc. US 1983

Note: This manual does not cover the XJ650 Turbo models

ISBN 978 1 85010 353 0

Haynes
THE BOOK

British Library Cataloguing in Publication Data
A catalogue record for this book is available from the British Library

Library of Congress Control Number 86-82634

ABCDE
FG

3

Haynes Publishing
Sparkford, Yeovil,
Somerset BA22 7JJ, England

Haynes North America, Inc
859 Lawrence Drive,
Newbury Park, California 91320, USA

Printed using NORBRITE BOOK 48.8gsm (CODE: 40N6533) from NORPAC; procurement system certified under Sustainable Forestry Initiative standard. Paper produced is certified to the SFI Certified Fiber Sourcing Standard (CERT - 0094271)

Acknowledgements

Our thanks are due to Motorcycle City of Farnborough, who supplied the Yamaha XJ650 featured in the photographs throughout this manual, and to Mitsui Machinery Sales (UK) Ltd who supplied the necessary service information and gave permission to reproduce many of the original line drawings used. Yeovil Motorcycle Services too, gave valuable technical assistance. The XJ750 shown on the front cover was supplied by Mark Pike of Chiselborough, Somerset.

The Avon Rubber Company supplied information and technical assistance on tyre care and fitting, and NGK Spark Plugs (UK) Ltd provided information on plug maintenance and electrode conditions.

About this manual

The purpose of this manual is to present the owner with a concise and graphic guide which will enable him to tackle any operation from basic routine maintenance to a major overhaul. It has been assumed that any work would be undertaken without the luxury of a well-equipped workshop and a range of manufacturer's service tools.

To this end, the machine featured in the manual was stripped and rebuilt in our own workshop, by a team comprising a mechanic, a photographer and the author. The resulting photographic sequence depicts events as they took place, the hands shown being those of the author and the mechanic.

The use of specialised, and expensive, service tools was avoided unless their use was considered to be essential due to risk of breakage or injury. There is usually some way of improvising a method of removing a stubborn component, providing that a suitable degree of care is exercised.

The author learnt his motorcycle mechanics over a number of years, faced with the same difficulties and using similar facilities to those encountered by most owners. It is hoped that this practical experience can be passed on through the pages of this manual.

Where possible, a well-used example of the machine is chosen for the workshop project, as this highlights any areas which might be particularly prone to giving rise to problems. In this way, any such difficulties are encountered and resolved before the text is written, and the techniques used to deal with them can be incorporated in the relevant section. Armed with a working knowledge of the machine, the author undertakes a considerable amount of research in order that the maximum amount of data can be included in the manual.

A comprehensive section, preceding the main part of the manual, describes procedures for carrying out the routine maintenance of the machine at intervals of time and mileage. This section is included particularly for those owners who wish to ensure the efficient day-to-day running of their motorcycle, but who choose not to undertake overhaul or renovation work.

Each Chapter is divided into numbered sections. Within these sections are numbered paragraphs. Cross reference throughout the manual is quite straightforward and logical. When reference is made 'See Section 6.10' it means Section 6, paragraph 10 in the same Chapter. If another Chapter were intended, the reference would read, for example, 'See Chapter 2, Section 6.10'. All the photographs are captioned with a section/paragraph number to which they refer and are relevant to the Chapter text adjacent.

Figures (usually line illustrations) appear in a logical but numerical order, within a given Chapter. Fig. 1.1 therefore refers to the first figure in Chapter 1.

Left-hand and right-hand descriptions of the machines and their components refer to the left and right of a given machine when the rider is seated normally.

Motorcycle manufacturers continually make changes to specifications and recommendations, and these, when notified, are incorporated into our manuals at the earliest opportunity.

We take great pride in the accuracy of information given in this manual, but motorcycle manufacturers make alterations and design changes during the production run of a particular motorcycle of which they do not inform us. No liability can be accepted by the authors or publishers for loss, damage or injury caused by any errors in, or omissions from, the information given.

Contents

Yamaha XJ650 right-hand view

Yamaha XJ750 left-hand view

Engine/gearbox unit of Yamaha XJ650

Introduction to the Yamaha XJ650 and 750 Fours

The ancestry of the Yamaha XJ range can be traced back to 1976 when the three cylinder XS750 D was introduced. This was Yamaha's first large capacity four-stroke and it was noteworthy as being the first widely accepted shaft-drive motorcycle to come from that manufacturer. Prior to its introduction, shaft drive conferred an image of strength, reliability and medium performance. Fine on a BMW or Gold Wing, but something of a shock on a DOHC triple.

The popularity of the XS750 models and the later 850cc versions had an obvious effect when Yahama decided to join in the race amongst the 'big four' to produce the biggest road-going motorcycle in the late 1970s. To Yamaha's credit, they were not drawn into the six-cylinder behemoth battle that two of the other factories had indulged in, but brought out a comparatively usable four-cylinder 1100 cc sports tourer, the XS1100 E. This, and its subsequent derivatives, were quite well received, and bore many similarities to the 750 triples, including the now proven shaft drive.

By the end of the 1970s sanity once again prevailed in the industry and the spiralling cost of fuel encouraged the manufacturers to look for a new wave of mid-capacity machines. These were to be smaller, lighter and more economical, but were to have a level of performance close to that of the 1000 cc machines.

Yamaha's answer was to be the XJ series, which retained and developed the shaft drive arrangement to a point where it was almost competitive with chain systems in terms of rideability, with few of the legendary shaft drive vices being obtrusive. The XJ series has also heralded a period of serious refinement which can be seen in almost every area of the various models.

The engine is astonishingly narrow for a four, this being achieved by relocating the alternator from the usual crankshaft end mounting to the end of a secondary shaft above and behind the crankshaft. Recent developments in suspension and braking technology are also evident on various models. The XJ models also mark one of the first serious attempts at applying computer technology on a motorcycle. Although the microprocessor-based monitor system is really little more than impressive gadgetry, it does indicate the way in which enhanced precision at relatively low cost may shape the development of the motorcycle during the next few years.

The various models described in this manual are identified in one of two ways. Where necessary, the UK models are identified by their Yamaha production code, eg XJ650 (11N), but all US models are identified by their Yamaha model code and the suffix letter indicating the year of production, eg XJ650 RJ. To assist the owner with model identification, given below are the engine/frame numbers with which each machine's production run commenced and the approximate date of import.

Model	Frame number	Date
XJ650 (4KO)	4KO – 000101 on	1980 to 1982
XJ650 (11N)	4KO – 055101 on	1982 to 1984
XJ650 G	4H7 – 000101 on	1980
XJ650 H	4H7 – 100101 on	1981
XJ650 LH	4W5 – 000001 on	1981
XJ650 J	5N8 – 000101 on	1982
XJ650 RJ	5V2 – 000101 on	1982
XJ750 (11M)	11M – 000101 on	1981 to 1984
XJ750 RH	5G2 – 000101 on	1981
XJ750 J	15R – 000101 on	1982
XJ750 RJ	5G2 – 100101 on	1982

Model dimensions and weights

	XJ650 G	XJ650 H, LH,J	XJ650(UK) XJ650 RJ
Overall length	2165 mm (85.2 in)	2165 mm (85.2 in)	2170 mm (85.4 in)
Overall width	860 mm (33.9 in)	855 mm (33.7 in)	730 mm (28.7 in)
Overall height	1180 mm (46.5 in)	1170 mm (46.1 in)	1130 mm (44.5 in)
Wheelbase	1445 mm (56.9 in)	1445 mm (56.9 in)	1435 mm (56.5 in)
Seat height	750 mm (29.5 in)	770 mm (30.3 in)	780 mm (30.7 in)
Ground clearance	150 mm (5.9 in)	145 mm (5.7 in)	140 mm (5.5 in)
Kerb weight	217 kg (478 lb)	217 kg (478 lb)	227 kg (500 lb)

	XJ750 RH, RJ	XJ750 J	XJ750(UK)
Overall length	2110 mm (83.1 in)	2145 mm (84.4 in)	2195 mm (86.4 in)
Overall width	860 mm (33.9 in)	830 mm (32.7 in)	860 mm (33.9 in)
Overall height	1120 mm (44.1 in)	1175 mm (46.3 in)	1130 mm (44.5 in)
Wheelbase	1445 mm (56.9 in)	1445 mm (56.9 in)	1445 mm (56.9 in)
Seat height	775 mm (30.5 in)	780 mm (30.7 in)	780 mm (30.7 in)
Ground clearance	140 mm (5.5 in)	135 mm (5.3 in)	145 mm (5.7 in)
Kerb weight	237 kg (522 lb)	236 kg (520 lb)	238 kg (525 lb)

Ordering spare parts

Before attempting any overhaul or maintenance work it is important to ensure that any parts likely to be required are to hand. Many of the more common parts such as gaskets and seals will be available off the shelf from the local Yamaha dealer, but often it will prove necessary to order more specialised parts well in advance. It is worthwhile running through the operation to be undertaken, referring to the appropriate Chapter and section of this book, so that a note can be made of the items most likely to be required. In some instances it will of course be necessary to dismantle the assembly in question so that the various components can be examined and measured for wear and in these instances, it must be remembered that the machine may have to be left dismantled while the replacement parts are obtained.

It is advisable to purchase almost all new parts from an official Yamaha dealer. Almost any motorcycle dealer should be able to obtain the parts in time, but this may take longer than it would through the official factory spares arrangement. It is quite in order to purchase expendable items such as spark plugs, bulbs, tyres, oil and grease from the nearest convenient source.

Owners should be very wary of some of the pattern parts that might be offered at a lower price than the Yamaha originals. Whilst in most cases these will be of an adequate standard, some of the more important parts have been known to fail suddenly and cause extensive damage in the process. A particular danger in recent years is the growing number of counterfeit parts from Taiwan. These include items such as oil filters and brake pads and are often sold in packaging which is almost indistinguishable from the manufacturer's own. Again, these are often quite serviceable parts, but can sometimes be dangerously inadequate in materials or construction. Apart from rendering the manufacturer's warranty invalid, use of sub-standard parts may put the life of the rider (or the machine) at risk. In short, where there are any doubts on safety grounds purchase parts **only** from a reputable Yamaha dealer. The extra cost involved pays for a high standard of quality and the parts will be guaranteed to work effectively.

Most machines are subject to continuous detail modifications throughout their production run, and in addition to annual model changes. In most cases these changes will be known to the dealer but not to the general public, so it is essential to quote the engine and frame numbers in full when ordering parts. The engine number is embossed in a rectangular section of the crankcase close to the engine rear right-hand upper mounting, and the frame number is stamped on the right-hand side of the steering head.

Engine number is stamped on crankcase

Frame number is stamped on steering head

Safety first!

Professional motor mechanics are trained in safe working procedures. However enthusiastic you may be about getting on with the job in hand, do take the time to ensure that your safety is not put at risk. A moment's lack of attention can result in an accident, as can failure to observe certain elementary precautions.

There will always be new ways of having accidents, and the following points do not pretend to be a comprehensive list of all dangers; they are intended rather to make you aware of the risks and to encourage a safety-conscious approach to all work you carry out on your vehicle.

Essential DOs and DON'Ts

DON'T start the engine without first ascertaining that the transmission is in neutral.

DON'T suddenly remove the filler cap from a hot cooling system – cover it with a cloth and release the pressure gradually first, or you may get scalded by escaping coolant.

DON'T attempt to drain oil until you are sure it has cooled sufficiently to avoid scalding you.

DON'T grasp any part of the engine, exhaust or silencer without first ascertaining that it is sufficiently cool to avoid burning you.

DON'T allow brake fluid or antifreeze to contact the machine's paintwork or plastic components.

DON'T syphon toxic liquids such as fuel, brake fluid or antifreeze by mouth, or allow them to remain on your skin.

DON'T inhale dust – it may be injurious to health (see *Asbestos* heading).

DON'T allow any spilt oil or grease to remain on the floor – wipe it up straight away, before someone slips on it.

DON'T use ill-fitting spanners or other tools which may slip and cause injury.

DON'T attempt to lift a heavy component which may be beyond your capability – get assistance.

DON'T rush to finish a job, or take unverified short cuts.

DON'T allow children or animals in or around an unattended vehicle.

DON'T inflate a tyre to a pressure above the recommended maximum. Apart from overstressing the carcase and wheel rim, in extreme cases the tyre may blow off forcibly.

DO ensure that the machine is supported securely at all times. This is especially important when the machine is blocked up to aid wheel or fork removal.

DO take care when attempting to slacken a stubborn nut or bolt. It is generally better to pull on a spanner, rather than push, so that if slippage occurs you fall away from the machine rather than on to it.

DO wear eye protection when using power tools such as drill, sander, bench grinder etc.

DO use a barrier cream on your hands prior to undertaking dirty jobs – it will protect your skin from infection as well as making the dirt easier to remove afterwards; but make sure your hands aren't left slippery. Note that long-term contact with used engine oil can be a health hazard.

DO keep loose clothing (cuffs, tie etc) and long hair well out of the way of moving mechanical parts.

DO remove rings, wristwatch etc, before working on the vehicle – especially the electrical system.

DO keep your work area tidy – it is only too easy to fall over articles left lying around.

DO exercise caution when compressing springs for removal or installation. Ensure that the tension is applied and released in a controlled manner, using suitable tools which preclude the possibility of the spring escaping violently.

DO ensure that any lifting tackle used has a safe working load rating adequate for the job.

DO get someone to check periodically that all is well, when working alone on the vehicle.

DO carry out work in a logical sequence and check that everything is correctly assembled and tightened afterwards.

DO remember that your vehicle's safety affects that of yourself and others. If in doubt on any point, get specialist advice.

IF, in spite of following these precautions, you are unfortunate enough to injure yourself, seek medical attention as soon as possible.

Asbestos

Certain friction, insulating, sealing, and other products – such as brake linings, clutch linings, gaskets, etc – contain asbestos. *Extreme care must be taken to avoid inhalation of dust from such products since it is hazardous to health*. If in doubt, assume that they *do* contain asbestos.

Fire

Remember at all times that petrol (gasoline) is highly flammable. Never smoke, or have any kind of naked flame around, when working on the vehicle. But the risk does not end there – a spark caused by an electrical short-circuit, by two metal surfaces contacting each other, by careless use of tools, or even by static electricity built up in your body under certain conditions, can ignite petrol vapour, which in a confined space is highly explosive.

Always disconnect the battery earth (ground) terminal before working on any part of the fuel or electrical system, and never risk spilling fuel on to a hot engine or exhaust.

It is recommended that a fire extinguisher of a type suitable for fuel and electrical fires is kept handy in the garage or workplace at all times. Never try to extinguish a fuel or electrical fire with water.

Note: *Any reference to a 'torch' appearing in this manual should always be taken to mean a hand-held battery-operated electric lamp or flashlight. It does **not** mean a welding/gas torch or blowlamp.*

Fumes

Certain fumes are highly toxic and can quickly cause unconsciousness and even death if inhaled to any extent. Petrol (gasoline) vapour comes into this category, as do the vapours from certain solvents such as trichloroethylene. Any draining or pouring of such volatile fluids should be done in a well ventilated area.

When using cleaning fluids and solvents, read the instructions carefully. Never use materials from unmarked containers – they may give off poisonous vapours.

Never run the engine of a motor vehicle in an enclosed space such as a garage. Exhaust fumes contain carbon monoxide which is extremely poisonous; if you need to run the engine, always do so in the open air or at least have the rear of the vehicle outside the workplace.

The battery

Never cause a spark, or allow a naked light, near the vehicle's battery. It will normally be giving off a certain amount of hydrogen gas, which is highly explosive.

Always disconnect the battery earth (ground) terminal before working on the fuel or electrical systems.

If possible, loosen the filler plugs or cover when charging the battery from an external source. Do not charge at an excessive rate or the battery may burst.

Take care when topping up and when carrying the battery. The acid electrolyte, even when diluted, is very corrosive and should not be allowed to contact the eyes or skin.

If you ever need to prepare electrolyte yourself, always add the acid slowly to the water, and never the other way round. Protect against splashes by wearing rubber gloves and goggles.

Mains electricity and electrical equipment

When using an electric power tool, inspection light etc, always ensure that the appliance is correctly connected to its plug and that, where necessary, it is properly earthed (grounded). Do not use such appliances in damp conditions and, again, beware of creating a spark or applying excessive heat in the vicinity of fuel or fuel vapour. Also ensure that the appliances meet the relevant national safety standards.

Ignition HT voltage

A severe electric shock can result from touching certain parts of the ignition system, such as the HT leads, when the engine is running or being cranked, particularly if components are damp or the insulation is defective. Where an electronic ignition system is fitted, the HT voltage is much higher and could prove fatal.

Tools and working facilities

The first priority when undertaking maintenance or repair work of any sort on a motorcycle is to have a clean, dry, well-lit working area. Work carried out in peace and quiet in the well-ordered atmosphere of a good workshop will give more satisfaction and much better results than can usually be achieved in poor working conditions. A good workshop must have a clean flat workbench or a solidly constructed table of convenient working height. The workbench or table should be equipped with a vice which has a jaw opening of at least 4 in (100 mm). A set of jaw covers should be made from soft metal such as aluminium alloy or copper, or from wood. These covers will minimise the marking or damaging of soft or delicate components which may be clamped in the vice. Some clean, dry, storage space will be required for tools, lubricants and dismantled components. It will be necessary during a major overhaul to lay out engine/gearbox components for examination and to keep them where they will remain undisturbed for as long as is necessary. To this end it is recommended that a supply of metal or plastic containers of suitable size is collected. A supply of clean, lint-free, rags for cleaning purposes and some newspapers, other rags, or paper towels for mopping up spillages should also be kept. If working on a hard concrete floor note that both the floor and one's knees can be protected from oil spillages and wear by cutting open a large cardboard box and spreading it flat on the floor under the machine or workbench. This also helps to provide some warmth in winter and to prevent the loss of nuts, washers, and other tiny components which have a tendency to disappear when dropped on anything other than a perfectly clean, flat, surface.

Unfortunately, such working conditions are not always available to the home mechanic. When working in poor conditions it is essential to take extra time and care to ensure that the components being worked on are kept scrupulously clean and to ensure that no components or tools are lost or damaged.

A selection of good tools is a fundamental requirement for anyone contemplating the maintenance and repair of a motor vehicle. For the owner who does not possess any, their purchase will prove a considerable expense, offsetting some of the savings made by doing-it-yourself. However, provided that the tools purchased meet the relevant national safety standards and are of good quality, they will last for many years and prove an extremely worthwhile investment.

To help the average owner to decide which tools are needed to carry out the various tasks detailed in this manual, we have compiled three lists of tools under the following headings: *Maintenance and minor repair, Repair and overhaul,* and *Specialized*. The newcomer to practical mechanics should start off with the simpler jobs around the vehicle. Then, as his confidence and experience grow, he can undertake more difficult tasks, buying extra tools as and when they are needed.

In this way, a *Maintenance and minor repair* tool kit can be built-up into a *Repair and overhaul* tool kit over a considerable period of time without any major cash outlays. The experienced home mechanic will have a tool kit good enough for most repair and overhaul procedures and will add tools from the specialized category when he feels the expense is justified by the amount of use these tools will be put to.

It is obviously not possible to cover the subject of tools fully here. For those who wish to learn more about tools and their use there is a book entitled *Motorcycle Workshop Practice Manual* (Bk No 1454) available from the publishers of this manual.

As a general rule, it is better to buy the more expensive, good quality tools. Given reasonable use, such tools will last for a very long time, whereas the cheaper, poor quality, item will wear out faster and need to be renewed more often, thus nullifying the original saving. There is also the risk of a poor quality tool breaking while in use, causing personal injury or expensive damage to the component being worked on.

For practically all tools, a tool factor is the best source since he will have a very comprehensive range compared with the average garage or accessory shop. Having said that, accessory shops often offer excellent quality tools at discount prices, so it pays to shop around. There are plenty of tools around at reasonable prices, but always aim to purchase items which meet the relevant national safety standards. If in doubt, seek the advice of the shop proprietor or manager before making a purchase.

The basis of any toolkit is a set of spanners. While open-ended spanners with their slim jaws, are useful for working on awkwardly-positioned nuts, ring spanners have advantages in that they grip the nut far more positively. There is less risk of the spanner slipping off the nut and damaging it, for this reason alone ring spanners are to be preferred. Ideally, the home mechanic should acquire a set of each, but if expense rules this out a set of combination spanners (open-ended at one end and with a ring of the same size at the other) will provide a good compromise. Another item which is so useful it should be considered an essential requirement for any home mechanic is a set of socket spanners. These are available in a variety of drive sizes. It is recommended that the $\frac{1}{2}$-inch drive type is purchased to begin with as although bulkier and more expensive than the $\frac{3}{8}$-inch type, the larger size is far more common and will accept a greater variety of torque wrenches, extension pieces and socket sizes. The socket set should comprise sockets of sizes between 8 and 24 mm, a reversible ratchet drive, an extension bar of about 10 inches in length, a spark plug socket with a rubber insert, and a universal joint. Other attachments can be added to the set at a later date.

Maintenance and minor repair tool kit

Set of spanners 8 – 24 mm
Set of sockets and attachments
Spark plug spanner with rubber insert – 10, 12, or 14 mm
as appropriate
Adjustable spanner
C-spanner/pin spanner
Torque wrench (same size drive as sockets)
Set of screwdrivers (flat blade)
Set of screwdrivers (cross-head)
Set of Allen keys 4 – 10 mm
Impact screwdriver and bits
Ball pein hammer – 2 lb
Hacksaw (junior)
Self-locking pliers – Mole grips or vice grips
Pliers – combination
Pliers – needle nose
Wire brush (small)
Soft-bristled brush
Tyre pump
Tyre pressure gauge
Tyre tread depth gauge
Oil can
Fine emery cloth
Funnel (medium size)
Drip tray
Grease gun
Set of feeler gauges
Brake bleeding kit
Strobe timing light
Continuity tester (dry battery and bulb)
Soldering iron and solder
Wire stripper or craft knife
PVC insulating tape
Assortment of split pins, nuts, bolts, and washers

Repair and overhaul toolkit

The tools in this list are virtually essential for anyone undertaking major repairs to a motorcycle and are additional to the tools listed above. Concerning Torx driver bits, Torx screws are encountered on some of the more modern machines where their use is restricted to fastening certain components inside the engine/gearbox unit. It is therefore recommended that if Torx bits cannot be borrowed from a local dealer, they are purchased individually as the need arises. They are not in regular use in the motor trade and will therefore only be available in specialist tool shops.

Plastic or rubber soft-faced mallet
Torx driver bits
Pliers – electrician's side cutters
Circlip pliers – internal (straight or right-angled tips are available)
Circlip pliers – external
Cold chisel
Centre punch
Pin punch
Scriber
Scraper (made from soft metal such as aluminium or copper)
Soft metal drift
Steel rule/straight edge
Assortment of files
Electric drill and bits
Wire brush (large)
Soft wire brush (similar to those used for cleaning suede shoes)
Sheet of plate glass
Hacksaw (large)
Valve grinding tool
Valve grinding compound (coarse and fine)
Stud extractor set (E-Z out)

Specialized tools

This is not a list of the tools made by the machine's manufacturer to carry out a specific task on a limited range of models. Occasional references are made to such tools in the text of this manual and, in general, an alternative method of carrying out the task without the manufacturer's tool is given where possible. The tools mentioned in this list are those which are not used regularly and are expensive to buy in view of their infrequent use. Where this is the case it may be possible to hire or borrow the tools against a deposit from a local dealer or tool hire shop. An alternative is for a group of friends or a motorcycle club to join in the purchase.

Valve spring compressor
Piston ring compressor
Universal bearing puller
Cylinder bore honing attachment (for electric drill)
Micrometer set
Vernier calipers
Dial gauge set
Cylinder compression gauge
Vacuum gauge set
Multimeter
Dwell meter/tachometer

Care and maintenance of tools

Whatever the quality of the tools purchased, they will last much longer if cared for. This means in practice ensuring that a tool is used for its intended purpose; for example screwdrivers should not be used as a substitute for a centre punch, or as chisels. Always remove dirt or grease and any metal particles but remember that a light film of oil will prevent rusting if the tools are infrequently used. The common tools can be kept together in a large box or tray but the more delicate, and more expensive, items should be stored separately where they cannot be damaged. When a tool is damaged or worn out, be sure to renew it immediately. It is false economy to continue to use a worn spanner or screwdriver which may slip and cause expensive damage to the component being worked on.

Fastening systems

Fasteners, basically, are nuts, bolts and screws used to hold two or more parts together. There are a few things to keep in mind when working with fasteners. Almost all of them use a locking device of some type; either a lock washer, lock nut, locking tab or thread adhesive. All threaded fasteners should be clean, straight, have undamaged threads and undamaged corners on the hexagon head where the spanner fits. Develop the habit of replacing all damaged nuts and bolts with new ones.

Rusted nuts and bolts should be treated with a rust penetrating fluid to ease removal and prevent breakage. After applying the rust penetrant, let it 'work' for a few minutes before trying to loosen the nut or bolt. Badly rusted fasteners may have to be chiseled off or removed with a special nut breaker, available at tool shops.

Flat washers and lock washers, when removed from an assembly should always be replaced exactly as removed. Replace any damaged washers with new ones. Always use a flat washer between a lock washer and any soft metal surface (such as aluminium), thin sheet metal or plastic. Special lock nuts can only be used once or twice before they lose their locking ability and must be renewed.

If a bolt or stud breaks off in an assembly, it can be drilled out and removed with a special tool called an E-Z out. Most dealer service departments and motorcycle repair shops can perform this task, as well as others (such as the repair of threaded holes that have been stripped out).

Spanner size comparison

Jaw gap (in)	Spanner size		Jaw gap (in)	Spanner size
0.250	$\frac{1}{4}$ in AF		0.945	24 mm
0.276	7 mm		1.000	1 in AF
0.313	$\frac{5}{16}$ in AF		1.010	$\frac{9}{16}$ in Whitworth; $\frac{5}{8}$ in BSF
0.315	8 mm		1.024	26 mm
0.344	$\frac{11}{32}$ in AF; $\frac{1}{8}$ in Whitworth		1.063	$1\frac{1}{16}$ in AF; 27 mm
0.354	9 mm		1.100	$\frac{5}{16}$ in Whitworth; $\frac{11}{16}$ in BSF
0.375	$\frac{3}{8}$ in AF		1.125	$1\frac{1}{8}$ in AF
0.394	10 mm		1.181	30 mm
0.433	11 mm		1.200	$\frac{11}{16}$ in Whitworth; $\frac{3}{4}$ in BSF
0.438	$\frac{7}{16}$ in AF		1.250	$1\frac{1}{4}$ in AF
0.445	$\frac{3}{16}$ in Whitworth; $\frac{1}{4}$ in BSF		1.260	32 mm
0.472	12 mm		1.300	$\frac{3}{4}$ in Whitworth; $\frac{7}{8}$ in BSF
0.500	$\frac{1}{2}$ in AF		1.313	$1\frac{5}{16}$ in AF
0.512	13 mm		1.390	$\frac{13}{16}$ in Whitworth; $\frac{15}{16}$ in BSF
0.525	$\frac{1}{4}$ in Whitworth; $\frac{5}{16}$ in BSF		1.417	36 mm
0.551	14 mm		1.438	$1\frac{7}{16}$ in AF
0.563	$\frac{9}{16}$ in AF		1.480	$\frac{7}{8}$ in Whitworth; 1 in BSF
0.591	15 mm		1.500	$1\frac{1}{2}$ in AF
0.600	$\frac{5}{16}$ in Whitworth; $\frac{3}{8}$ in BSF		1.575	40 mm; $\frac{15}{16}$ in Whitworth
0.625	$\frac{5}{8}$ in AF		1.614	41 mm
0.630	16 mm		1.625	$1\frac{5}{8}$ in AF
0.669	17 mm		1.670	1 in Whitworth; $1\frac{1}{8}$ in BSF
0.686	$\frac{11}{16}$ in AF		1.688	$1\frac{11}{16}$ in AF
0.709	18 mm		1.811	46 mm
0.710	$\frac{3}{8}$ in Whitworth; $\frac{7}{16}$ in BSF		1.813	$1\frac{13}{16}$ in AF
0.748	19 mm		1.860	$1\frac{1}{8}$ in Whitworth; $1\frac{1}{4}$ in BSF
0.750	$\frac{3}{4}$ in AF		1.875	$1\frac{7}{8}$ in AF
0.813	$\frac{13}{16}$ in AF		1.969	50 mm
0.820	$\frac{7}{16}$ in Whitworth; $\frac{1}{2}$ in BSF		2.000	2 in AF
0.866	22 mm		2.050	$1\frac{1}{4}$ in Whitworth; $1\frac{3}{8}$ in BSF
0.875	$\frac{7}{8}$ in AF		2.165	55 mm
0.920	$\frac{1}{2}$ in Whitworth; $\frac{9}{16}$ in BSF		2.362	60 mm
0.938	$\frac{15}{16}$ in AF			

Standard torque settings

Specific torque settings will be found at the end of the specifications section of each chapter. Where no figure is given, bolts should be secured according to the table below.

Fastener type (thread diameter)	kgf m	lbf ft
5mm bolt or nut	0.45 – 0.6	3.5 – 4.5
6 mm bolt or nut	0.8 – 1.2	6 – 9
8 mm bolt or nut	1.8 – 2.5	13 – 18
10 mm bolt or nut	3.0 – 4.0	22 – 29
12 mm bolt or nut	5.0 – 6.0	36 – 43
5 mm screw	0.35 – 0.5	2.5 – 3.6
6 mm screw	0.7 – 1.1	5 – 8
6 mm flange bolt	1.0 – 1.4	7 – 10
8 mm flange bolt	2.4 – 3.0	17 – 22
10 mm flange bolt	3.0 – 4.0	22 – 29

Choosing and fitting accessories

The range of accessories available to the modern motor-cyclist is almost as varied and bewildering as the range of motorcycles. This Section is intended to help the owner in choosing the correct equipment for his needs and to avoid some of the mistakes made by many riders when adding accessories to their machines. It will be evident that the Section can only cover the subject in the most general terms and so it is recommended that the owner, having decided that he wants to fit, for example, a luggage rack or carrier, seeks the advice of several local dealers and the owners of similar machines. This will give a good idea of what makes of carrier are easily available, and at what price. Talking to other owners will give some insight into the drawbacks or good points of any one make. A walk round the motorcycles in car parks or outside a dealer will often reveal the same sort of information.

The first priority when choosing accessories is to assess exactly what one needs. It is, for example, pointless to buy a large heavy-duty carrier which is designed to take the weight of fully laden panniers and topbox when all you need is a place to strap on a set of waterproofs and a lunchbox when going to work. Many accessory manufacturers have ranges of equipment to cater for the individual needs of different riders and this point should be borne in mind when looking through a dealer's catalogues. Having decided exactly what is required and the use to which the accessories are going to be put, the owner will need a few hints on what to look for when making the final choice. To this end the Section is now sub-divided to cover the more popular accessories fitted. Note that it is in no way a customizing guide, but merely seeks to outline the practical considerations to be taken into account when adding aftermarket equipment to a motorcycle.

Fairings and windscreens

A fairing is possibly the single, most expensive, aftermarket item to be fitted to any motorcycle and, therefore, requires the most thought before purchase. Fairings can be divided into two main groups: front fork mounted handlebar fairings and wind-screens, and frame mounted fairings.

The first group, the front fork mounted fairings, are becoming far more popular than was once the case, as they offer several advantages over the second group. Front fork mounted fairings generally are much easier and quicker to fit, involve less modification to the motorcycle, do not as a rule restrict the steering lock, permit a wider selection of handlebar styles to be used, and offer adequate protection for much less money than the frame mounted type. They are also lighter, can be swapped easily between different motorcycles, and are available in a much greater variety of styles. Their main disadvantages are that they do not offer as much weather protection as the frame mounted types, rarely offer any storage space, and, if poorly fitted or naturally incompatible, can have an adverse effect on the stability of the motorcycle.

The second group, the frame mounted fairings, are secured so rigidly to the main frame of the motorcycle that they can offer a substantial amount of protection to motorcycle and rider in the event of a crash. They offer almost complete protection from the weather and, if double-skinned in construction, can provide a great deal of useful storage space. The feeling of peace, quiet and complete relaxation encountered when riding behind a good full fairing has to be experienced to be believed. For this reason full fairings are considered essential by most touring motorcyclists and by many people who ride all year round. The main disadvantages of this type are that fitting can take a long time, often involving removal or modification of standard motorcycle components, they restrict the steering lock and they can add up to about 40 lb to the weight of the machine. They do not usually affect the stability of the machine to any great extent once the front tyre pressure and suspension have been adjusted to compensate for the extra weight, but can be affected by sidewinds.

The first thing to look for when purchasing a fairing is the quality of the fittings. A good fairing will have strong, substantial brackets constructed from heavy-gauge tubing; the brackets must be shaped to fit the frame or forks evenly so that the minimum of stress is imposed on the assembly when it is bolted down. The brackets should be properly painted or finished — a nylon coating being the favourite of the better manufacturers — the nuts and bolts provided should be of the same thread and size standard as is used on the motorcycle and be properly plated. Look also for shakeproof locking nuts or locking washers to ensure that everything remains securely tightened down. The fairing shell is generally made from one of two materials: fibreglass or ABS plastic. Both have their advantages and disadvantages, but the main consideration for the owner is that fibreglass is much easier to repair in the event of damage occurring to the fairing. Whichever material is used, check that it is properly finished inside as well as out, that the edges are protected by beading and that the fairing shell is insulated from vibration by the use of rubber grommets at all mounting points. Also be careful to check that the windscreen is retained by plastic bolts which will snap on impact so that the windscreen will break away and not cause personal injury in the event of an accident.

Having purchased your fairing or windscreen, read the manufacturer's fitting instructions very carefully and check that you have all the necessary brackets and fittings. Ensure that the mounting brackets are located correctly and bolted down securely. Note that some manufacturers use hose clamps to retain the mounting brackets; these should be discarded as they are convenient to use but not strong enough for the task. Stronger clamps should be substituted; car exhaust pipe clamps of suitable size would be a good alternative. Ensure that the front forks can turn through the full steering lock available without fouling the fairing. With many types of frame-mounted fairing the handlebars will have to be altered or a different type fitted and the steering lock will be restricted by stops provided with the fittings. Also check that the fairing does not foul the front wheel or mudguard, in any steering position, under full fork compression. Re-route any cables, brake pipes or electrical wiring which may snag on the fairing and take great care to protect all electrical connections, using insulating tape. If the manufacturer's instructions are followed carefully at every stage no serious problems should be encountered. Remember that hydraulic pipes that have been disconnected must be carefully re-tightened and the hydraulic system purged of air bubbles by bleeding.

Two things will become immediately apparent when taking a motorcycle on the road for the first time with a fairing – the first is the tendency to underestimate the road speed because of the lack of wind pressure on the body. This must be very carefully watched until one has grown accustomed to riding behind the fairing. The second thing is the alarming increase in engine noise which is an unfortunate but inevitable by-product of fitting any type of fairing or windscreen, and is caused by normal engine noise being reflected, and in some cases amplified, by the flat surface of the fairing.

Luggage racks or carriers

Carriers are possibly the commonest item to be fitted to modern motorcycles. They vary enormously in size, carrying capacity, and durability. When selecting a carrier, always look for one which is made specifically for your machine and which is bolted on with as few separate brackets as possible. The universal-type carrier, with its mass of brackets and adaptor pieces, will generally prove too weak to be of any real use. A good carrier should bolt to the main frame, generally using the two suspension unit top mountings and a mudguard mounting bolt as attachment points, and have its luggage platform as low and as far forward as possible to minimise the effect of any load on the machine's stability. Look for good quality, heavy gauge tubing, good welding and good finish. Also ensure that the carrier does not prevent opening of the seat, sidepanels or tail compartment, as appropriate. When using a carrier, be very careful not to overload it. Excessive weight placed so high and so far to the rear of any motorcycle will have an adverse effect on the machine's steering and stability.

Luggage

Motorcycle luggage can be grouped under two headings: soft and hard. Both types are available in many sizes and styles and have advantages and disadvantages in use.

Soft luggage is now becoming very popular because of its lower cost and its versatility. Whether in the form of tankbags, panniers, or strap-on bags, soft luggage requires in general no brackets and no modification to the motorcycle. Equipment can be swapped easily from one motorcycle to another and can be fitted and removed in seconds. Awkwardly shaped loads can easily be carried. The disadvantages of soft luggage are that the contents cannot be secure against the casual thief, very little protection is afforded in the event of a crash, and waterproofing is generally poor. Also, in the case of panniers, carrying capacity is restricted to approximately 10 lb, although this amount will vary considerably depending on the manufacturer's recommendation. When purchasing soft luggage, look for good quality material, generally vinyl or nylon, with strong, well-stitched attachment points. It is always useful to have separate pockets, especially on tank bags, for items which will be needed on the journey. When purchasing a tank bag, look for one which has a separate, well-padded, base. This will protect the tank's paintwork and permit easy access to the filler cap at petrol stations.

Hard luggage is confined to two types: panniers, and top boxes or tail trunks. Most hard luggage manufacturers produce matching sets of these items, the basis of which is generally that manufacturer's own heavy-duty luggage rack. Variations on this theme occur in the form of separate frames for the better quality panniers, fixed or quickly-detachable luggage, and in size and carrying capacity. Hard luggage offers a reasonable degree of security against theft and good protection against weather and accident damage. Carrying capacity is greater than that of soft luggage, around 15 – 20 lb in the case of panniers, although top boxes should never be loaded as much as their

apparent capacity might imply. A top box should only be used for lightweight items, because one that is heavily laden can have a serious effect on the stability of the machine. When purchasing hard luggage look for the same good points as mentioned under fairings and windscreens, ie good quality mounting brackets and fittings, and well-finished fibreglass or ABS plastic cases. Again as with fairings, always purchase luggage made specifically for your motorcycle, using as few separate brackets as possible, to ensure that everything remains securely bolted in place. When fitting hard luggage, be careful to check that the rear suspension and brake operation will not be impaired in any way and remember that many pannier kits require re-siting of the indicators. Remember also that a non-standard exhaust system may make fitting extremely difficult.

Handlebars

The occupation of fitting alternative types of handlebar is extremely popular with modern motorcyclists, whose motives may vary from the purely practical, wishing to improve the comfort of their machines, to the purely aesthetic, where form is more important than function. Whatever the reason, there are several considerations to be borne in mind when changing the handlebars of your machine. If fitting lower bars, check carefully that the switches and cables do not foul the petrol tank on full lock and that the surplus length of cable, brake pipe, and electrical wiring are smoothly and tidily disposed of. Avoid tight kinks in cable or brake pipes which will produce stiff controls or the premature and disastrous failure of an overstressed component. If necessary, remove the petrol tank and re-route the cable from the engine/gearbox unit upwards, ensuring smooth gentle curves are produced. In extreme cases, it will be necessary to purchase a shorter brake pipe to overcome this problem. In the case of higher handlebars than standard it will almost certainly be necessary to purchase extended cables and brake pipes. Fortunately, many standard motorcycles have a custom version which will be equipped with higher handlebars and, therefore, factory-built extended components will be available from your local dealer. It is not usually necessary to extend electrical wiring, as switch clusters may be used on several different motorcycles, some being custom versions. This point should be borne in mind however when fitting extremely high or wide handlebars.

When fitting different types of handlebar, ensure that the mounting clamps are correctly tightened to the manufacturer's specifications and that cables and wiring, as previously mentioned, have smooth easy runs and do not snag on any part of the motorcycle throughout the full steering lock. Ensure that the fluid level in the front brake master cylinder remains level to avoid any chance of air entering the hydraulic system. Also check that the cables are adjusted correctly and that all handlebar controls operate correctly and can be easily reached when riding.

Crashbars

Crashbars, also known as engine protector bars, engine guards, or case savers, are extremely useful items of equipment which can contribute protection to the machine's structure if a crash occurs. They do not, as has been inferred in the US, prevent the rider from crashing, or necessarily prevent rider injury should a crash occur.

It is recommended that only the smaller, neater, engine protector type of crashbar is considered. This type will offer protection while restricting, as little as is possible, access to the engine and the machine's ground clearance. The crashbars should be designed for use specifically on your machine, and should be constructed of heavy-gauge tubing with strong, integral mounting brackets. Where possible, they should bolt to a strong lug on the frame, usually at the engine mounting bolts.

The alternative type of crashbar is the larger cage type. This type is not recommended in spite of their appearance which promises some protection to the rider as well as to the machine. The larger amount of leverage imposed by the size of this type of crashbar increases the risk of severe frame damage in the event of an accident. This type also decreases the machine's ground clearance and restricts access to the engine. The amount of protection afforded the rider is open to some doubt as the design is based on the premise that the rider will stay in the normally seated position during an accident, and the crash bar structure will not itself fail. Neither result can in any way be guaranteed.

As a general rule, always purchase the best, ie usually the most expensive, set of crashbars you an afford. The investment will be repaid by minimising the amount of damage incurred, should the machine be involved in an accident. Finally, avoid the universal type of crashbar. This should be regarded only as a last resort to be used if no alternative exists. With its usual multitude of separate brackets and spacers, the universal crashbar is far too weak in design and construction to be of any practical value.

Exhaust systems

The fitting of aftermarket exhaust systems is another extremely popular pastime amongst motorcyclists. The usual motive is to gain more performance from the engine but other considerations are to gain more ground clearance, to lose weight from the motorcycle, to obtain a more distinctive exhaust note or to find a cheaper alternative to the manufacturer's original equipment exhaust system. Original equipment exhaust systems often cost more and may well have a relatively short life. It should be noted that it is rare for an aftermarket exhaust system alone to give a noticeable increase in the engine's power output. Modern motorcycles are designed to give the highest power output possible allowing for factors such as quietness, fuel economy, spread of power, and long-term reliability. If there were a magic formula which allowed the exhaust system to produce more power without affecting these other considerations you can be sure that the manufacturers, with their large research and development facilities, would have found it and made use of it. Performance increases of a worthwhile and noticeable nature only come from well-tried and properly matched modifications to the entire engine, from the air filter, through the carburettors, port timing or camshaft and valve design, combustion chamber shape, compression ratio, and the exhaust system. Such modifications are well outside the scope of this manual but interested owners might refer to the 'Piper Tuning Manual' produced by the publisher of this manual; this book goes into the whole subject in great detail.

Whatever your motive for wishing to fit an alternative exhaust system, be sure to seek expert advice before doing so. Changes to the carburettor jetting will almost certainly be required for which you must consult the exhaust system manufacturer. If he cannot supply adequately specific information it is reasonable to assume that insufficient development work has been carried out, and that particular make should be avoided. Other factors to be borne in mind are whether the exhaust system allows the use of both centre and side stands, whether it allows sufficient access to permit oil and filter changing and whether modifications are necessary to the standard exhaust system. Many two-stroke expansion chamber systems require the use of the standard exhaust pipe; this is all very well if the standard exhaust pipe and silencer are separate units but can cause problems if the two, as with so many modern two-strokes, are a one-piece unit. While the exhaust pipe can be removed easily by means of a hacksaw it is not so easy to refit the original silencer should you at any time wish to return the machine to standard trim. The same applies to several four-stroke systems.

On the subject of the finish of aftermarket exhausts, avoid black-painted systems unless you enjoy painting. As any trail-bike owner will tell you, rust has a great affinity for black exhausts and re-painting or rust removal becomes a task which must be carried out with monotonous regularity. A bright chrome finish is, as a general rule, a far better proposition as it is much easier to keep clean and to prevent rusting. Although the general finish of aftermarket exhaust systems is not always

up to the standard of the original equipment the lower cost of such systems does at least reflect this fact.

When fitting an alternative system always purchase a full set of new exhaust gaskets, to prevent leaks. Fit the exhaust first to the cylinder head or barrel, as appropriate, tightening the retaining nuts or bolts by hand only and then line up the exhaust rear mountings. If the new system is a one-piece unit and the rear mountings do not line up exactly, spacers must be fabricated to take up the difference. Do not force the system into place as the stress thus imposed will rapidly cause cracks and splits to appear. Once all the mountings are loosely fixed, tighten the retaining nuts or bolts securely, being careful not to overtighten them. Where the motorcycle manufacturer's torque settings are available, these should be used. Do not forget to carry out any carburation changes recommended by the exhaust system's manufacturer.

Electrical equipment

The vast range of electrical equipment available to motorcyclists is so large and so diverse that only the most general outline can be given here. Electrical accessories vary from electric ignition kits fitted to replace contact breaker points, to additional lighting at the front and rear, more powerful horns, various instruments and gauges, clocks, anti-theft systems, heated clothing, CB radios, radio-cassette players, and intercom systems, to name but a few of the more popular items of equipment.

As will be evident, it would require a separate manual to cover this subject alone and this section is therefore restricted to outlining a few basic rules which must be borne in mind when fitting electrical equipment. The first consideration is whether your machine's electrical system has enough reserve capacity to cope with the added demand of the accessories you wish to fit. The motorcycle's manufacturer or importer should be able to furnish this sort of information and may also be able to offer advice on uprating the electrical system. Failing this, a good dealer or the accessory manufacturer may be able to help. In some cases, more powerful generator components may be available, perhaps from another motorcycle in the manufacturer's range. The second consideration is the legal requirements in force in your area. The local police may be prepared to help with this point. In the UK for example, there are strict regulations governing the position and use of auxiliary riding lamps and fog lamps.

When fitting electrical equipment always disconnect the battery first to prevent the risk of a short-circuit, and be careful to ensure that all connections are properly made and that they are waterproof. Remember that many electrical accesories are designed primarily for use in cars and that they cannot easily withstand the exposure to vibration and to the weather. Delicate components must be rubber-mounted to insulate them from vibration, and sealed carefully to prevent the entry of rainwater and dirt. Be careful to follow exactly the accessory manufacturer's instructions in conjunction with the wiring diagram at the back of this manual.

Accessories – general

Accessories fitted to your motorcycle will rapidly deteriorate if not cared for. Regular washing and polishing will maintain the finish and will provide an opportunity to check that all mounting bolts and nuts are securely fastened. Any signs of chafing or wear should be watched for, and the cause cured as soon as possible before serious damage occurs.

As a general rule, do not expect the re-sale value of your motorcycle to increase by an amount proportional to the amount of money and effort put into fitting accessories. It is usually the case that an absolutely standard motorcycle will sell more easily at a better price than one that has been modified. If you are in the habit of exchanging your machine for another at frequent intervals, this factor should be borne in mind to avoid loss of money.

Fault diagnosis

Contents

1 Introduction

This Section provides an easy reference-guide to the more common ailments that are likely to afflict your machine. Obviously, the opportunities are almost limitless for faults to occur as a result of obscure failures, and to try and cover all eventualities would require a book. Indeed, a number have been written on the subject.

Successful fault diagnosis is not a mysterious 'black art' but the application of a bit of knowledge combined with a systematic and logical approach to the problem. Approach any fault diagnosis by first accurately identifying the symptom and then checking through the list of possible causes, starting with the simplest or most obvious and progressing in stages to the most complex. Take nothing for granted, but above all apply liberal quantities of common sense.

The main symptom of a fault is given in the text as a major heading below which are listed, as Section headings, the various systems or areas which may contain the fault. Details of each possible cause for a fault and the remedial action to be taken are given, in brief, in the paragraphs below each Section heading. Further information should be sought in the relevant Chapter.

In some cases reference will be made in the singular to a component, for example a carburettor, where in fact more than one item is fitted to the machine. The particular reference should be applied to all those components.

Starter motor problems

2 Starter motor not rotating

Engine stop switch off or starter interlock switches (where fitted) not operated correctly.

Fuse blown. Check the main fuse located behind the battery side cover.

Battery voltage low. Switching on the headlamp and operating the horn will give a good indication of the charge level. If necessary recharge the battery from an external source.

Neutral gear not selected. Where a neutral indicator switch is fitted.

Faulty neutral indicator switch, clutch interlock switch, or other starter interlock switch (where fitted). Check the switch wiring and switches for correct operation. See Chapter 6 for a description of the starter interlock switch system.

Ignition switch defective. Check switch for continuity and connections for security.

Engine stop switch defective. Check switch for continuity in 'Run' position. Fault will be caused by broken, wet or corroded switch contacts. Clean or renew as necessary.

Starter button switch faulty. Check continuity of switch. Faults as for engine stop switch.

Starter relay (solenoid) faulty. If the switch is functioning correctly a pronounced click should be heard when the starter button is depressed. This presupposes that current is flowing to the solenoid when the button is depressed.

Wiring open or shorted. Check first that the battery terminal connections are tight and corrosion free. Follow this by checking that all wiring connections are dry, tight and corrosion free. Check also for frayed or broken wiring. Occasionally a wire may become trapped between two moving components, particularly in the vicinity of the steering head, leading to breakage of the internal core but leaving the softer but more resilient outer cover intact. This can cause mysterious intermittent or total power loss.

Starter motor defective. A badly worn starter motor may cause high current drain from a battery without the motor rotating. If current is found to be reaching the motor, after checking the starter button and starter relay, suspect a damaged motor. The motor should be removed for inspection.

On all but XJ650 G, H and LH models a starter interlock system is fitted to prevent engine starting unless the machine's controls are so positioned to ensure a safe and foolproof start. Included in the interlock circuitry is a diode block fitted to prevent cross-feeding of current in the circuit. In very rare cases it may be found that one or more diodes have failed, thus causing complete interlock system failure. See Chapter 6 for further information.

3 Starter motor rotates but engine does not turn over

Starter motor clutch defective. Suspect jammed or worn engagement rollers, plungers and springs.

Damaged starter motor drive train. Inspect and renew component where necessary. Failure in this area is unlikely.

4 Starter motor and clutch function but engine will not turn over

Engine seized. Seizure of the engine is always a result of damage to internal components due to lubrication failure, or component breakage resulting from abuse, neglect or old age. A seizing or partially seized component may go un-noticed until the engine has cooled down and an attempt is made to restart the engine. Suspect first seizure of the valves, valve gear and the pistons. Instantaneous seizure whilst the engine is running indicates component breakage. In either case major dismantling and inspection will be required.

Engine does not start when turned over

5 No fuel flow to carburettor

No fuel or insufficient fuel in tank.

Fuel tap lever position incorrectly selected.

Float chambers require priming after running dry (vacuum taps only).

Tank filler cap air vent obstructed. Usually caused by dirt or water. Clean the vent orifice.

Fuel tap or filter blocked. Blockage may be due to accumulation of rust or paint flakes from the tank's inner surface or of foreign matter from contaminated fuel. Remove the tap and clean it and the filter. Look also for water droplets in the fuel.

Fuel line blocked. Blockage of the fuel line is more likely to result from a kink in the line rather than the accumulation of debris.

6 Fuel not reaching cylinder

Float chamber not filling. Caused by float needle or floats sticking in up position. This may occur after the machine has been left standing for an extended length of time allowing the fuel to evaporate. When this occurs a gummy residue is often left which hardens to a varnish-like substance. This condition may be worsened by corrosion and crystaline deposits produced prior to the total evaporation of contaminated fuel. Sticking of the float needle may also be caused by wear. In any case removal of the float chamber will be necessary for inspection and cleaning.

Blockage in starting circuit, slow running circuit or jets. Blockage of these items may be attributable to debris from the fuel tank by-passing the filter system or to gumming up as described in paragraph 1. Water droplets in the fuel will also block jets and passages. The carburettor should be dismantled for cleaning.

Fuel level too low. The fuel level in the float chamber is controlled by float height. The float height may increase with wear or damage but will never reduce, thus a low float height is an inherent rather than developing condition. Check the float height and make any necessary adjustment.

7 Engine flooding

Float valve needle worn or stuck open. A piece of rust or other debris can prevent correct seating of the needle against the valve seat thereby permitting an uncontrolled flow of fuel. Similarly, a worn needle or needle seat will prevent valve closure. Dismantle the carburettor float bowl for cleaning and, if necessary, renewal of the worn components.

Fuel level too high. The fuel level is controlled by the float height which may increase due to wear of the float needle, pivot pin or operating tang. Check the float height, and make any necessary adjustment. A leaking float will cause an increase in fuel level, and thus should be renewed.

Accelerator pump. On those models so equipped, repeated operation of the throttle prior to starting will cause flooding due to too much raw fuel being injected into the venturi.

Cold starting mechanism. Check the choke (starter mechanism) for correct operation. If the mechanism jams in the 'On' position subsequent starting of a hot engine will be difficult.

Blocked air filter. A badly restricted air filter will cause flooding. Check the filter and clean or renew as required. A collapsed inlet hose will have a similar effect.

8 No spark at plug

Ignition switch not on.

Engine stop switch off.

Fuse blown. Check fuse for ignition circuit. See wiring diagram.

Battery voltage low. The current draw required by a starter motor is sufficiently high that an under-charged battery may not have enough spare capacity to provide power for the ignition circuit during starting.

Starter motor inefficient. A starter motor with worn brushes and a worn or dirty commutator will draw excessive amounts of current causing power starvation in the ignition system. See the preceding paragraph. Starter motor overhaul will be required.

Spark plug failure. Clean the spark plug thoroughly and reset the electrode gap. Refer to the spark plug section and the condition guide in Chapter 3. If the spark plug shorts internally or has sustained visible damage to the electrodes, core or ceramic insulator it should be renewed. On rare occasions a plug that appears to spark vigorously will fail to do so when refitted to the engine and subjected to the compression pressure in the cylinder.

Spark plug cap or high tension (HT) lead faulty. Check condition and security. Replace if deterioration is evident.

Spark plug cap loose. Check that the spark plug cap fits securely over the plug and, where fitted, the screwed terminal on the plug end is secure.

Shorting due to moisture. Certain parts of the ignition system are susceptible to shorting when the machine is ridden or parked in wet weather. Check particularly the area from the spark plug cap back to the ignition coil. A water dispersant spray may be used to dry out waterlogged components. Recurrence of the problem can be prevented by using an ignition sealant spray after drying out and cleaning.

Ignition or stop switch shorted. May be caused by water, corrosion or wear. Water dispersant and contact cleaning sprays may be used. If this fails to overcome the problem dismantling and visual inspection of the switches will be required.

Shorting or open circuit in wiring. Failure in any wire connecting any of the ignition components will cause ignition malfunction. Check also that all connections are clean, dry and tight.

Ignition coil failure. Check the coil, referring to Chapter 3.

Pickup coil failure. Check the two coils, referring to Chapter 3.

TCI (igniter) unit failure. See Chapter 3.

9 Weak spark at plug

Feeble sparking at the plug may be caused by any of the faults mentioned in the preceding Section other than those items in paragraphs 1 and 2.

10 Compression low

Spark plug loose. This will be self-evident on inspection, and may be accompanied by a hissing noise when the engine is turned over. Remove the plug and check that the threads in the cylinder head are not damaged. Check also that the plug sealing washer is in good condition.

Cylinder head gasket leaking. This condition is often accompanied by a high pitched squeak from around the cylinder head and oil loss, and may be caused by insufficiently tightened cylinder head fasteners, a warped cylinder head or mechanical failure of the gasket material. Re-torqueing the fasteners to the correct specification may seal the leak in some instances but if damage has occurred this course of action will provide, at best, only a temporary cure.

Valve not seating correctly. The failure of a valve to seat may be caused by insufficient valve clearance, pitting of the valve seat or face, carbon deposits on the valve seat or seizure of the valve stem or valve gear components. Valve spring breakage will also prevent correct valve closure. The valve clearances should be checked first and then, if these are found to be in order, further dismantling will be required to inspect the relevant components for failure.

Cylinder, piston and ring wear. Compression pressure will be lost if any of these components are badly worn. Wear in one component is invariably accompanied by wear in another. A top end overhaul will be required.

Piston rings sticking or broken. Sticking of the piston rings may be caused by seizure due to lack of lubrication or heating as a result of poor carburation or incorrect fuel type. Gumming of the rings may result from lack of use, or carbon deposits in the ring grooves. Broken rings result from over-revving, over-heating or general wear. In either case a top-end overhaul will be required.

Engine stalls after starting

11 General causes

Improper cold start mechanism operation. Check that the operating controls function smoothly and, where applicable, are correctly adjusted. A cold engine may not require application of an enriched mixture to start initially but may baulk without choke once firing. Likewise a hot engine may start with an enriched mixture but will stop almost immediately if the choke is inadvertently in operation.

Ignition malfunction. See Section 9, 'Weak spark at plug'.

Carburettor incorrectly adjusted. Maladjustment of the idle speed may cause the engine to stop immediately after starting. See Chapter 2.

Fuel contamination. Check for filter blockage by debris or water which reduces, but does not completely stop, fuel flow or blockage of the slow speed circuit in the carburettor by the same agents. If water is present it can often be seen as droplets in the bottom of the float bowl. Clean the filter and, where water is in evidence, drain and flush the fuel tank and float bowl.

Intake air leak. Check for security of the carburettor mounting and hose connections, and for cracks or splits in the hoses. Check also that the carburettor top is secure and that the vacuum gauge adaptor plug (where fitted) is tight.

Air filter blocked or omitted. A blocked filter will cause an over-rich mixture; the omission of a filter will cause an excessively weak mixture. Both conditions will have a detrimental effect on carburation. Clean or renew the filter as necessary.

Fuel filler cap air vent blocked. Usually caused by dirt or water. Clean the vent orifice.

Poor running at idle and low speed

12 Weak spark at plug or erratic firing

Battery voltage low. In certain conditions low battery charge, especially when coupled with a badly sulphated battery, may result in misfiring. If the battery is in good general condition it should be recharged; an old battery suffering from sulphated plates should be renewed.

Spark plug fouled, faulty or incorrectly adjusted. See Section 8 or refer to Chapter 3.

Spark plug cap or high tension lead shorting. Check the condition of both these items ensuring that they are in good condition and dry and that the cap is fitted correctly.

Spark plug type incorrect. Fit plug of correct type and heat range as given in Specifications. In certain conditions a plug of hotter or colder type may be required for normal running.

Faulty ignition coil. Partial failure of the coil internal insulation will diminish the performance of the coil. No repair is possible, a new component must be fitted.

Pickup coil faulty. Partial failure of a pickup coil internal insulation will diminish the performance of the coil. No repair is possible, a new component must be fitted.

TCI (igniter) unit malfunction. See Chapter 6 for details.

Ignition timing reluctor loose on crankshaft end. Re-secure the reluctor in the correct position.

13 Fuel/air mixture incorrect

Intake air leak. See Section 11.

Mixture strength incorrect. Adjust slow running mixture strength using pilot adjustment screw.

Carburettor synchronisation.

Pilot jet or slow running circuit blocked. The carburettor should be removed and dismantled for thorough cleaning. Blow through all jets and air passages with compressed air to clear obstructions.

Air cleaner clogged or omitted. Clean or fit air cleaner element as necessary. Check also that the element and air filter cover are correctly seated.

Cold start mechanism in operation. Check that the choke has not been left on inadvertently and the operation is correct. Where applicable check the operating cable free play.

Fuel level too high or too low. Check the float height and adjust as necessary. See Section 7.

Fuel tank air vent obstructed. Obstruction usually caused by dirt or water. Clean vent orifice.

Valve clearance incorrect. Check, and if necessary, adjust, the clearances.

14 Compression low

See Section 10.

Acceleration poor

15 General causes

All items as for previous Section.

Accelerator pump defective. Where so equipped, check that the accelerator pump injects raw fuel into the carburettor venturi, when the throttle is open fully. If this does not occur check the condition of the pump components and that the feed passage to the pump is not obstructed.

Timing incorrect or not advancing. Other than as a result of the pickup coils becoming loose incorrect timing is only likely to be caused by TCI unit failure. TCI unit failure will also prevent correct ignition advance as the engine speed increases.

Sticking throttle vacuum piston. CD carburettors only.

Brakes binding. Usually caused by maladjustment or partial seizure of the operating mechanism due to poor maintenance. Check brake adjustment (where applicable). A bent wheel spindle or warped brake disc can produce similar symptoms.

Poor running or lack of power at high speeds

16 Weak spark at plug or erratic firing

All items as for Section 12.

HT lead insulation failure. Insulation failure of the HT lead and spark plug cap due to old age or damage can cause shorting when the engine is driven hard. This condition may be less noticeable, or not noticeable at all at lower engine speeds.

17 Fuel/air mixture incorrect

All items as for Section 13, with the exception of items 2 and 4.

Main jet blocked. Debris from contaminated fuel, or from the fuel tank, and water in the fuel can block the main jet. Clean the fuel filter, the float bowl area, and if water is present, flush and refill the fuel tank.

Main jet is the wrong size. The standard carburettor jetting is for sea level atmospheric pressure. For high altitudes, usually above 5000 ft, a smaller main jet will be required.

Jet needle and needle jet worn. These can be renewed individually but should be renewed as a pair. Renewal of both items requires partial dismantling of the carburettor.

Air bleed holes blocked. Dismantle carburettor and use compressed air to blow out all air passages.

Reduced fuel flow. A reduction in the maximum fuel flow from the fuel tank to the carburettor will cause fuel starvation, proportionate to the engine speed. Check for blockages through debris or a kinked fuel line.

Vacuum diaphragm split. Renew.

18 Compression low

See Section 10.

Knocking or pinking

19 General causes

Carbon build-up in combustion chamber. After high mileages have been covered large accumulation of carbon may occur. This may glow red hot and cause premature ignition of the fuel/air mixture, in advance of normal firing by the spark plug. Cylinder head removal will be required to allow inspection and cleaning.

Fuel incorrect. A low grade fuel, or one of poor quality may result in compression induced detonation of the fuel resulting in knocking and pinking noises. Old fuel can cause similar problems. A too highly leaded fuel will reduce detonation but will accelerate deposit formation in the combustion chamber and may lead to early pre-ignition as described in item 1.

Spark plug heat range incorrect. Uncontrolled pre-ignition can result from the use of a spark plug the heat range of which is too hot.

Weak mixture. Overheating of the engine due to a weak mixture can result in pre-ignition occurring where it would not occur when engine temperature was within normal limits. Maladjustment, blocked jets or passages and air leaks can cause this condition.

Overheating

20 Firing incorrect

Spark plug fouled, defective or maladjusted. See Section 6.
Spark plug type incorrect. Refer to the Specifications and ensure that the correct plug type is fitted.
Incorrect ignition timing. Timing that is far too much advanced or far too much retarded will cause overheating. Check the ignition timing is correct and that the advance mechanism is functioning.

21 Fuel/air mixture incorrect

Slow speed mixture strength incorrect. Adjust pilot air screw.
Main jet wrong size. The carburettor is jetted for sea level atmospheric conditions. For high altitudes, usually above 5000 ft, a smaller main jet will be required.
Air filter badly fitted or omitted. Check that the filter element is in place and that it and the air filter box cover are sealing correctly. Any leaks will cause a weak mixture.
Induction air leaks. Check the security of the carburettor mountings and hose connections, and for cracks and splits in the hoses. Check also that the carburettor top is secure and that the vacuum gauge adaptor plug (where fitted) is tight.
Fuel level too low. See Section 6.
Fuel tank filler cap air vent obstructed. Clear blockage.

22 Lubrication inadequate

Engine oil too low. Not only does the oil serve as a lubricant by preventing friction between moving components, but it also acts as a coolant. Check the oil level and replenish.
Engine oil overworked. The lubricating properties of oil are lost slowly during use as a result of changes resulting from heat and also contamination. Always change the oil at the recommended interval.
Engine oil of incorrect viscosity or poor quality. Always use the recommended viscosity and type of oil.
Oil filter and filter by-pass valve blocked. Renew filter and clean the by-pass valve.

23 Miscellaneous causes

Engine fins clogged. A build-up of mud in the cylinder head and cylinder barrel cooling fins will decrease the cooling capabilities of the fins. Clean the fins as required.

Clutch operating problems

24 Clutch slip

No clutch lever play. Adjust clutch lever end play according to the procedure in Routine Maintenance.
Friction plates worn or warped. Overhaul clutch assembly, replacing plates out of specification.
Steel plates worn or warped. Overhaul clutch assembly, replacing plates out of specification.
Clutch springs broken or wear. Old or heat-damaged (from slipping clutch) springs should be replaced with new ones.
Clutch release not adjusted properly.
Clutch inner cable snagging. Caused by a frayed cable or kinked outer cable. Replace the cable with a new one. Repair of a frayed cable is not advised.
Clutch release mechanism defective. Worn or damaged parts in the clutch release mechanism could include the shaft, actuating arm or pivot. Replace parts as necessary.
Clutch hub and outer drum worn. Severe indentation by the clutch plate tangs of the channels in the hub and drum will cause snagging of the plates preventing correct engagement. If this damage occurs, renewal of the worn components is required.
Lubricant incorrect. Use of a transmission lubricant other than that specified may allow the plates to slip.

25 Clutch drag

Clutch lever play excessive. Adjust lever at bars or at cable end if necessary.
Clutch plates warped or damaged. This will cause a drag on the clutch, causing the machine to creep. Overhaul clutch assembly.
Clutch spring tension uneven. Usually caused by a sagged or broken spring. Check and replace springs.
Engine oil deteriorated. Badly contaminated engine oil and a heavy deposit of oil sludge and carbon on the plates will cause plate sticking. The oil recommended for this machine is of the detergent type, therefore it is unlikely that this problem will arise unless regular oil changes are neglected.
Engine oil viscosity too high. Drag in the plates will result from the use of an oil with too high a viscosity. In very cold weather clutch drag may occur until the engine has reached operating temperature.
Clutch hub and outer drum worn. Indentation by the clutch plate tangs of the channels in the hub and drum will prevent easy plate disengagement. If the damage is light the affected areas may be dressed with a fine file. More pronounced damage will necessitate renewal of the components.
Clutch housing seized to shaft. Lack of lubrication, severe wear or damage can cause the housing to seize to the shaft. Overhaul of the clutch, and perhaps the transmission, may be necessary to repair damage.
Clutch release mechanism defective. Worn or damaged release mechanism parts can stick and fail to provide leverage. Overhaul clutch cover components.
Loose clutch hub nut. Causes drum and hub misalignment, putting a drag on the engine. Engagement adjustment continually varies. Overhaul clutch assembly.

Gear selection problems

26 Gear lever does not return

Weak or broken centraliser spring. Renew the spring.
Gearchange shaft bent or seized. Distortion of the gearchange shaft often occurs if the machine is dropped heavily on the gear lever. Provided that damage is not severe straightening of the shaft is permissible.

Sticking gearchange external linkage. As a result of corrosion or lack of lubrication. Clean and lubricate the relevant components.

27 Gear selection difficult or impossible

Clutch not disengaging fully. See Section 25.

Gearchange shaft bent. This often occurs if the machine is dropped heavily on the gear lever. Straightening of the shaft is permissible if the damage is not too great.

Gearchange arms, pawls or pins worn or damaged. Wear or breakage of any of these items may cause difficulty in selecting one or more gears. Overhaul the selector mechanism.

Gearchange arm spring broken. Renew spring.

Gearchange drum stopper cam or detent plunger damage. Failure, rather than wear, of these items may jam the drum thereby preventing gearchanging. The damaged items must be renewed.

Selector forks bent or seized. This can be caused by dropping the machine heavily on the gearchange lever or as a result of lack of lubrication. Though rare, bending of a shaft can result from a missed gearchange or false selection at high speed.

Selector fork end and pin wear. Pronounced wear of these items and the grooves in the gearchange drum can lead to imprecise selection and, eventually, no selection. Renewal of the worn components will be required.

Structural failure. Failure of any one component of the selector rod and change mechanism will result in improper or fouled gear selection.

28 Jumping out of gear

Detent plunger assembly worn or damaged. Wear of the plunger and the cam with which it locates and breakage of the detent spring can cause imprecise gear selection resulting in jumping out of gear. Renew the damaged components.

Gear pinion dogs worn or damaged. Rounding off the dog edges and the mating recesses in adjacent pinion can lead to jumping out of gear when under load. The gears should be inspected and renewed. Attempting to reprofile the dogs is not recommended.

Selector forks, gearchange drum and pinion grooves worn. Extreme wear of these interconnected items can occur after high mileages especially when lubrication has been neglected. The worn components must be renewed.

Gear pinions, bushes and shafts worn. Renew the worn components.

Bent gearchange shaft. Often caused by dropping the machine on the gear lever.

Gear pinion tooth broken. Chipped teeth are unlikely to cause jumping out of gear once the gear has been selected fully; a tooth which is completely broken off, however, may cause problems in this respect and in any event will cause transmission noise.

29 Overselection

Pawl spring weak or broken. Renew the spring.

Detent plunger worn or broken. Renew the damaged items.

Stopper arm spring worn or broken. Renew the spring.

Gearchange arm stop pads worn. Repairs can be made by welding and reprofiling with a file.

Abnormal engine noise.

30 Knocking or pinking

See Section 19.

31 Piston slap or rattling from cylinder

Cylinder bore/piston clearance excessive. Resulting from wear, partial seizure or improper boring during overhaul. This condition can often be heard as a high, rapid tapping noise when the engine is under little or no load, particularly when power is just beginning to be applied. Reboring to the next correct oversize should be carried out and a new oversize piston fitted.

Connecting rod bent. This can be caused by over-revving, trying to start a very badly flooded engine (resulting in a hydraulic lock in the cylinder) or by earlier mechanical failure such as a dropped valve. Attempts at straightening a bent connecting rod from a high performance engine are not recommended. Careful inspection of the crankshaft should be made before renewing the damaged connecting rod.

Gudgeon pin, piston boss bore or small-end bearing wear or seizure. Excess clearance or partial seizure between normal moving parts of these items can cause continuous or intermittent tapping noises. Rapid wear or seizure is caused by lubrication starvation resulting from an insufficient engine oil level or oilway blockage.

Piston rings worn, broken or sticking. Renew the rings after careful inspection of the piston and bore.

32 Valve noise or tapping from the cylinder head

Valve clearance incorrect. Adjust the clearances with the engine cold.

Valve spring broken or weak. Renew the spring set.

Camshaft or cylinder head worn or damaged. The camshaft lobes are the most highly stressed of all components in the engine and are subject to high wear if lubrication becomes inadequate. The bearing surfaces on the camshaft and cylinder head are also sensitive to a lack of lubrication. Lubrication failure due to blocked oilways can occur, but over-enthusiastic revving before engine warm-up is complete is the usual cause.

Worn camshaft drive components. A rustling noise or light tapping which is not improved (where appropriate) by correct re-adjustment of the cam chain tension can be emitted by a worn cam chain or worn sprockets and chain. If uncorrected, subsequent cam chain breakage may cause extensive damage. The worn components must be renewed before wear becomes too far advanced.

Malfunction of the cam chain tensioner. Seizure of the chain tensioner (semi-automatic type or fully-automatic type, depending on the model) will prevent correct tensioning of the chain.

33 Other noises

Big-end bearing wear. A pronounced knock from within the crankcase which worsens rapidly is indicative of big-end bearing failure as a result of extreme normal wear or lubrication failure. Remedial action in the form of a bottom end overhaul should be taken; continuing to run the engine will lead to further damage including the possibility of connecting rod breakage.

Main bearing failure. Extreme normal wear or failure of the main bearings is characteristically accompanied by a rumble from the crankcase and vibration felt through the frame and footrests. Renew the worn bearings and carry out a very careful examination of the crankshaft.

Crankshaft excessively out of true. A bent crank may result from over-revving or damage from an upper cylinder component or gearbox failure. Damage can also result from dropping the machine on either crankshaft end. Straightening of the crankshaft is not possible in normal circumstances; a replacement item should be fitted.

Engine mounting loose. Tighten all the engine mounting nuts and bolts.

Cylinder head gasket leaking. The noise most often associated with a leaking head gasket is a high pitched squeaking, although any other noise consistent with gas being forced out under pressure from a small orifice can also be emitted. Gasket leakage is often accompanied by oil seepage from around the mating joint or from the cylinder head holding down bolts and nuts. Leakage into the cam chain tunnel or oil return passages will increase crankcase pressure and may cause oil leakage at joints and oil seals. Also, oil contamination will be accelerated. Leakage results from insufficient or uneven tightening of the cylinder head fasteners, or from random mechanical failure. Retightening to the correct torque figure will, at best, only provide a temporary cure. The gasket should be renewed at the earliest opportunity.

Exhaust system leakage. Popping or crackling in the exhaust system, particularly when it occurs with the engine on the overrun, indicates a poor joint either at the cylinder port or at the exhaust pipe/silencer connection. Failure of the gasket or looseness of the clamp should be looked for.

Abnormal transmission noise

34 Clutch noise

Clutch outer drum/friction plate tang clearance excessive.
Clutch outer drum/spacer clearance excessive.
Clutch outer drum/thrust washer clearance excessive.
Primary drive gear teeth worn or damaged.
Clutch shock absorber assembly worn or damaged.

35 Transmission noise

Bearing or bushes worn or damaged. Renew the affected components.

Gear pinions worn or chipped. Renew the gear pinions.

Metal chips jam in gear teeth. This can occur when pieces of metal from any failed component are picked up by a meshing pinion. The condition will lead to rapid bearing wear or early gear failure.

Engine/transmission oil level too low. Top up immediately to prevent damage to gearbox and engine.

Gearchange mechanism worn or damaged. Wear or failure of certain items in the selection and change components can induce mis-selection of gears (see Section 27) where incipient engagement of more than one gear set is promoted. Remedial action, by the overhaul of the gearbox, should be taken without delay.

Worn or damaged bevel gear sets. A whine emitted from either bevel gear set is indicative of improper meshing. This may increase progressively as wear develops or suddenly due to mechanical failure. Drain the lubricant and inspect for metal chips prior to dismantling.

Output shaft joint failure. This can cause vibration and noise. Renew the affected component.

Exhaust smokes excessively

36 White/blue smoke (caused by oil burning)

Piston rings worn or broken. Breakage or wear of any ring, but particularly the oil control ring, will allow engine oil past the

piston into the combustion chamber. Overhaul the cylinder barrel and piston.

Cylinder cracked, worn or scored. These conditions may be caused by overheating, lack of lubrication, component failure or advanced normal wear. The cylinder barrel should be renewed or rebored and the next oversize piston fitted.

Valve oil seal damaged or worn. This can occur as a result of valve guide failure or old age. The emission of smoke is likely to occur when the throttle is closed rapidly after acceleration, for instance, when changing gear. Renew the valve oil seals and, if necessary, the valve guides.

Valve guides worn. See the preceding paragraph.

Engine oil level too high. This increases the crankcase pressure and allows oil to be forced past the piston rings. Often accompanied by seepage of oil at joints and oil seals.

Cylinder head gasket blown between cam chain tunnel or oil return passage. Renew the cylinder head gasket.

Abnormal crankcase pressure. This may be caused by blocked breather passages or hoses causing back-pressure at high engine revolutions.

37 Black smoke (caused by over-rich mixture)

Air filter element clogged. Clean or renew the element.

Main jet loose or too large. Remove the float chamber to check for tightness of the jet. If the machine is used at high altitudes rejetting will be required to compensate for the lower atmospheric pressure.

Cold start mechanism jammed on. Check that the mechanism works smoothly and correctly and that, where fitted, the operating cable is lubricated and not snagged.

Fuel level too high. The fuel level is controlled by the float height which can increase as a result of wear or damage. Remove the float bowl and check the float height. Check also that floats have not punctured; a punctured float will loose buoyancy and allow an increased fuel level.

Float valve needle stuck open. Caused by dirt or a worn valve. Clean the float chamber or renew the needle and, if necessary, the valve seat.

Oil pressure indicator lamp goes on

38 Engine lubrication system failure

Engine oil defective. Oil pump shaft or locating pin sheared off from ingesting debris or seizing from lack of lubrication (low oil level).

Engine oil screen clogged. Change oil and filter and service pickup screen.

Engine oil level too low. Inspect for leak or other problem causing low oil level and add recommended lubricant.

Engine oil viscosity too low. Very old, thin oil, or an improper weight of oil used in engine. Change to correct lubricant.

Camshaft or journals worn. High wear causing drop in oil pressure. Replace cam and/or head. Abnormal wear could be caused by oil starvation at high rpm from low oil level, improper oil weight or type, or loose oil fitting on upper cylinder oil line.

Crankshaft and/or bearings worn. Same problems as paragraph 5. Overhaul lower end.

Relief valve stuck open. This causes the oil to be dumped back into the sump. Repair or replace.

39 Electrical system failure

Oil level switch defective. Check switch according to the procedures in Chapter 6. Replace if defective.

Oil level indicator lamp wiring system defective. Check for pinched, shorted, disconnected or damaged wiring.

Poor handling or roadholding

40 Directional instability

Steering head bearing adjustment too tight. This will cause rolling or weaving at low speeds. Re-adjust the bearings.

Steering head bearing worn or damaged. Correct adjustment of the bearing will prove impossible to achieve if wear or damage has occurred. Inconsistent handling will occur including rolling or weaving at low speed and poor directional control at indeterminate higher speeds. The steering head bearing should be dismantled for inspection and renewed if required. Lubrication should also be carried out.

Bearing races pitted or dented. Impact damage caused, perhaps, by an accident or riding over a pot-hole can cause indentation of the bearing, usually in one position. This should be noted as notchiness when the handlebars are turned. Renew and lubricate the bearings.

Steering stem bent. This will occur only if the machine is subjected to a high impact such as hitting a curb or a pot-hole. The lower yoke/stem should be renewed; do not attempt to straighten the stem.

Front or rear tyre pressures too low.

Front or rear tyre worn. General instability, high speed wobbles and skipping over white lines indicates that tyre renewal may be required. Tyre induced problems, in some machine/tyre combinations, can occur even when the tyre in question is by no means fully worn.

Swinging arm bearings worn or badly adjusted. Difficulties in holding line, particularly when cornering or when changing power settings indicates wear in the swinging arm bearings. The swinging arm bearings should be adjusted; if this fails to effect a cure the swinging arm should be removed from the machine and the bearings renewed.

Swinging arm flexing. The symptoms given in the preceding paragraph will also occur if the swinging arm fork flexes badly. This can be caused by structural weakness as a result of corrosion, fatigue or impact damage, or because the rear wheel spindle is slack.

Wheel bearings worn. Renew the worn bearings.

Tyres unsuitable for machine. Not all available tyres will suit the characteristics of the frame and suspension, indeed, some tyres or tyre combinations may cause a transformation in the handling characteristics. If handling problems occur immediately after changing to a new tyre type or make, revert to the original tyres to see whether an improvement can be noted. In some instances a change to what are, in fact, suitable tyres may give rise to handling deficiences. In this case a thorough check should be made of all frame and suspension items which affect stability.

Improperly balanced suspension components. Where the machine is fitted with variable spring rates and/or damping in the front and rear suspension systems ensure that the variables are set within the ranges indicated by the manufacturer.

41 Steering bias to left or right

Rear wheel out of alignment. A bent rear wheel spindle will misalign the wheel in the swinging arm.

Wheels out of alignment. This can be caused by impact damage to the frame, swinging arm, wheel spindles or front forks. Although occasionally a result of material failure or corrosion it is usually as a result of a crash.

Front forks twisted in the steering yokes. A light impact, for instance with a pot-hole or low curb, can twist the fork legs in the steering yokes without causing structural damage to the fork legs or the yokes themselves. Re-alignment can be made by loosening the yoke pinch bolts, wheel spindle and mudguard bolts. Re-align the wheel with the handlebars and tighten the bolts working upwards from the wheel spindle. This action should be carried out only when there is no chance that structural damage has occurred.

42 Handlebar vibrates or oscillates

Tyres worn or out of balance. Either condition, particularly in the front tyre, will promote shaking of the fork assembly and thus the handlebars. A sudden onset of shaking can result if a balance weight is displaced during use.

Tyres badly positioned on the wheel rims. A moulded line on each wall of a tyre is provided to allow visual verification that the tyre is correctly positioned on the rim. A check can be made by rotating the tyre; any misalignment will be immediately obvious.

Wheel rims warped or damaged. Inspect the wheels for runout as described in Chapter 5.

Swinging arm bearings worn. Renew the bearings.

Wheel bearings worn. Renew the bearings.

Steering head bearings incorrectly adjusted. Vibration is more likely to result from bearings which are too loose rather than too tight. Re-adjust the bearings.

Loose fork component fasteners. Loose nuts and bolts holding the fork legs, wheel spindle, mudguards or steering stem can promote shaking at the handlebars. Fasteners on running gear such as the forks and suspension should be check tightened occasionally to prevent dangerous looseness of components occurring.

Engine mounting bolts loose. Tighten all fasteners.

43 Poor front fork performance

Damping fluid level incorrect. If the fluid level is too low poor suspension control will occur resulting in a general impairment of roadholding and early loss of tyre adhesion when cornering and braking. Too much oil is unlikely to change the fork characteristics unless severe overfilling occurs when the fork action will become stiffer and oil seal failure may occur.

Damping oil viscosity incorrect. The damping action of the fork is directly related to the viscosity of the damping oil. The lighter the oil used, the less will be the damping action imparted. For general use, use the recommended viscosity of oil, changing to a slightly higher or heavier oil only when a change in damping characteristic is required. Overworked oil, or oil contaminated with water which has found its way past the seals, should be renewed to restore the correct damping performance and to prevent bottoming of the forks.

Air pressure incorrect. On models with air assisted forks an imbalance in the pressure between the fork legs can give rise to poor fork performance. Similarly, if the air pressure is outside the recommended range problems can occur.

Damping components worn or corroded. Advanced normal wear of the fork internals is unlikely to occur until a very high mileage has been covered. Continual use of the machine with damaged oil seals which allows the ingress of water, or neglect, will lead to rapid corrosion and wear. Dismantle the forks for inspection and overhaul.

Weak fork springs. Progressive fatigue of the fork springs, resulting in a reduced spring free length, will occur after extensive use. This condition will promote excessive fork dive under braking, and in its advanced form will reduce the at-rest extended length of the forks and thus the fork geometry. Renewal of the springs as a pair is the only satisfactory course of action.

Bent stanchions or corroded stanchions. Both conditions

will prevent correct telescoping of the fork legs, and in an advanced state can cause sticking of the fork in one position. In a mild form corrosion will cause stiction of the fork thereby increasing the time the suspension takes to react to an uneven road surface. Bent fork stanchions should be attended to immediately because they indicate that impact damage has occurred, and there is a danger that the forks will fail with disastrous consequences.

44 Front fork judder when braking (see also Section 56)

Wear between the fork stanchions and the fork legs. Renewal of the affected components is required.
Slack steering head bearings. Re-adjust the bearings.
Warped brake disc. If irregular braking action occurs fork judder can be induced in what are normally serviceable forks. Renew the damaged brake components.

45 Poor rear suspension performance

Rear suspension unit damper worn out or leaking. The damping performance of most rear suspension units falls off with age. This is a gradual process, and thus may not be immediately obvious. Indications of poor damping include hopping of the rear end when cornering or braking, and a general loss of positive stability. See Chapter 4.
Weak rear springs. If the suspension unit springs fatigue they will promote excessive pitching of the machine and reduce the ground clearance when cornering. Although replacement springs are available separately from the rear suspension damper unit it is probable that if spring fatigue has occurred the damper units will also require renewal.
Swinging arm flexing or bearings worn. See Sections 40 and 41.
Bent suspension unit damper rod. This is likely to occur only if the machine is dropped or if seizure of the piston occurs. If either happens the suspension units should be renewed as a pair.

Abnormal frame and suspension noise

46 Front end noise

Oil level low or too thin. This can cause a 'spurting' sound and is usually accompanied by irregular fork action.
Spring weak or broken. Makes a clicking or scraping sound. Fork oil will have a lot of metal particles in it.
Steering head bearings loose or damaged. Clicks when braking. Check, adjust or replace.
Fork clamps loose. Make sure all fork clamp pinch bolts are tight.
Fork stanchion bent. Good possibility if machine has been dropped. Repair or replace tube.

47 Rear suspension noise

Fluid level too low. Leakage of a suspension unit, usually evident by oil on the outer surfaces, can cause a spurting noise. The suspension units should be renewed as a pair.
Defective rear suspension unit with internal damage. Renew the suspension units as a pair.

Brake problems

48 Brakes are spongy or ineffective

Air in brake circuit. This is only likely to happen in service due to neglect in checking the fluid level or because a leak has developed. The problem should be identified and the brake system bled of air.
Pad worn. Check the pad wear against the wear lines provided and renew the pads if necessary.
Contaminated pads. Cleaning pads which have been contaminated with oil, grease or brake fluid is unlikely to prove successful; the pads should be renewed.
Pads glazed. This is usually caused by overheating. The surface of the pads may be roughened using glass-paper or a fine file.
Brake fluid deterioration. A brake which on initial operation is firm but rapidly becomes spongy in use may be failing due to water contamination of the fluid. The fluid should be drained and then the system refilled and bled.
Master cylinder seal failure. Wear or damage of master cylinder internal parts will prevent pressurisation of the brake fluid. Overhaul the master cylinder unit.
Caliper seal failure. This will almost certainly be obvious by loss of fluid, a lowering of fluid in the master cylinder reservoir and contamination of the brake pads and caliper. Overhaul the caliper assembly.
Brake lever or pedal improperly adjusted. Adjust the clearance between the lever end and master cylinder plunger to take up lost motion, as recommended in Routine maintenance.

49 Brakes drag

Disc warped. The disc must be renewed.
Caliper piston, caliper or pads corroded. The brake caliper assembly is vulnerable to corrosion due to water and dirt, and unless cleaned at regular intervals and lubricated in the recommended manner, will become sticky in operation.
Piston seal deteriorated. The seal is designed to return the piston in the caliper to the retracted position when the brake is released. Wear or old age can affect this function. The caliper should be overhauled if this occurs.
Brake pad damaged. Pad material separating from the backing plate due to wear or faulty manufacture. Renew the pads. Faulty installation of a pad also will cause dragging.
Wheel spindle bent. The spindle may be straightened if no structural damage has occurred.
Brake lever or pedal not returning. Check that the lever or pedal works smoothly throughout its operating range and does not snag on any adjacent cycle parts. Lubricate the pivot if necessary.
Twisted caliper support bracket. This is likely to occur only after impact in an accident. No attempt should be made to re-align the caliper; the bracket should be renewed.

50 Brake lever or pedal pulsates in operation

Disc warped or irregularly worn. The disc must be renewed.
Wheel spindle bent. The spindle may be straightened provided no structural damage has occurred.

51 Brake noise

Brake squeal. This can be caused by the omission or incorrect installation of the anti-squeal shim fitted to the rear of one pad. The arrow on the shim should face the direction of wheel normal rotation. Squealing can also be caused by dust on the pads, usually in combination with glazed pads, or other contamination from oil, grease, brake fluid or corrosion. Persistent squealing which cannot be traced to any of the normal causes can often be cured by applying a thin layer of high temperature silicone grease to the rear of the pads. Make absolutely certain that no grease is allowed to contaminate the braking surface of the pads.

Glazed pads. This is usually caused by high temperatures or contamination. The pad surfaces may be roughened using glass-paper or a fine file. If this approach does not effect a cure the pads should be renewed.

Disc warped. This can cause a chattering, clicking or intermittent squeal and is usually accompanied by a pulsating brake lever or pedal or uneven braking. The disc must be renewed.

Brake pads fitted incorrectly or undersize. Longitudinal play in the pads due to omission of the locating springs (where fitted) or because pads of the wrong size have been fitted will cause a single tapping noise every time the brake is operated. Inspect the pads for correct installation and security.

52 Brake induced fork judder

Worn front fork stanchions and legs, or worn or badly adjusted steering head bearings. These conditions, combined with uneven or pulsating braking as described in Sections 50 and 51 will induce more or less judder when the brakes are applied, dependent on the degree of wear and poor brake operation. Attention should be given to both areas of malfunction. See the relevant Sections.

Electrical problems

53 Battery dead or weak

Battery faulty. Battery life should not be expected to exceed 3 to 4 years, particularly where a starter motor is used regularly. Gradual sulphation of the plates and sediment deposits will reduce the battery performance. Plate and insulator damage can often occur as a result of vibration. Complete power failure, or intermittent failure, may be due to a broken battery terminal. Lack of electrolyte will prevent the battery maintaining charge.

Battery leads making poor contact. Remove the battery leads and clean them and the terminals, removing all traces of corrosion and tarnish. Reconnect the leads and apply a coating of petroleum jelly to the terminals.

Load excessive. If additional items such as spot lamps, are fitted, which increase the total electrical load above the maximum alternator output, the battery will fail to maintain full charge. Reduce the electrical load to suit the electrical capacity.

Regulator/rectifier failure.

Alternator generating coils open-circuit or shorted.

Charging circuit shorting or open circuit. This may be caused by frayed or broken wiring, dirty connectors or a faulty ignition switch. The system should be tested in a logical manner. See Section 60.

54 Battery overcharged

Rectifier/regulator faulty. Overcharging is indicated if the battery becomes hot or it is noticed that the electrolyte level falls repeatedly between checks. In extreme cases the battery will boil causing corrosive gases and electrolyte to be emitted through the vent pipes.

Battery wrongly matched to the electrical circuit. Ensure that the specified battery is fitted to the machine.

55 Total electrical failure

Fuse blown. Check the main fuse. If a fault has occurred, it must be rectified before a new fuse is fitted.

Battery faulty. See Section 53.

Earth failure. Check that the frame main earth strap from the battery is securely affixed to the frame and is making a good contact.

Ignition switch or power circuit failure. Check for current flow through the battery positive lead (red) to the ignition switch. Check the ignition switch for continuity.

56 Circuit failure

Cable failure. Refer to the machine's wiring diagram and check the circuit for continuity. Open circuits are a result of loose or corroded connections, either at terminals or in-line connectors, or because of broken wires. Occasionally, the core of a wire will break without there being any apparent damage to the outer plastic cover.

Switch failure. All switches may be checked for continuity in each switch position, after referring to the switch position boxes incorporated in the wiring diagram for the machine. Switch failure may be a result of mechanical breakage, corrosion or water.

Fuse blown. Refer to the wiring diagram to check whether or not a circuit fuse is fitted. Replace the fuse, if blown, only after the fault has been identified and rectified.

57 Bulbs blowing repeatedly

Vibration failure. This is often an inherent fault related to the natural vibration characteristics of the engine and frame and is, thus, difficult to resolve. Modifications of the lamp mounting, to change the damping characteristics may help.

Intermittent earth. Repeated failure of one bulb, particularly where the bulb is fed directly from the generator, indicates that a poor earth exists somewhere in the circuit. Check that a good contact is available at each earthing point in the circuit.

Reduced voltage. Where a quartz-halogen bulb is fitted the voltage to the bulb should be maintained or early failure of the bulb will occur. Do not overload the system with additional electrical equipment in excess of the system's power capacity and ensure that all circuit connections are maintained clean and tight.

YAMAHA XJ 650/750

Check list

UK models

Weekly or every 400 miles (600 km)

1 Check the engine/transmission oil level
2 Check control and cable adjustment
3 Check the tyre pressures and condition
4 Make a general safety check of the machine
5 Check the lights, horn(s), indicators and speedometer for correct operation

Monthly or every 1000 miles (1500 km)

1 Lubricate the control cables, controls and pivot points
2 Check the battery electrolyte level
3 Check the brake pads for wear, and where applicable fluid level and brake adjustment
4 Check the operation of the steering and suspension
5 Clean the air filter element

2 monthly or every 2000 miles (3000 km)

1 Clean and adjust the spark plugs
2 Check carburettor adjustment and synchronisation
3 Check the final drive oil level

3 monthly or every 3000 miles (4500 km)

1 Check and adjust the cam chain tension*
2 Change the engine/transmission oil

4 monthly or every 4000 miles (6000 km)

1 Check the cylinder compression
2 Check the ignition timing

6 monthly or every 6000 miles (10 000 km)

1 Check and adjust the valve clearances
2 Change the engine/transmission oil
3 Change the final drive oil

Yearly or every 10 000 miles (15 000 km)

1 Change the front fork oil
2 Dismantle, check and regrease the steering head bearings
3 Dismantle, check and regrease the swinging arm bearings

*Applies only to machines with a manually adjusted cam chain tensioner XJ650 4KO – (UK)

US models

Weekly or every 100 miles (150 km)

1 Check the engine/transmission oil level
2 Check control and cable adjustment
3 Check the tyre pressures and condition
4 Make a general safety check of the machine
5 Check the lights, horn(s), indicators and speedometer for correct operation

6 monthly, or every 2500 miles (4000 km)

1 Change the engine/transmission oil
2 Check the brake pads for wear, and where applicable fluid level and brake adjustment
3 Check and adjust the clutch cable
4 Lubricate the control cables, controls and pivot points
5 Check the operation of the steering and suspension
6 Check the condition of the wheel bearings
7 Check the battery electrolyte level
8 Clean the fuel filters and check the fuel system
9 Clean and adjust the spark plugs
10 Inspect the exhaust system
11 Check carburettor adjustment and synchronisation

Yearly or every 5000 miles (8000 km)

1 Check and adjust the cam chain tension*
2 Check and adjust the valve clearances
3 Check the crankcase ventilation system
4 Check the condition of the fuel pipes
5 Renew the oil filter
6 Change the final drive oil
7 Clean the air filter element

18 monthly or every 7500 miles (12 000 km)

1 Renew the spark plugs

2 yearly or every 10 000 miles (16 000 km)

1 Remove, check and regrease the wheel bearings
2 Change the front fork oil
3 Dismantle, check and regrease the steering head bearings
4 Dismantle, check and regrease the swinging arm bearings

*Applies only to models fitted with a manually adjusted cam chain tensioner (XJ650 G, H, LH and RJ models)

Adjustment data

Tyre pressures – tyres cold	Front – psi (kg/cm²)	Rear – psi (kg/cm²)
XJ650 (UK), XJ650 RJ:		
Up to 198lb (90kg)	26 (1.8)	28 (2.0)
198 – 331lb (90 – 150kg)	28 (2.0)	33 (2.3)
331 – 478lb (150 – 217kg)	28 (2.0)	40 (2.8)
High speed riding	33 (2.3)	36 (2.5)
XJ650 G, H, LH, J:		
Up to 198lb (90kg)	26 (1.8)	28 (2.0)
198 – 353lb (90 – 160kg)	28 (2.0)	33 (2.3)
353 – 507lb (160 – 230kg)	28 (2.0)	40 (2.8)
High speed riding	33 (2.3)	36 (2.5)
XJ750 (UK), XJ750 RH, RJ:		
Up to 198lb (90kg)	26 (1.8)	28 (2.0)
198 – 474lb (90 – 215kg)	28 (2.0)	33 (2.3)
High speed riding	33 (2.3)	36 (2.5)
XJ750 J:		
Up to 198lb (90kg)	26 (1.8)	28 (2.0)
198 – 507lb (90 – 230kg)	28 (2.0)	33 (2.3)
High speed riding	28 (2.0)	33 (2.3)

Valve clearances (cold)
XJ650(UK) and
XJ650 J

Inlet	0.16-0.20 mm (0.006-0.008 in)
Exhaust	0.16-0.20 mm (0.006-0.008 in)

All other models

Inlet	0.11-0.15 mm (0.004-0.006 in)
Exhaust	0.16-0.20 mm (0.006-0.008 in)

Spark plug gap	0.7-0.8 mm (0.028-0.032 in)
Spark plug type	NGK BP7ES or ND W22EP
Idle speed	1050 ± 50 rpm

Recommended lubricants

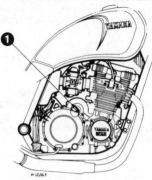

Component	Quantity	Type/viscosity
1 Engine/ transmission		
XJ650 (11N) XJ650 RJ	2.65 lit (5.6/47 US/Imp pint) – at oil change 2.95 lit (6.2/5.1 US/Imp pint) – at oil and filter change	Not below 5°C (40°F); SAE 20W/40 'SE' engine oil
All other XJ650 models	2.35 lit (5.0/4.1 US/Imp pint) – at oil change 2.65 lit (5.6/4.7 US/Imp pint) – at oil and filter change	Not above 15°C (59°F); SAE 10W/30 'SE' engine oil
All XJ750 models	2.50 lit (5.2/4.4 US/Imp pint) – at oil change 2.80 lit (5.9/4.9 US/Imp pint) – at oil and filter change	
2 Final drive box	0.20 lit (0.42/0.35 US/Imp pint)	SAE 80 API 'GL-4' or SAE 80/90 Hypoid gear oil
3 Front forks (per leg)		
XJ650(UK), XJ650RH, RJ XJ650J	236cc (7.98/8.31 US/Imp fl oz) 278cc (9.40 US fl oz)	SAE 10W fork oil or SAE 10W/30 motor oil
All other XJ650 models	262cc (9.24 US fl oz)	
XJ750J XJ750(UK) XJ750RH and RJ	257cc (8.7 US fl oz) 312cc (11.0 Imp fl oz) 309cc (10.5 US fl oz)	SAE 20W fork oil
4 Swinging arm	As required	High melting-point grease
5 Wheel bearings	As required	As above
6 Steering head bearings	As required	As above
7 General lubrication	As required	Light machine oil
8 Disc brakes	As required	Hydraulic brake fluid DOT 3 (US) or SAE J1703 (UK)

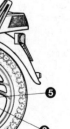

ROUTINE MAINTENANCE GUIDE

Routine maintenance

Refer to Chapter 7 for information on the 1983 US models

Periodic routine maintenance is a continuous process which should commence immediately the machine is used. The object is to maintain all adjustments and to diagnose and rectify minor defects before they develop into more extensive, and often more expensive, problems.

It follows that if the machine is maintained properly, it will both run and perform with optimum efficiency, and be less prone to unexpected breakdowns. Regular inspection of the machine will show up any parts which are wearing, and with a little experience, it is possible to obtain the maximum life from any one component, renewing it when it becomes so worn that it is liable to fail.

Regular cleaning can be considered as important as mechanical maintenance. This will ensure that all the cycle parts are inspected regularly and are kept free from accumulations of road dirt and grime.

The various maintenance tasks are described under their respective mileage and calendar headings, and are accompanied by diagrams and photographs where pertinent.

It should be noted that the intervals between each maintenance task serve only as a guide. As the machine gets older, or if it is used under particularly arduous conditions, it is advisable to reduce the period between each check.

For ease of reference, most service operations are described in detail under the relevant heading. However, if further general information is required, this can be found under the pertinent Section heading and Chapter in the main text.

Other than the valve shim removal tool mentioned in Section 11 no special tools are required for routine maintenance; a good selection of general workshop tools is, however, essential. Included in the tools must be a range of metric ring or combination spanners, a selection of crosshead screwdrivers, and two pairs of circlip pliers, one external opening and the other internal opening. Additionally, owing to the extreme tightness of most casing screws on Japanese machines, an impact screwdriver, together with a choice of large or small cross-head screw bits, is absolutely indispensable. This is particularly so if the engine has not been dismantled since leaving the factory.

Cleaning the machine

Keeping the motorcycle clean should be considered as an important part of the routine maintenance, to be carried out whenever the need arises. A machine cleaned regularly will not only succumb less speedily to the inevitable corrosion of external surfaces, and hence maintain its market value, but will be far more approachable when the time comes for maintenance or service work. Furthermore, loose or failing components are more readily spotted when not partially obscured by a mantle of road grime and oil.

Surface dirt should be removed using a sponge and warm, soapy water; the latter being applied copiously to remove the particles of grit which might otherwise cause damage to the paintwork and polished surfaces.

Oil and grease is removed most easily by the application of a cleaning solvent such as 'Gunk' or 'Jizer'. The solvent should be applied when the parts are still dry and worked in with a stiff brush. Large quantities of water should be used when rinsing off, taking care that water does not enter the carburettors, air cleaners or electrics.

Application of a wax polish to the cycle parts and a good chrome cleaner to the chrome parts will give a good finish. Always wipe the machine down if used in the wet. There is less chance of water getting into control cables if they are regularly lubricated, which will prevent stiffness of action.

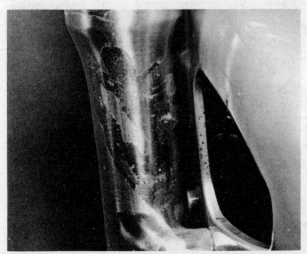

Peeling lacquer on alloy components requires stripping and relacquering to prevent corrosion

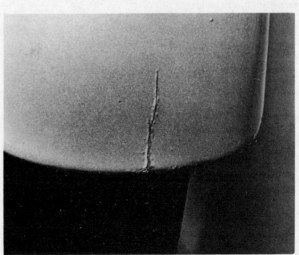

This split has been caused by vibration and will spread unless it is repaired

Routine Maintenance schedule – US models

Weekly, or every 100 miles (150 km)

	See Section
Check engine/transmission oil level	1
Check control and cable adjustments	2,3,4,5,9
Check tyre pressures and condition	6
Safety check	7
Legal check	8

6 monthly, or every 2500 miles (4000 km)

Change engine/transmission oil	14
Check front disc brake pads	15
Check front brake fluid level	16,17
Check rear brake adjustment	2
Check front brake lever adjustment – handlebar mounted master cylinder	3
Check front brake cable adjustment – remote mounted master cylinder	4
Check and adjust the clutch cable	5
Lubricate control cables	18
Lubricate controls and pivots	19
Check operation of steering and suspension	20
Check wheel bearing condition	21
Check battery electrolyte level	22
Checking and cleaning the fuel system	23
Check, clean and adjust spark plugs	24
Check the exhaust system condition	25
Check carburettor adjustment and synchronisation	26

Yearly, or every 5000 miles (8000 km)

Check and adjust cam chain tension*	10
Check and adjust valve clearances	11
Check crankcase ventilation system	12
Check condition of fuel pipes	13
Renew oil filter	27
Change final drive oil	28
Clean air filter element	29

18 monthly or every 7500 miles (12 000 km)

Renew the spark plugs	–

2 yearly, or every 10 000 miles (16 000 km)

Remove, check and grease wheel bearings	33
Change front fork oil	34,35,36,37
Dismantle, check and regrease steering head bearings	38
Dismantle, check and regrease swinging arm bearings	39

Applies only to models fitted with manually adjusted cam chain tensioner (XJ650 G, H, LH and RJ models).

Routine Maintenance schedule – UK models

Weekly or every 400 miles (600 km)

	See section
Check engine/transmission oil level	1
Check control and cable adjustments	2,3,4 and 5
Check tyre condition and pressures	6
Safety check	7
Legal check	8

Monthly, or every 1000 miles (1500 km)

Lubricate control cables	18
Lubricate controls and pivots	19
Check battery condition	22

Check front brake pad condition	15
Check front brake fluid level	16,17
Check front brake lever adjustment	3,4
Check rear brake adjustment	2
Check operation of steering and suspension	20
Check and clean the air filter	29

2 monthly, or every 2000 miles (3000 km)

Check, clean and adjust spark plugs	24
Check carburettors adjustment and synchronisation	26
Check final drive oil level	30

3 monthly, or every 3000 miles (4500 km)

Check and adjust cam chain tension*	10
Change engine/transmission oil	14

4 monthly, or every 4000 miles (6000 km)

Check cylinder compression	31
Check the ignition timing	32

6 monthly, or every 6000 miles (10 000 km)

Check and adjust valve clearances	11
Change engine/transmission oil and filter	27
Change final drive oil	28

Yearly, or every 10 000 miles (15 000 km)

Change the front fork oil	34–37
Dismantle, check and regrease swinging arm bearings	39
Dismantle, check and regrease steering head bearings	38

Applies only to machines with manually adjusted cam chain tensioner – XJ650 – 4KO (UK)

1 Checking the engine/transmission oil level

The oil level in the sump is monitored by a float-type level switch which controls a warning lamp or LCD, depending on the model concerned. If the oil level drops far below maximum, the lamp or LCD will tend to flash on under acceleration or braking because the oil in the sump is forced away from the switch giving a temporary low level reading. This can prove annoying, and can be minimised by keeping the oil level to maximum. This is indicated by the upper line in the oil sight glass on the side of the engine unit.

Top up the crankcase using oil of the specified grade

2 Rear brake adjustment – except XJ750 J

The brake pedal height can be adjusted in relation to that of the footrest by altering the position of the stop bolt. Slacken the adjuster lock nut and screw the stop bolt in or out until the desired position has been found, then, tighten the locknut. As a guide, Yamaha recommend that the pedal height is set about 1.6 in (40 mm) below the top of the footrest. Note that brake pedal free play must be checked after this adjustment.

Brake pedal free play is adjusted by moving the adjuster nut at the rear of the brake rod to give about 0.8 – 1.2 in (20 – 30 mm) of movement before the brake operates. Once set, check that the brake light comes on as the brake begins to operate. If necessary the point at which the light comes on can be adjusted by altering the position of the switch body.

3 Front brake lever: free play adjustment – all models except XJ750(UK), XJ750 RH and RJ

All models equipped with handlebar-mounted master cylinders have provision for lever free play adjustment. The adjustment screw and locknut are accessible after the rubber gaiter has been slid clear of the lever stock. Lever free play is measured at the tip of the lever and should be 5 – 8 mm (0.2 – 0.3 in) before the lever starts to move the master cylinder piston. Adjustment is effected by slackening the locknut, moving the adjustment screw by the required amount, and tightening the locknut to secure it.

4 Front brake cable: free play adjustment – XJ750(UK), XJ750 RH and RJ

On the above machines, a cable is fitted between the front brake lever and the remote master cylinder. A knurled adjuster is incorporated in the lever stock to facilitate adjustment for free play. This is measured between the lever stock and blade, and should be 1 – 2 mm (0.04 – 0.08 in). Note that the adjuster has a detent device which prevents the adjuster from moving during normal riding. Check that this has clicked home during changes to the setting.

5 Clutch adjustment

Periodic clutch cable adjustment will be necessary to compensate for stretch in the cable and for the gradual wear of the clutch friction plates. Start by slackening the lever adjuster fully to give maximum free play. Using the adjuster at the lower end of the cable, set the cable free play to 2 – 3 mm (0.08 – 0.12 in) measured between the lever stock and blade. Subsequent adjustments can now be made at the handlebar end of the cable.

6 Checking tyre pressures and condition

The tyre pressures should always be checked when cold, preferably after the machine has stood overnight. Whenever the machine is ridden the temperature inside the tyre, and thus the air pressure, increases. It follows that any pressure measurement made after a run will be inaccurate by several psi. It is recommended that a small pocket type pressure gauge is purchased and carried on the machine. There is inevitably some variation between gauges found on garage forecourts, and to be sure of consistent readings it is better to stick to a known gauge. Remember that apparently small discrepancies in pressure can have a tangible or even dangerous effect on handling and roadholding. The pressure settings shown below are those given for the original tyre fitments. If different brands or sizes of tyre have been fitted, check the tyre supplier or manufacturer for the correct pressure recommendation for your machine, and mark the owners handbook and this manual for future reference. The figures quoted refer to the machine with various loadings. These include the rider, passenger and any luggage carried.

The rear brake adjuster

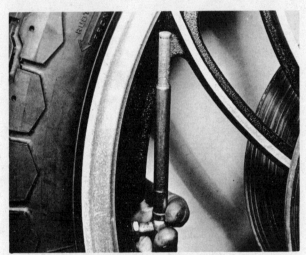
Check tyre pressures using an accurate gauge

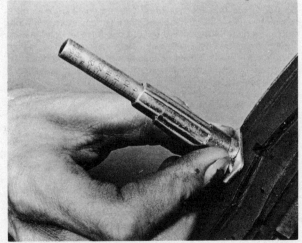
Tread depth can be checked as shown

In the case of this tyre, a tread depth gauge was unnecessary. This tyre is dangerous, and in many areas, illegal

Tyre pressures – tyres cold XJ650 (UK), XJ650 RJ:	Front – psi (kg/cm²)	Rear – psi (kg/cm²)
Up to 198lb (90kg)	26 (1.8)	28 (2.0)
198 – 331lb (90 – 150kg)	28 (2.0)	33 (2.3)
331 – 478lb (150 – 217kg)	28 (2.0)	40 (2.8)
High speed riding	33 (2.3)	36 (2.5)
XJ650 G, H, LH, J:		
Up to 198lb (90kg)	26 (1.8)	28 (2.0)
198 – 353lb (90 – 160kg)	28 (2.0)	33 (2.3)
353 – 507lb (160 – 230kg)	28 (2.0)	40 (2.8)
High speed riding	33 (2.3)	36 (2.5)
XJ750 (UK), XJ750 RH, RJ:		
Up to 198lb (90kg)	26 (1.8)	28 (2.0)
198 – 474lb (90 – 215kg)	28 (2.0)	33 (2.3)
High speed riding	33 (2.3)	36 (2.5)
XJ750 J:		
Up to 198lb (90kg)	26 (1.8)	28 (2.0)
198 – 507lb (90 – 230kg)	28 (2.0)	33 (2.3)
High speed riding	28 (2.0)	33 (2.3)

Each time that the tyre pressures are checked, give each tyre a close visual examination and remove any stones or other objects trapped between the tread. Look for cuts or abrasions on the tread and tyre sidewalls. Note that any such defect warrants full investigation and may require tyre renewal. It cannot be stressed too highly that the rider's and passenger's lives depend on the tyres being in a safe condition. It is possible for serious damage to the tyre carcass to show no outward signs, so if an impact with a kerb or an object in the road has occurred **make sure** that the tyre is safe. This may necessitate tyre removal and recourse to a tyre expert, but do it. Internally damaged tyres can and do fail suddenly, and the results can be fatal.

Note that the tyres are of the tubeless type on some models and this will minimise the risk of sudden deflation. The drawback of tubeless tyres is that they are difficult to remove and fit at home (see Chapter 5, Section 26). If a nail or other sharp object can be seen to have punctured the tyre it must be investigated immediately, even if no pressure loss has been noticed. For most people, this will entail removal of the wheel and a trip to a tyre specialist or a Yamaha dealer. If so, leave the object in the tyre to identify the position of the suspected puncture.

There are a number of tyre sealers or puncture preventatives on the market, but their use should be considered with caution. Opinion is divided as to whether they are safe, and a number of tyre manufacturers have issued clear warnings against their use. Although the tyre companies cannot be considered unbiased in this matter it is generally agreed that no sealer can be considered a safe permanent repair. Use your discretion and check local laws before using any of these products.

Finally, spin each wheel and check for bearing play by rocking the wheel against the spindle axis. If movement is discovered it is likely that the bearing will be in need of renewal, though very slight movement is almost always evident. If in doubt remove the bearings for cleaning and examination as described in Chapter 5, Sections 21 or 23.

7 Safety check

Give the machine a close visual inspection, checking for loose nuts and fittings, frayed control cables etc. Check the tyres for damage, especially splitting on the sidewalls. Remove any stones or other objects caught between the treads. This is particularly important on the front tyre, where rapid deflation due to penetration of the inner tube will almost certainly cause total loss of control.

8 Legal check

Ensure that the lights, horn and trafficators function correctly, also the speedometer.

9 Control adjustments – XJ750 J only

The various controls on the XJ750 J are adjustable to suit various riding positions. The relevant adjustment procedures are described in detail in the following Sections of Chapter 4:

Footrest adjustment	Section 29
Rear brake pedal	Section 31
Gearchange linkage	Section 34
Handlebar adjustment	Section 35

10 Cam chain tensioner adjustment

Important note: This section applies *only* to those models which have non-automatic chain tensioners. These are listed below. Do not disturb the automatic tensioner assembly on the remaining models, since, once it has been disturbed, it must be removed completely and refitted from scratch.

The models fitted with non-automatic tensioners are the UK model XJ650 (4KO), and the US models XJ650 G, H, LH and RJ.

The camshaft drive chain tension should be checked at the intervals specified above and whenever the camshaft drive mechanism appears to be unusually noisy. It should be noted that the engine cannot run properly if there is excessive play in the camshaft drive, because the valve timing accuracy will be lost.

Remove the inspection cover at the left-hand crankshaft end to expose the ignition rotor assembly. Locate the 'C' mark on the rotor periphery, and turn the crankshaft anti-clockwise by means of the large square at the centre until the mark aligns with the fixed pointer. Moving to the tensioner mechanism at the rear of the engine unit, slacken the adjuster locknut and bolt to free the spring loaded plunger. The chain will be tensioned automatically as the plunger finds its own level. Secure the bolt and locknut to hold the setting. On no account over-tighten the bolt or locknut because this will only succeed in shearing the bolt. The appropriate torque setting is 4.3 lbf ft (0.6 kgf m) for the bolt and 6.5 lbf ft (0.9 kgf m) for the locknut. Leave the inspection cover off until the ignition timing has been checked as described below.

Prise adjustment shim out of cam follower ...

... to reveal size marking on underside

11 Valve clearance adjustment

The valve clearances on the XJ650 and 750 models are set by fitting hardened steel pads of various thicknesses between the cam followers and lobes. Whilst this results in a rather complicated adjustment sequence, it does allow the engine to run for long periods without the need for adjustment. The valve clearances should be checked at the intervals specified above or whenever the valve gear appears to be unusually noisy.

To gain access to the cylinder head area a certain amount of preliminary dismantling will be necessary. Start by lifting the dualseat to gain access to the rear of the fuel tank. Check that the fuel tap is set to the 'On' or 'Res' position and prise off the fuel feed pipe and vacuum pipe at the carburettors. Where appropriate, disconnect the fuel level sender wire.

The rear of the tank is secured by a single bolt or a flat retaining clip, depending on the model. Release the bolt or pull off the clip, then lift the rear of the tank and pull it back to free the buffer rubbers at the front. Remove the tank and place it to one side, preferably on a soft surface to protect the paintwork.

Disconnect the cold start (choke) cable and route it clear of the cylinder head cover. Pull off the spark plug leads and lodge them clear of the engine. Remove the cylinder head cover holding bolts and remove the cover. If it sticks to the gasket face, tap around the joint using a soft-faced mallet. Do not use excessive force; the object is to break the seal, not the alloy castings. Remove the inspection cover from the left-hand end of the crankcase to expose the timing plate assembly. Note that the timing plate incorporates a large square at its centre. This is provided to allow the crankshaft to be rotated without risk of damage.

Before checking the valve clearances, make a rough plan sketch of the cylinder head so that a note of each clearance can be made against the relevant valve. The clearance of each valve should be checked with the appropriate cam lobe at 180° from the cam follower. Check the gap between the cam's base circle and the adjustment shim and note the reading. Repeat this sequence with the remaining valves. The valve clearances are as shown below.

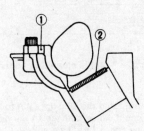

Valve adjustment shim removal

1 *Holding tool*   2 *Shim*

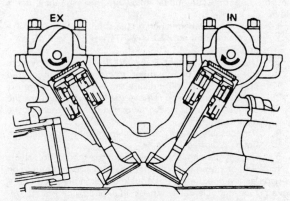

Correct direction of camshaft rotation with holding tool in place

thickness which will restore the correct clearance. This operation entails the use of a special Yamaha tool, part number 90890-01245. It is possible to fabricate a suitable substitute, but in view of the fact that it will be required fairly frequently it is probably as well to obtain the correct item. It is useful to note that an identical tool is used on the XS750, 850 and 1100 models.

Turn the crankshaft until the valve in question is fully open, having first positioned the slot in the cam follower to point away from the holding tool. Fit the tool in position and retain it with one of the cylinder head cover screws. The crankshaft can now be rotated so that the cam lobe moves clear of the cam

XJ650(UK) and XJ650 J

Inlet	0.16 – 0.20 mm (0.006 – 0.008 in)
Exhaust	0.16 – 0.20 mm (0.006 – 0.008 in)

All other models

Inlet	0.11 – 0.15 mm (0.004 – 0.006 in)
Exhaust	0.16 – 0.20 mm (0.006 – 0.008 in)

If any of the clearances are outside these limits it will be necessary to remove the old pad and fit a new one of a

MEASURED CLEARANCE	INTAKE/EXHAUST — INSTALLED PAD NUMBER*																								
---	200	205	210	215	220	225	230	235	240	245	250	255	260	265	270	275	280	285	290	295	300	305	310	315	320
0.00 ~ 0.05				200	205	210	215	220	225	230	235	240	245	250	255	260	265	270	275	280	285	290	295	300	305
0.06 ~ 0.10			200	205	210	215	220	225	230	235	240	245	250	255	260	265	270	275	280	285	290	295	300	305	310
0.11 ~ 0.15		200	205	210	215	220	225	230	235	240	245	250	255	260	265	270	275	280	285	290	295	300	305	310	315
0.16 ~ 0.20																									
0.21 ~ 0.25	205	210	215	220	225	230	235	240	245	250	255	260	265	270	275	280	285	290	295	300	305	310	315	320	
0.26 ~ 0.30	210	215	220	225	230	235	240	245	250	255	260	265	270	275	280	285	290	295	300	305	310	315	320		
0.31 ~ 0.35	215	220	225	230	235	240	245	250	255	260	265	270	275	280	285	290	295	300	305	310	315	320			
0.36 ~ 0.40	220	225	230	235	240	245	250	255	260	265	270	275	280	285	290	295	300	305	310	315	320				
0.41 ~ 0.45	225	230	235	240	245	250	255	260	265	270	275	280	285	290	295	300	305	310	315	320					
0.46 ~ 0.50	230	235	240	245	250	255	260	265	270	275	280	285	290	295	300	305	310	315	320						
0.51 ~ 0.55	235	240	245	250	255	260	265	270	275	280	285	290	295	300	305	310	315	320							
0.56 ~ 0.60	240	245	250	255	260	265	270	275	280	285	290	295	300	305	310	315	320								
0.61 ~ 0.65	245	250	255	260	265	270	275	280	285	290	295	300	305	310	315	320									
0.66 ~ 0.70	250	255	260	265	270	275	280	285	290	295	300	305	310	315	320										
0.71 ~ 0.75	255	260	265	270	275	280	285	290	295	300	305	310	315	320											
0.76 ~ 0.80	260	265	270	275	280	285	290	295	300	305	310	315	320												
0.81 ~ 0.85	265	270	275	280	285	290	295	300	305	310	315	320													
0.86 ~ 0.90	270	275	280	285	290	295	300	305	310	315	320														
0.91 ~ 0.95	275	280	285	290	295	300	305	310	315	320															
0.96 ~ 1.00	280	285	290	295	300	305	310	315	320																
1.01 ~ 1.05	285	290	295	300	305	310	315	320																	
1.06 ~ 1.10	290	295	300	305	310	315	320																		
1.11 ~ 1.15	295	300	305	310	315	320																			
1.16 ~ 1.20	300	305	310	315	320																				
1.21 ~ 1.25	305	310	315	320																					
1.26 ~ 1.30	310	315	320																						
1.31 ~ 1.35	315	320																							
1.36 ~ 1.40	320																								

Valve clearance adjustment shim selection chart – XJ650(UK) and XJ650 J

follower. It is important that the lobe does not touch the holding tool because this could cause damage to the cylinder head or the camshaft. To this end, rotate the crankshaft so that the inlet camshaft turns clockwise and the exhaust camshaft turns anti-clockwise, as viewed from the left-hand side of the machine.

Remove the adjustment shim by prising it clear with a small screwdriver inserted in the slot or by using a magnet. Pad (shim) selection can now be made following the sequence given below.

1   Note the number etched on the shim. This gives its size in millimetres; eg 270 is 2.70 mm, 245 is 2.45 mm, and so on.

2   Using the accompanying chart, trace the point where the installed pad number intersects the measured clearance. This will indicate the new shim required.

3   Fit the new pad and remove the holding tool. Turn the crankshaft through several revolutions and recheck the clearance.

4   If necessary, repeat the sequence until the clearance figure is correct.

5   Repeat the above sequence with the remaining valves.

It should be noted that Yamaha treat the size numbering on the pads as a guide, hence the double check in stages 3 and 4 above. Once the clearance check has been completed, re-assemble the cylinder head cover, fuel tank and seat in the reverse order of that given for removal.

12 Checking the crankcase ventilation system

The oil-laden gases in the crankcase are expelled through a baffle arrangement and then fed through a breather pipe into the air cleaner casing. This arrangement prevents the emission of unburnt hydrocarbons by the simple expedient of burning them along with the fuel/air mixture. The condition of the breather hose should be checked periodically, and if damaged or split should be renewed.

13 Checking the fuel pipes

Examine the fuel pipes and connections for signs of leakage or imminent failure. The synthetic rubber material used is strong and lasts well in service, but will inevitably deteriorate in time. Apart from environmental considerations, fuel leakage can be extremely hazardous, and warrants immediate attention.

14 Changing the engine/transmission oil

Oil changes should always be made with the engine warm, to assist in draining fully the old oil, and to ensure that any contaminants are in suspension in it. Before starting work, find a suitable drain bowl. This should be shallow enough to slide under the crankcase and, of course, of sufficient capacity to catch the old oil. The amount varies between models but an old one gallon can with one side cut out will prove ideal.

Remove the crankcase filler cap, then slacken the sump drain plug which is located at the front edge of the sump. Place

Drain plug has magnetic tip to trap ferrous particles

Intake

MEASURED CLEARANCE	INSTALLED PAD NUMBER*

MEASURED CLEARANCE	200	205	210	215	220	225	230	235	240	245	250	255	260	265	270	275	280	285	290	295	300	305	310	315	320
0.00 ~ 0.05			200	205	210	215	220	225	230	235	240	245	250	255	260	265	270	275	280	285	290	295	300	305	310
0.06 ~ 0.10		200	205	210	215	220	225	230	235	240	245	250	255	260	265	270	275	280	285	290	295	300	305	310	315
0.11 ~ 0.15																									
0.16 ~ 0.20	205	210	215	220	225	230	235	240	245	250	255	260	265	270	275	280	285	290	295	300	305	310	315	320	
0.21 ~ 0.25	210	215	220	225	230	235	240	245	250	255	260	265	270	275	280	285	290	295	300	305	310	315	320		
0.26 ~ 0.30	215	220	225	230	235	240	245	250	255	260	265	270	275	280	285	290	295	300	305	310	315	320			
0.31 ~ 0.35	220	225	230	235	240	245	250	255	260	265	270	275	280	285	290	295	300	305	310	315	320				
0.36 ~ 0.40	225	230	235	240	245	250	255	260	265	270	275	280	285	290	295	300	305	310	315	320					
0.41 ~ 0.45	230	235	240	245	250	255	260	265	270	275	280	285	290	295	300	305	310	315	320						
0.46 ~ 0.50	235	240	245	250	255	260	265	270	275	280	285	290	295	300	305	310	315	320							
0.51 ~ 0.55	240	245	250	255	260	265	270	275	280	285	290	295	300	305	310	315	320								
0.56 ~ 0.60	245	250	255	260	265	270	275	280	285	290	295	300	305	310	315	320									
0.61 ~ 0.65	250	255	260	265	270	275	280	285	290	295	300	305	310	315	320										
0.66 ~ 0.70	255	260	265	270	275	280	285	290	295	300	305	310	315	320											
0.71 ~ 0.75	260	265	270	275	280	285	290	295	300	305	310	315	320												
0.76 ~ 0.80	265	270	275	280	285	290	295	300	305	310	315	320													
0.81 ~ 0.85	270	275	280	285	290	295	300	305	310	315	320														
0.86 ~ 0.90	275	280	285	290	295	300	305	310	315	320															
0.91 ~ 0.95	280	285	290	295	300	305	310	315	320																
0.96 ~ 1.00	285	290	295	300	305	310	315	320																	
1.01 ~ 1.05	290	295	300	305	310	315	320																		
1.06 ~ 1.10	295	300	305	310	315	320																			
1.11 ~ 1.15	300	305	310	315	320																				
1.16 ~ 1.20	305	310	315	320																					
1.21 ~ 1.25	310	315	320																						
1.26 ~ 1.30	315	320																							
1.31 ~ 1.35	320																								

VALVE CLEARANCE (engine cold) 0.11~ 0.15mm

Example: Installed is 250
Measured clearance is 0.32 mm
Replace 250 pad with 270

*Pad number (example):
Pad No. 250 = 2.50 mm
Pad No. 255 = 2.55 mm
Always install pad with number down.

Exhaust

MEASURED CLEARANCE	200	205	210	215	220	225	230	235	240	245	250	255	260	265	270	275	280	285	290	295	300	305	310	315	320
0.00 ~ 0.05				200	205	210	215	220	225	230	235	240	245	250	255	260	265	270	275	280	285	290	295	300	305
0.06 ~ 0.10			200	205	210	215	220	225	230	235	240	245	250	255	260	265	270	275	280	285	290	295	300	305	310
0.11 ~ 0.15		200	205	210	215	220	225	230	235	240	245	250	255	260	265	270	275	280	285	290	295	300	305	310	315
0.16 ~ 0.20																									
0.21 ~ 0.25	205	210	215	220	225	230	235	240	245	250	255	260	265	270	275	280	285	290	295	300	305	310	315	320	
0.26 ~ 0.30	210	215	220	225	230	235	240	245	250	255	260	265	270	275	280	285	290	295	300	305	310	315	320		
0.31 ~ 0.35	215	220	225	230	235	240	245	250	255	260	265	270	275	280	285	290	295	300	305	310	315	320			
0.36 ~ 0.40	220	225	230	235	240	245	250	255	260	265	270	275	280	285	290	295	300	305	310	315	320				
0.41 ~ 0.45	225	230	235	240	245	250	255	260	265	270	275	280	285	290	295	300	305	310	315	320					
0.46 ~ 0.50	230	235	240	245	250	255	260	265	270	275	280	285	290	295	300	305	310	315	320						
0.51 ~ 0.55	235	240	245	250	255	260	265	270	275	280	285	290	295	300	305	310	315	320							
0.56 ~ 0.60	240	245	250	255	260	265	270	275	280	285	290	295	300	305	310	315	320								
0.61 ~ 0.65	245	250	255	260	265	270	275	280	285	290	295	300	305	310	315	320									
0.66 ~ 0.70	250	255	260	265	270	275	280	285	290	295	300	305	310	315	320										
0.71 ~ 0.75	255	260	265	270	275	280	285	290	295	300	305	310	315	320											
0.76 ~ 0.80	260	265	270	275	280	285	290	295	300	305	310	315	320												
0.81 ~ 0.85	265	270	275	280	285	290	295	300	305	310	315	320													
0.86 ~ 0.90	270	275	280	285	290	295	300	305	310	315	320														
0.91 ~ 0.95	275	280	285	290	295	300	305	310	315	320															
0.96 ~ 1.00	280	285	290	295	300	305	310	315	320																
1.01 ~ 1.05	285	290	295	300	305	310	315	320																	
1.06 ~ 1.10	290	295	300	305	310	315	320																		
1.11 ~ 1.15	295	300	305	310	315	320																			
1.16 ~ 1.20	300	305	310	315	320																				
1.21 ~ 1.25	305	310	315	320																					
1.26 ~ 1.30	310	315	320																						
1.31 ~ 1.35	315	320																							
1.36 ~ 1.40	320																								

VALVE CLEARANCE (engine cold) 0.16 ~ 0.20 mm

Example: Installed is 250
Measured clearance is 0.32 mm
Replace 250 pad with 265

*Pad number (example):
Pad No. 250 = 2.50 mm
Pad No. 255 = 2.55 mm
Always install pad with number down.

Valve clearance adjustment shim selection chart – All models except XJ650(UK) and XJ650 J

the drain bowl beneath the sump, remove the plug, and allow the oil to drain. The middle gear assembly is lubricated by immersion in a well of engine/transmission oil held in a pocket in the crankcase casting. This has its own small drain plug situated just to the rear of the sump and slightly to the left of the engine centre line. This too should be removed and the oil allowed to drain. Once all the oil has drained, clean the drain plugs and the orifices and refit them using new sealing washers where necessary. Tighten to the following torque figures.

Sump drain plug   4.3 kgf m (31.0 lbf ft)
Middle gear drain plug   2.4 kgf m (17.5 lbf ft)

Add the appropriate quantity of oil via the crankcase filler, noting that Yamaha specify SAE 10W/30 'SE' motor oil for ambient temperatures up to 15°C (59°F) or SAE 20W/40 'SE' motor oil for ambient temperatures down to 5°C (41°F). After filling run the engine for a few minutes to distribute the oil, the let the machine stand for a while before checking the level in the sight window. This should be at the maximum mark to lessen the tendency for the low level warning to operate under acceleration or braking.

Engine/transmission oil quantities – oil change only
XJ650 – 11N (UK) and XJ650 RJ
2.65 litre (5.6/4.7 US/Imp pint)
All other XJ650 models
2.35 litre (5.0/4.1 US/Imp pint)
All XJ750 models
2.50 litre (5.2/4.4 US/Imp pint)

15 Front disc brake – pad checking and renewal

The condition of the front brake pads is of great importance, and must be checked regularly to ensure that the braking system continues to function reliably. In addition, regular inspection will bring to light any other potential faults which might otherwise go unnoticed. It is particularly important that the pads are renewed as soon as the wear limit is reached. Despite the apparent abundance of friction material left it should be noted that this provides an essential thermal barrier between the mechanical and hydraulic components. If the pads are allowed to wear beyond the prescribed limit, excessive heat may be transferred to the caliper, causing the fluid in it to boil. The resulting bubbles will render the brake ineffective.

A variety of calipers are used on the XJ range, each requiring a somewhat different approach to pad checking and renewal. Reference should be made to the appropriate Section of Chapter 5 for details.

16 Front disc brake: checking the hydraulic fluid – All models except XJ750(UK), XJ750 RH and RJ

Place the machine on its centre stand, then turn the handlebar until the fluid reservoir is level. Check that the fluid is at or above the lower level mark. This is indicated by the centre line through the small inspection window at the rear of the reservoir. If topping up is necessary, check first that the drop in level is not due to leakage of the master cylinder, hydraulic pipes and hoses or the caliper. A sudden drop in fluid level warrants a careful check of the system, though a gradual decrease is normally due to pad wear.

Remove the rectangular cover from the reservoir after releasing the four retaining screws. The cover should be lifted away carefully, taking care not to allow fluid to drip onto paintwork or plastic components; it will cause damage to either. Lift out the rubber diaphragm beneath the cap and top up the reservoir using a DOT 3 or SAE J1703 hydraulic fluid. On no account use any other type of fluid in the system or rapid seal failure or complete brake failure may result. Refit the diaphragm, then place the cover in position and refit the retaining screws.

Note: Hydraulic fluid is hygroscopic, which means that it will absorb moisture from the air if exposed to it for long periods. To guard against this ensure that fluid is kept in a sealed container, and make sure that the rubber diaphragm is refitted correctly to isolate the contents of the reservoir from possible contamination. If moisture or water contamination is suspected, drain and flush the system by bleeding it through as described in Chapter 5. Contaminated fluid may boil in service causing vapour locks in the system.

17 Front disc brakes: checking the hydraulic fluid – XJ750(UK), XJ750 RH and RJ

The procedure for topping up the master cylinder on the above models is broadly similar to that outlined in the preceding Section, and the remarks concerning the specification and characteristics of the hydraulic fluid should be noted. The remote mounting, however, precludes removal of the top of the reservoir under normal circumstances. To allow topping up to take place an Allen-headed filler plug is fitted to the rear of the reservoir. Access to it can be gained by turning the handlebar fully to the left. The rear of the master cylinder and the filler plug is now visible on the right-hand side of the steering head, between the fork yokes.

No sight glass arrangement is provided, but the reservoir incorporates a sensor which operates a warning panel on the LCD unit when the fluid level is low. The reservoir should be topped up as soon as possible if the warning panel comes on.

Topping up is not easy, given the small filler hole, and by far the best way of approaching the problem is to obtain a syringe to which a length of small-bore clear plastic tubing can be attached. This arrangement will allow the fluid to be injected into the reservoir without spillage, and also permits excess fluid to be drawn off where necessary. It should be possible to obtain a suitable syringe from a local doctor or veterinary practice, and

Fluid lower level is visible through sight glass

the tubing from a motor accessory shop (screen washer tubing) or pet shop (aquarium air pump tubing).

18 Control cable lubrication

Use motor oil or a good general purpose oil to lubricate the control cables. A good method of lubricating cables is shown in the accompanying illustration, using a plasticine funnel. This method has the disadvantage that the cables usually need removing from the machine to allow the oil to drip through. A hydraulic cable oiler which pressurises the lubricant overcomes this problem. Nylon lined cables, which may have been fitted as replacements, should not be lubricated; in some cases the oil will cause the lining to swell leading to cable seizure.

19 General lubrication – controls and pivots

It is sound practice to clean and lubricate the various control levers, the centre and side stands and other linkages regularly. As well as ensuring smooth operation and preventing corrosion, this routine will highlight any impending failures which might otherwise have gone unnoticed. Engine oil or general lubricating oil can be used, or as an alternative, one of the general purpose maintenance aerosols such as WD40.

20 Checking steering and suspension operation

Deficiences in the steering or suspension components will often make themselves obvious by deteriorating handling characteristics, though some faults develop slowly and may not be noticed until they are dangerously advanced. It is sound practice to check for wear in these components regularly.

With the machine on its centre stand, ask an assistant to press down or sit on the back of the seat to raise the front wheel clear of the ground. If necessary place a wooden crate or similar item under the crankcase to hold this position. Turn the handlebar from lock to lock noting any sign of roughness or tightness in its movement. Next, grasp one fork leg in each hand and push and pull the forks so that any play can be felt. Check, where play is evident, whether it is due to wear in the forks or steering head bearings. These are dealt with in detail in the main body of the Manual and should be investigated promptly.

With the rear wheel raised, push and pull from one side of the swinging arm to check for play in the pivot bearings. If play is found it can probably be corrected by adjustment as described in Chapter 4.

With the machine on its wheels, sit astride it and bounce up and down to ensure that the suspension is working smoothly and progressively. Note that most of the models covered have a number of possible suspension setting combinations which are described in Chapter 4, Section 24.

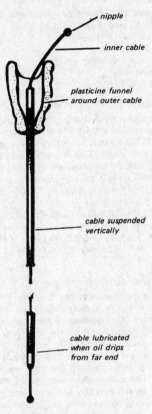

Electrolyte level can be seen through battery case

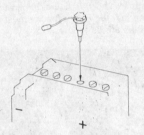

Oiling a control cable

Position of electrolyte level sensor in battery – XJ750(UK), RH, RJ and J

21 Checking wheel bearing condition

Raise each wheel clear of the ground in turn and spin the wheel to check for roughness in the bearings. Note that this may not prove easy since the front wheel tends to suffer from brake drag and the rear wheel from resistance from the transmission components, but serious pitting should show up as distinct notchiness.

Worn bearings can be checked by pushing and pulling the wheel rim from one side. A small amount of movement is almost inevitable at the wheel rim, but significant wear, say 2 or 3 mm movement, indicates that further examination is required. Refer to Chapter 5 for details.

22 Checking the battery electrolyte level

The XJ650 models are equipped with a 12 Ah (Ampere-hour) GS battery, whilst the XJ750 variants have a Yuasa battery rated at 14 Ah. In all cases the battery is housed behind the right-hand side panel, access being gained by removing the side panel and lifting the dualseat.

Maintenance is normally limited to keeping the electrolyte level between the prescribed upper and lower limits and by making sure that the vent pipe is not blocked.

Unless acid is spilt, as may occur if the machine falls over, the electrolyte should always be topped up with distilled water, to restore the correct level. If acid is spilt on any part of the machine, it should be neutralised with an alkali such as washing soda and washed away with plenty of water, otherwise serious corrosion will occur. Top up with sulphuric acid of the correct specific gravity (1.260 – 1.280) only when spillage has occurred. Check that the vent pipe is well clear of the frame tubes or any of the other cycle parts, for obvious reasons.

Note that machines fitted with the computer monitor system (CMS) are equipped with a battery level sensor which takes the place of the fourth cell cap from the negative (-) terminal. This is described in detail in Chapter 6, but note that it must **always** be fitted to this cell or damage to the microprocessor unit will occur.

23 Checking and cleaning the fuel system

It is difficult to give any recommendation for regular cleaning of the system though Yamaha advise this at 6 monthly intervals. Trouble with contaminated fuel is, fortunately, rare these days, but it can happen. Dirt or paint particles are usually trapped by the filter at the fuel tap inlet, and this can be removed at any convenient time and checked (see Chapter 2). Water contamination, whether from contaminated fuel from a garage or caused by condensation or the ingress of rain, can work its way into the carburettors and cause endless trouble with obstructed jets. If this happens it will be necessary to remove the tank for flushing and to clean the carburettor jets with compressed air. Again, see Chapter 2 for more information.

24 Cleaning and adjusting spark plugs

Remove the spark plugs and clean them using a wire brush. Clean the electrode points using emery paper or a fine file and then reset the gaps. To reset the gap, bend the outer electrode to bring it closer to or further from the central electrode, until a feeler gauge of the correct size can just be slid between the gap. Never bend the central electrode or the insulator will crack, causing engine damage if the particles fall in whilst the engine is running. The correct plug gap is 0.7 – 0.8 mm (0.028 – 0.032 in). Before refitting the plugs, smear the threads with a small quantity of graphite grease to aid subsequent removal.

25 Checking the exhaust system

Examine the exhaust system for signs of damage or corrosion. The latter is an eventual inevitability, but can be kept at bay by regular cleaning. Most exhaust systems are eventually eaten away from the inside by the acidic exhaust gases, and signs of this should be checked for. Apart from this, check the security of all clamps and mountings, tightening or renewing them as required.

26 Carburettors – checking and adjustment

If carburettor malfunction has been indicated by rough running or poor fuel consumption, attention to the carburettor adjustment and synchronisation is likely to be required. This is a somewhat lengthy procedure and requires the use of a set of vacuum gauges to ensure accurate synchronisation. Reference should be made to Chapter 2 for further details.

27 Changing the oil filter

The oil filter should be renewed at every second oil change. Carry out the oil change as described in Section 14, then proceed as follows prior to adding the new oil.

Move the drain tray beneath the oil filter housing and remove the filter retaining bolt. Lift away the filter housing assembly and discard the old filter element. Prise out the O-ring from the edge of the housing. Clean the groove out, then fit a new O-ring ensuring that it locates evenly in the groove. Offer up the filter assembly and tighten the holding bolt to 1.5 kgf m (11.0 lbf ft). **Do not** exceed this figure. When topping up with new engine oil, note that the filter capacity must be allowed for. The correct quantities are as follows.

Fitting a new oil filter element

Engine/transmission oil quantities – oil and filter change
XJ650 – 11N (UK) and XJ650 RJ
   2.95 litre (6.2/5.1 US/Imp pint)
All other XJ650 models
   2.65 litre (5.6/4.7 US/Imp pint)
All XJ750 models
   2.80 litre (5.9/4.9 US/Imp pint)

28 Changing the final drive oil

The final drive oil is best changed after a long run, when the oil has had a chance to warm up a little to assist in draining, and contaminants will have been stirred up into suspension in the oil. Place a bowl of about 3 or 4 pints capacity beneath the drain plug, which should be removed to allow the oil to drain. This may take time, so it is best to leave the machine, coming back when the oil has finished dripping. Clean and refit the drain plug, tightening it securely.

The casing holds 0.20 litre (0.42/0.35 US/Imp pint), and this should bring the oil level to the edge of the filler hole. The recommended oil grade is SAE 80 AP1 'GL-4' Hypoid gear oil. Alternatively an SAE 80W/90 Hypoid gear oil may be used where available.

29 Cleaning the air filter element

Lift the dualseat and remove the tool tray. The top of the air cleaner casing doubles as a mounting board for the fuse box, and reasonable care must be taken when removing it. Remove the three securing screws and lift the top clear of the casing. The element can now be withdrawn.

Start by getting rid of the loose dust by tapping the element on a hard surface, or by flicking the dust away using a soft brush. The remaining dust will be trapped by the porous filter surface and can be removed by blowing through the element from the inside, using compressed air. It is important to use a dry air supply for this purpose, because any moisture will tend to cause the dust to stick.

No exact interval for element renewal can be given, because this will depend entirely upon the usual operating conditions. Experience will show when the element will require removal. It will be noted that each time the element is cleaned it will become progressively more difficult to dislodge all of the dust. Eventually, after 6 – 10 cleaning operations, the element will require renewal.

Any obvious damage, such as holes or oil contamination, will necessitate immediate renewal of the element, because the entry of unfiltered air into the engine will cause accelerated wear, and the reduced resistance will result in a weak mixture. Conversely, a badly contaminated filter will obstruct the passage of air through the engine and will lead to an over-rich mixture.

Final drive oil level should just reach the edge of the filler hole

Renewing the air cleaner element

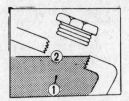

Final drive oil level

1 Oil   2 Correct oil level

30 Checking the final drive oil level

Unless something untoward occurs, the oil level in the final drive casing should not vary. It is good practice to check the level from time to time, however, since the final drive components will wear rapidly if not properly lubricated. Start by placing the machine on its centre stand on level ground. The check should be made when the machine has been stood for some hours, preferably overnight. Unscrew the level plug and check that the oil is up to the edge of the filler hole. If the level is noticeably below this, look for signs of leakage, such as oil staining on the underside of the casing. If leakage is evident it is vital that the problem is rectified before it becomes worse. Apart from the risk of damage to the final drive, oil leakage onto the rear tyre can be extremely hazardous.

31 Cylinder compression check – UK models only

Yamaha recommend cylinder compression is checked at 4000 mile (6000 km) intervals to ensure that the engine is working at full efficiency. In practice, unless a drop in performance has been noticed, the operation is a somewhat academic exercise. Few owners will possess a compression tester, so any suspected compression loss is probably best left to a dealer to diagnose. For owners having access to this equipment, and who wish to make the test, proceed as follows.

Check that the engine is in good general adjustment and that the valve clearances are set correctly. Start the engine and allow it to reach normal operating temperature, preferably by riding it for a few miles. Stop the engine, place the machine on the centre stand and then remove all of the spark plugs. Following the gauge manufacturer's directions, connect the instrument to each cylinder in turn, cranking the engine on the starter motor until the highest possible reading is shown. Note that the HT leads should be earthed during this operation and that the battery must be in good condition to attain full cranking speed. Make a note of each reading and compare it with the figures shown below.

Cylinder compression pressure (at sea level)

Standard	156 psi (11.0 kg cm^2)
Minimum	128 psi (9.0 kg cm^2)
Maximum	171 psi (12.0 kg cm^2)
Maximum difference between cylinders	14 psi (1.0 kg cm^2)

A marked difference in pressure of one or more cylinders usually indicates a mechanical failure rather than general wear. Try introducing a few drops of oil into the offending cylinder and repeat the test. If the reading improves, worn or broken piston rings should be suspected whilst no change of pressure indicates valve rather than bore problems. Refer to Chapter 1 for more information.

32 Checking the ignition timing

The ignition timing must be checked using a strobosopic timing lamp or 'strobe'. Of the two types available, a good quality xenon tube type is preferable to the cheaper neon types, because it will give a more defined image in use, although the latter will usually prove adequate if used in a shaded position. Where a strobe is not available, the timing check must be entrusted to a Yamaha service agent.

Run the engine until it reaches normal operating temperature, then stop it and remove the inspection cover at the left-hand end of the crankshaft. Connect the strobe to the spark plug lead of cylinder No 1 (left-hand cylinder) following the manufacturer's instructions. Note that where a 12 volt supply is required it is preferable to use a separate battery rather than the machine's own, because it is possible that pulses in the electrical system may act as rogue trigger-pulses causing some degree of confusion when the test is made.

It will be noted that the timing plate carries a number of marks. The 'T' mark indicates top dead centre (tdc) and a few degrees from it is the ignition timing mark which takes the form of an inverted U. This indicates the limits of the timing accuracy at the specified idle speed of 1050 ± 50 rpm.

Start the engine and set the idle speed to the above figure. It should be noted that a test tachometer is recommended by Yamaha for this purpose, but under normal circumstances the instrument fitted to the machine will give a reasonably accurate reading. Direct the light at the area near the fixed pointer and check that the timing mark 'freezes' with the fixed pointer mark within the limits indicated by the arms of the inverted U.

The timing is not adjustable, so a discrepancy can indicate only two things. It is possible that the fixed pointer has been bent or that the pickup backplate is loose. Alternatively, there is a fault in the TCI unit or the pickup assembly. In either case reference should be made to Chapter 3, Section 8 for more information. All being well, however, the fixed mark should fall between the two arms of the inverted U mark.

33 Checking and greasing the wheel bearings

The useful life of the wheel bearings will be greatly extended if they are removed from the hub for cleaning and greasing at the prescribed interval. This also affords the opportunity to inspect the bearings for signs of wear, allowing the bearing to be renewed if required rather than have it fail unexpectedly. The procedure is described in detail in Chapter 5, Sections 21 and 23.

34 Changing the front fork oil – XJ650(UK), XJ650 G, H, LH, RH and RJ

Place the machine on its centre stand so that weight is taken off the front wheel. This can be achieved by placing a support under the crankcase (a wooden crate is ideal), or by removing the front wheel. Of the two methods, it is preferable to remove the wheel because this will lessen the risk of the fork oil contaminating the brake components and will facilitate draining of the fork oil.

It will now be necessary to remove the fork top plug. This is held in place by a wire circlip in the top of the fork stanchion and is covered by a black plastic plug. To gain sufficient access it will usually be necessary to release the handlebar assembly.

The front fork oil drain screw

To avoid unnecessary work place a blanket or similar over the fuel tank so that the entire handlebar assembly can be freed and moved rearward to clear the top of the fork stanchions. Release the handlebar clamp screws and lift off the clamps to free the handlebar assembly. Place it on the protected fuel tank.

Prise off the plastic plug to reveal the metal plug which closes the fork. Removal of this item requires a large screwdriver, a small electrical screwdriver and considerable dexterity. The plug is under pressure from the fork spring and so it is necessary to depress it with the large screwdriver. Holding the plug down with one hand, use the electrical screwdriver to work the retaining clip out of its locating groove. It will be appreciated that the help of an assistant would prove very useful during this operation. Once the clip is free relax pressure on the plug until it is pushed out by the spring. Repeat the procedure on the remaining fork leg.

Place a drain tray under one fork leg and unscrew the small drain screw. Allow the fork oil to drain. If necessary, this can be speeded up by 'pumping' the fork leg up and down, but this is only really practicable if the wheel was removed and each fork leg is being dealt with separately. When all the oil has been expelled, clean and refit the drain plug, using a new sealing washer where necessary.

When refilling the fork, note that Yamaha recommend SAE 10W fork oil for all US models and SAE 10W/30 motor oil for the UK models. In practice, any reputable brand of fork oil of the appropriate grade can be used, this being less prone to cavitation (the formation of air bubbles with the attendant loss of damping effect). Fill each fork leg with the quantity of oil shown below, then complete reassembly by reversing the dismantling sequence.

Fork oil capacities (per leg)

XJ650 G, H, LH	262 cc (9.24 US fl oz)
XJ650(UK), XJ650 RH, RJ	236 cc (7.98/8.31 US/Imp fl oz)

35 Changing the front fork oil – XJ650 J

The procedure for changing the fork oil on the XJ650 J model is essentially the same as that described in Section 34, the only exception being that the above model employs air assistance and thus has Schrader-type valves in the top plugs. Before the plugs are removed the air pressure in the forks should be released by depressing the valve cores for a few seconds. The procedure described in Section 34 should then be followed, noting that the air valves may make removal of the retaining clip a little more awkward. Remember that the fork air pressure should be set up when the oil change is complete. Refer to Chapter 4, Section 24 for details.

The recommended fork oil grade is SAE 10W/30 motor oil or SAE 10W fork oil and the capacity 278 cc (9.40 US fl oz).

36 Changing the front fork oil – XJ750 J

The forks fitted to the XJ750 J are equipped with air-assisted forks with adjustable damping, and this will have a noticeable effect on the ease with which the fork oil can be changed. Start by placing the machine on its centre stand, then remove the front wheel so that no weight is on the forks. Place an old blanket or similar over the fuel tank to protect the paintwork.

Slacken the Allen-headed pinch bolt which retains each handlebar section to its splined mounting on the fork yoke. Lift each half of the handlebar upwards to disengage it from the splines, then lay it across the protected tank. This will allow access to the fork top bolts with minimal dismantling.

Remove the air valve cap from the union at the top of the left-hand stanchion and depressurise both forks by depressing the air valve insert for a few seconds. Remove the cross-head screw which secures the damping adjustment knob to the top of each stanchion and remove the knobs. Slacken fully each of the fork top bolts. Remove each one, noting that it incorporates a long damper adjustment rod. Take care not to drop or bend this assembly during removal and place each one in a safe place.

Place a drain bowl beneath each fork leg and remove the two drain screws. When most of the oil has drained, 'pump' each leg in turn to expel any residual oil. When draining has been completed, clean the drain holes and screws. Check, and if necessary renew the sealing washers, then refit and tighten the drain screws.

Top up each fork leg with 257 cc (8.7 US fl oz) of SAE 10W/30 type SE motor oil. Pump each fork leg up and down a few times to allow the damping oil to settle and any air pockets to be displaced.

Check the condition of the top bolt O-rings and renew them if in any doubt as to their condition. Lower the cap bolt assembly into the stanchion, noting that the flat on the rod must engage in the corresponding slot in the top of the damper rod. To check this, slowly turn the assembly until it can be felt to drop into engagement with the damper rod. Be absolutely certain of this and on no account force the cap bolt home. Once the rod is fully located the cap bolt can be secured. Repeat this process on the remaining fork leg.

Place the damping adjuster knob over the splined protrusion of the damping adjuster rod, but do not fit the retaining screw. Turn the knob fully clockwise until it stops, then back it off slightly until the No 1 position is indicated by a click. Lift off the knob and refit it so that the '1' aligns with the index mark to the rear of the top yoke, then fit the retaining screw. Complete reassembly by reversing the dismantling sequence. Remember to check and adjust the suspension settings as described in Chapter 4, Section 24.

37 Changing the front fork oil – all XJ750 models except XJ750 J

The procedure for changing the fork oil on the above models is generally similar to that described in Section 34. Follow the directions in that Section, noting the following points.

To remove the handlebars, prise off the rectangular trim blank from the centre of the handlebar cover to expose the cover retaining screws. Release the screws and lift the cover away. The handlebar assembly can now be removed by releasing the clamp halves.

Before removing the fork top plugs, unscrew the fork air valve caps and release the pressure from each fork by depressing the valve core for a few seconds.

The oil grades and quantities per leg are as shown below:

Fork oil capacities (per leg)

XJ750 RH and RJ	309 cc (10.5 US fl oz) SAE 20W fork oil
XJ750(UK)	312 cc (11.0 Imp, fl oz) SAE 10W/30 motor oil or SAE 10W fork oil

38 Lubricating and adjusting the steering head bearings

To obtain maximum life from the steering head bearings they should be removed for inspection, cleaning and re-greasing at the specified interval. Note also that excessive free play will cause poor handling, and this must be corrected by adjustment as soon as it becomes evident, or checked and set during overhaul. The relevant procedure varies according to the model, so reference should be made to the appropriate Section in Chapter 4.

39 Lubricating and adjusting the swinging arm bearings

The swinging arm assembly pivots on tapered roller bearings, and these can be adjusted to take up any free play which has developed. To prolong the life of the bearings they should be removed for examination and adjustment at the prescribed intervals. The procedure entails the removal of the swinging arm assembly and is described in detail in Chapter 4, Sections 18 and 19.

Castrol Lubricants

Castrol Engine Oils
Castrol Grand Prix

Castrol Grand Prix 10W/40 four stroke motorcycle oil is a superior quality lubricant designed for air or water cooled four stroke motorcycle engines, operating under all conditions.

Castrol Super TT Two Stroke Oil

Castrol Super TT Two Stroke Oil is a superior quality lubricant specially formulated for high powered Two Stroke engines. It is readily miscible with fuel and contains selective modern additives to provide excellent protection against deposit induced pre-ignition, high temperature ring sticking and scuffing, wear and corrosion.
Castrol Super TT Two Stroke Oil is recommended for use at petrol mixture ratios of up to 50:1.

Castrol R40

Castrol R40 is a castor-based lubricant specially designed for racing and high speed rallying, providing the ultimate in lubrication. Castrol R40 should never be mixed with mineral-based oils, and further additives are unnecessary and undesirable. A specialist oil for limited applications.

Castrol Gear Oils
Castrol Hypoy EP90

An SAE 90 mineral-based extreme pressure multi-purpose gear oil, primarily recommended for the lubrication of conventional hypoid differential units operating under moderate service conditions. Suitable also for some gearbox applications.

Castrol Hypoy Light EP 80W

A mineral-based extreme pressure multi-purpose gear oil with similar applications to Castrol Hypoy but an SAE rating of 80W and suitable where the average ambient temperatures are between 32°F and 10°F. Also recommended for manual transmissions where manufacturers specify an extreme pressure SAE 80 gear oil.

Castrol Hypoy B EP80 and B EP90

Are mineral-based extreme pressure multi-purpose gear oils with similar applications to Castrol Hypoy, operating in average ambient temperatures between 90°F and 32°F. The Castrol Hypoy B range provides added protection for gears operating under very stringent service conditions.

Castrol Greases

Castrol LM Grease

A multi-purpose high melting point lithium-based grease suitable for most automotive applications, including chassis and wheel bearing lubrication.

Castrol MS3 Grease

A high melting point lithium-based grease containing molybdenum disulphide. Suitable for heavy duty chassis application and some CV joints where a lithium-based grease is specified.

Castrol BNS Grease

A bentone-based non melting high temperature grease for ultra severe applications such as race and rally car front wheel bearings.

Other Castrol Products

Castrol Girling Universal Brake and Clutch Fluid

A special high performance brake and clutch fluid with an advanced vapour lock performance. It is the only fluid recommended by Girling Limited and surpasses the performance requirements of the current SAE J1703 Specification and the United States Federal Motor Vehicle Safety Standard No. 116 DOT 3 Specification.
In addition, Castrol Girling Universal Brake and Clutch fluid fully meets the requirements of the major vehicle manufacturers.

Castrol Fork Oil

A specially formulated fluid for the front forks of motorcycles, providing excellent damping and load carrying properties.

Castrol Chain Lubricant

A specially developed motorcycle chain lubricant containing nondrip, anti corrosion and water resistant additives which afford excellent penetration, lubrication and protection of exposed chains.

Castrol Everyman Oil

A light-bodied machine oil containing anti-corrosion additives for both household use and cycle lubrication.

Castrol DWF

A de-watering fluid which displaces moisture, lubricates and protects against corrosion of all metals. Innumerable uses in both car and home. Available in 400gm and 200gm aerosol cans.

Castrol Easing Fluid

A rust releasing fluid for corroded nuts, locks, hinges and all mechanical joints. Also available in 250ml tins.

Castrol Antifreeze

Contains anti-corrosion additives with ethylene glycol. Recommended for the cooling system of all petrol and diesel engines.

Chapter 1 Engine, clutch and gearbox

Refer to Chapter 7 for information on the 1983 US models

Contents

Specifications

Engine

	XJ650 models	XJ750 models
Type	Four cylinder, dohc air cooled, four stroke	Four cylinder, dohc air cooled, four stroke
Bore	63.0 mm (2.480 in)	65.0 mm (2.559 in)
Stroke	52.4 mm (2.063 in)	56.4 mm (2.220 in)
Capacity	653 cc (39.85 cu in)	748 cc (45.64 cu in)
Compression ratio	9.2:1	9.2:1

Cylinder block
Type .... Aluminium alloy, pressed in cast iron liners / Aluminium alloy, pressed in cast iron liners

Standard bore size	63.00 mm (2.4803 in)	65.00 mm (2.5591 in)
Service limit	63.10 mm (2.4843 in)	65.10 mm (2.5630 in)
Taper limit	0.05 mm (0.0020 in)	0.05 mm (0.0020 in)
Ovality limit	0.01 mm (0.0004 in)	0.01 mm (0.0004 in)
Piston/bore clearance	0.030 – 0.050 mm (0.0012 – 0.0020 in)	0.030 – 0.050 mm (0.0012 – 0.0020 in)
Service limit	0.1 mm (0.0039 in)	0.1 mm (0.0039 in)

Pistons
All models

Type	Light alloy
1st oversize	+0.25 mm (0.0098 in)
2nd oversize	+0.50 mm (0.0197 in)
3rd oversize	+0.75 mm (0.0295 in)
4th oversize	+1.00 mm (0.0394 in)

Piston rings
End gap (installed):

Top	0.15 – 0.35 mm (0.0059 – 0.0138 in)
Service limit	1.00 mm (0.0394 in)
2nd	0.15 – 0.35 mm (0.0059 – 0.0138 in)
Service limit	1.00 mm (0.0394 in)
Oil	0.3 – 0.9 mm (0.0118 – 0.035 in)
Service limit	1.5 mm (0.0591 in)

Piston ring to groove clearance:

Top	0.03 – 0.07 mm (0.0012 – 0.0028 in)
Service limit	0.15 mm (0.0059 in)
2nd	0.02 – 0.06 mm (0.008 – 0.0024 in)
Service limit	0.15 mm (0.0059 in)
Oil ring	Not applicable

Valves
Dimensions (refer to Fig. 1.13 for details)

'A' Overall valve head diameter:

Inlet	33 ± 0.1 mm (1.2992 ± 0.0039 in)
Exhaust	28 ± 0.1 mm (1.2205 ± 0.0039 in)

'B' Valve face width (overall):

Inlet	2.3 mm (0.0906 in)
Exhaust	2.3 mm (0.0906 in)

'C' Valve contact face width:

Inlet	1.0 ± 0.1 mm (0.0394 ± 0.0039 in)
Exhaust	1.0 ± 0.1 mm (0.0394 ± 0.0039 in)

'D' Margin thickness (min):

XJ650 RJ and XJ650(UK) models:	
Inlet and exhaust	0.7 mm (0.028 in)
All other models:	
Inlet	1.2 ± 0.2 mm (0.0472 ± 0.0079 in)
Exhaust	1.0 ± 0.2 mm (0.0394 ± 0.0079 in)
Valve seat width	1.0 mm (0.0394 in)
Service limit	2.0 mm (0.080 in)
Valve stem run-out (max)	0.03 mm (0.0012 in)

Valve stem diameter:

Inlet	6.975 – 6.990 mm (0.2746 – 0.2752 in)
Exhaust	6.960 – 6.975 mm (0.2740 – 0.2746 in)

Valve guide bore diameter:

Inlet	7.000 – 7.012 mm (0.2756 – 0.2761 in)
Exhaust	7.000 – 7.012 mm (0.2756 – 0.2761 in)

Valve stem to guide clearance:

Inlet	0.010 – 0.037 mm (0.0004 – 0.0015 in)
Exhaust	0.025 – 0.052 mm (0.0010 – 0.0020 in)

Valve clearances (engine cold)
Inlet:

XJ650 RJ and XJ650(UK)	0.16 – 0.20 mm (0.006 – 0.008 in)
All other models	0.11 – 0.15 mm (0.004 – 0.006 in)
Exhaust	0.16 – 0.20 mm (0.006 – 0.008 in)

Valve springs
Free length:

Inner	35.9 mm (1.413 in)
Outer	39.5 mm (1.555 in)

Installed length (valve closed):

    Inner .... 31.0 mm (1.220 in)

    Outer .... 34.0 mm (1.339 in)

Pressure at installed length:

    Inner .... 9.0 kg (19.8 lb)

    Outer .... 19.1 kg (42.1 lb)

Allowable warpage .... 1.6 mm (0.063 in) from vertical [2.5°]

Valve shims (pads)

Available sizes .... 2.00 – 3.20 mm (0.079 – 0.126 in).
in 0.05 mm (0.002 in) increments

Camshafts

	XJ650 RJ	All other models
Overall lobe height:		
Inlet	36.50 mm (1.437 in)	36.80 mm (1.449 in)
Service limit	Not available	36.65 mm (1.443 in)
Exhaust	35.80 mm (1.409 in)	
Service limit	35.65 mm (1.404 in)	
Base circle diameter:		
Inlet and exhaust	28.00 mm (1.102 in)	
Service limit	27.85 mm (1.096 in)	
Cam lift:		
Inlet:		
XJ650 RJ	8.50 mm (0.335 in)	
All other models	8.80 mm (0.347 in)	
Exhaust	7.80 mm (0.307 in)	
Bearing surface diameter	24.967 – 24.980 mm (0.9830 – 0.9835 in)	
Camshaft to cap clearance	0.020 – 0.054 mm (0.0008 – 0.0021 in)	
Service limit	0.160 mm (0.006 in)	
Camshaft runout (max):		
XJ650 RJ & XJ750 J	0.06 mm (0.0024 in)	
All other models	0.10 mm (0.0040 in)	

Crankshaft

Main bearing clearance:

  XJ650 RJ .... 0.020 – 0.044 mm (0.0008 – 0.0017 in)

  All other models .... 0.040 – 0.064 mm (0.0016 – 0.0025 in)

Big-end bearing clearance:

  XJ750 J and XJ750(UK) .... 0.016 – 0.040 mm (0.0006 – 0.0016 in)

  All other models .... 0.03 – 0.09 mm (0.0012 – 0.0035 in)

Main journal runout (max) .... 0.04 mm (0.0016 in)

Primary drive

Type .... Gear

Ratio 1.672:1 (97/58)

Clutch

Number of plates:

  Plain 7

  Friction 8

Number of springs 5

Friction plate thickness .... 3.0 mm (0.12 in)

Service limit .... 2.8 mm (0.11 in)

Plain plate thickness .... 1.6 mm (0.063 in)

Plain plate warpage (max) .... 0.05 mm (0.002 in)

Spring free length:

  XJ650 models .... 40.1 mm (1.579 in)

  Service limit .... 39.1 mm (1.539 in)

  XJ750 models .... 41.2 mm (1.622 in)

  Service limit .... 40.2 mm (1.583 in)

Gearbox

Type .... 5-speed, constant mesh

Ratios:

  1st 2.187:1 (35/16)

  2nd 1.500:1 (30/20)

  3rd 1.153:1 (30/26)

  4th 0.933:1 (28/30)

  Top 0.812:1 (26/32)

Secondary drive

Type .... Shaft

Output spur gear ratio:

XJ750 J 1.297:1 (48/37)

All other models 1.361:1 (49/36)

Middle gear ratio 1.055:1 (19/18)

Final drive gear ratio 2.909:1 (32/11)

Torque wrench settings

Component	kgf m	lbf ft
Cylinder head cap nuts – oil lightly	3.2	23.1
Cylinder head cover bolt	1.0	7.2
Spark plug	2.0	14.5
Cylinder head/block 8mm nuts	2.0	14.5
Cylinder head/block 6mm nuts – YICS model	N/Av	N/Av
Cylinder block/crankcase 8mm nut	2.0	14.5
Camshaft bearing cap bolt	1.0	7.2
Camshaft sprocket bolt	2.0	14.5
Non-automatic cam chain tensioner:		
Tensioner lock nut	0.9	6.5
Tensioner bolt	0.6	4.3
Camshaft chain tensioner plug	1.5	11.0
Connecting rod nut	2.5	18.1
Alternator rotor bolt	5.5	39.8
Crankcase drain plug	4.3	31.0
Middle gear case drain plug	2.4	17.5
Oil filter bolt	1.5	11.0
Oil cooler distributor block bolt – UK models	5.0	36.0
Oil pump cover screw	0.7	5.1
Oil strainer cover bolt	0.7	5.1
Crankcase bolt:		
8 mm	2.4	17.5
6 mm	1.2	8.7
Clutch centre nut	7.2	52.0
Clutch spring bolt	1.0	7.2
Gearchange pedal bolt	0.8	5.8
Neutral switch	2.0	14.5
Exhaust flange nut	0.8	5.4
Engine mounting:		
Front, upper nut	4.2	30.4
Front, lower bolt	4.2	30.4
Rear	7.0	50.6
Engine bracket nut	2.0	14.5

1 General description

The engine units fitted to the Yamaha XJ650 and XJ750 models are virtually identical, apart from the obvious differences in bore and stroke dimensions and the YICS system fitted to the later engine (see Chapter 2). The dismantling, overhaul and reassembly sequences are identical unless specific mention is made in the text.

The engine is of double overhead camshaft, transverse four-cylinder design and is built in unit with the primary transmission and gearbox. The cylinder head is of conventional light alloy construction and features two valves per cylinder operated directly from the camshafts via bucket type cam followers with shim adjustment pads. The camshafts run in plain bearing surfaces formed by detachable caps and by the cylinder head material itself. The camshafts are driven at their centres by a chain running from the crankshaft and through a cast-in tunnel between cylinders 2 and 3. The use of a cam tensioner assembly ensures constant cam chain tension.

The horizontally-split crankcase assembly houses a compact one-piece forged crankshaft supported by five re-newable plain main bearings. The connecting rods have split big-end eyes, also fitted with renewable bearing shells. The left-hand end of the crankshaft carries the ignition rotor and pickup coils. The right-hand end of the crankshaft does not carry any of the engine ancillaries and is covered by a blanking seal.

To the rear of the cylinder block, an extension of the crankcase houses a primary shaft running parallel to the crankshaft. The crankshaft incorporates a sprocket next to the

camshaft drive sprocket between the centre cylinders. Drive is taken from this to the primary shaft via a Hy-vo chain. The shaft carries the starter clutch assembly on its right-hand end, and the alternator on its left-hand end.

Primary drive to the clutch drum is taken from a large gear on the crankshaft, between cylinders 3 and 4. The clutch is of the conventional wet multiplate type and transmits power to the five-speed constant mesh gearbox. Drive from the gearbox output shaft is turned through 90° by a pair of bevel gears and is transmitted, at the rear of the casing, by means of a steel flange to which bolts the final drive shaft's universal joint.

2 Operations with engine/gearbox in frame

It is not necessary to remove the engine unit from the frame in order to dismantle the following items:

a) Right and left crankcase covers.
b) Clutch assembly and gear selector components (external).
c) Oil pump and filter.
d) Alternator and starter motor.
e) Cylinder head and cylinder head cover.
f) Cylinder block, pistons and rings*.
g) Ignition pickup and starter motor clutch.
h) Middle gear casing assembly

It should be noted that where a number of the above items require attention it can often be easier to remove the engine unit and carry out the dismantling work with the unit on a workbench.

* See Section 8 of this Chapter.

3 Operations with engine/gearbox unit removed from frame

As previously described the crankshaft and gearbox assemblies are housed within a common casing. Any work carried out on either of these two major assemblies will necessitate removal of the engine from the frame so that the crankcases can be separated.

4 Removing the engine/gearbox unit

1 Place the machine on its centre stand making sure that it is standing firmly. Although by no means essential it is useful to raise the machine a number of feet above floor level by placing it on a long bench or horizontal ramp. This will enable most of the work to be carried out in an upright position, which is eminently more comfortable than crouching or kneeling in a puddle of sump oil.

2 Place a suitable receptacle below the crankcase and drain off the engine oil. The sump plug lies just to the rear of the oil filter housing. The oil will drain at a higher rate if the engine has been warmed up previously, thereby heating amd thinning the oil. The engine holds a total of 5.8/7.0 Imp/US pint (3.3 litre) of engine oil, and a drain tray or bowl of suitable capacity should be placed beneath the drain plug prior to its removal. Leave the engine to drain whilst the middle gear drain plug is removed to allow the engine oil in this section of the crankcase to drain. The plug is located on the underside of the crankcase close the exhaust mounting.

3 It is now necessary to remove the oil filter, noting that the filter bowl will contain some residual oil and that provision must be made to catch this. The filter bowl is secured by a special bolt which has an unusually small hexagon to prevent overtightening. On the machine featured in this manual the bolt was extremely tight despite this, and it is recommended if there appears to be some risk of rounding the hexagon off, that removal is postponed until the exhaust system is dismantled and better access is available.

4 Pull off the plastic side panels which are retained by pegs held by frame-mounted grommets. Unlock and lift the dualseat. Check that the fuel tap is set to the 'On' position (do not set it on 'Pri' since this will allow the fuel to run out when the feed pipe is pulled off). The fuel tank is retained by a steel clip which pushes over a small stud below the rear edge of the tank, or by a single rubber-mounted bolt, depending on the model concerned. Release the rear of the tank and lift it slightly to allow the fuel and vacuum pipe to be pulled-off the fuel tap stubs. The spring clips securing each pipe should be released by squeezing the clip 'ears' when pulling the pipes from position. Keep the rear of the tank raised clear of its mounting and pull it rearwards to disengage the mounting rubbers at the front. Remove the tank and place it to one side to prevent damage to the paintwork or risk of fire. On machines equipped with a fuel level gauge, disconnect the sender lead as the tank is removed.

5 The exhaust system should be removed next noting that this operation is much easier if an assistant is available to help disentangle it from the underside of the machine. In the case of all but US XJ650 Maxim models, the four exhaust pipes are clamped to a collector box mounted below the crankcase. The collector box is retained by a pressed steel mounting, whilst each silencer is secured by a single bolt to its adjacent alloy support plate. The four exhaust pipes are held by finned alloy retainers which are secured by studs and nuts.

6 Remove the eight nuts and washers which hold the exhaust pipe retainers to the cylinder head. Unscrew the single bolt which secures the collector box to its mounting beneath the crankcase, leaving the system held in place by the silencer mounting bolts. Slacken the four clamps which hold the exhaust pipes to the collector box stubs. These will almost invariably be

rusty, and it will usually be necessary to apply some penetrating oil before attempting to slacken the clamp bolts. With an assistant supporting one side of the system, remove the silencer mounting bolts. Move the assembly forwards very slightly to allow the exhaust pipe retainers to be manoeuvred clear of the frame tubes, twisting the exhaust pipes as required to obtain clearance. The system can now be moved further forward and lowered to the ground. Rotate the system so that it can be pulled clear of the stand and the machine's underside.

7 The US XJ650 Maxim variants employ a slightly different arrangement in which four separate exhaust pipes are clamped to stubs on the silencer/balance pipe assembly beneath the crankcase. The latter assembly is in two parts, each of which comprises the silencer, two exhaust pipe stubs and one half of the balance pipe. The balance pipe is joined at the centre by a clamp. Removal of the system is accomplished by slackening the four exhaust pipe clamps and the balance pipe clamp. Remove the eight nuts which secure the exhaust pipe retainers to the cylinder head and release the silencer retaining bolts. The system can now be manoeuvred clear of the engine and frame and placed to one side.

8 Disconnect the battery negative (–) lead, then release the metal contact strip between the positive (+) terminal and the starter solenoid. Remove the rubber strap and lift the battery clear of its mounting. Disconnect the starter motor cable and switch leads at the solenoid and remove it. Unscrew the battery case holding bolts (2 off) and remove the battery case from its frame recess.

9 The CDI unit, regulator unit and, on certain models, an ignition cut-off relay are mounted on an injection moulded board behind the left-hand side panel. This is secured by a single mounting screw near the bottom, its top edge being clipped to the frame tube. Remove the screw and pull the board clear of the frame to give access to the wiring connectors. These should be separated and the assembly placed to one side. Note that the retaining screw also secures the engine earth strap. This should be pushed clear of the frame.

10 Slacken the clamps which retain the carburettors to the intake adapters and to the air cleaner connecting rubbers. Pull off the crankcase breather hose from its stub at the front of the air cleaner casing, having first displaced the wire clip which retains it. Remove the two air cleaner casing bolts to permit movement of the casing during carburettor removal. One of the screws will be found just forward of the fuel tank mounting, the other being on the left-hand side of the casing near the frame tube. Remove the air cleaner connecting rubbers by disengaging them from the carburettors and pulling them clear of the air cleaner casing.

11 Slacken the screw clamp which secures the choke cable outer to its support bracket and then disengage the inner cable from the choke arm. The throttle cable should be released in a similar fashion, the outer cable end being located by a socket end on its bracket. Note that if access to the cables proves difficult, it may prove preferable to postpone their removal until the carburettors have been partly removed. Grasp the carburettor bank and pull it rearwards to disengage the instruments from the mounting rubbers. Once free, remove the carburettors from the right-hand side.

12 Slacken the clutch cable adjuster assembly locknuts and screw in the adjuster to obtain maximum free play. Unhook the outer cable from its support bracket, then disengage the inner cable nipple from the clutch arm. Note that the arm has a small security tang which should be bent clear prior to removal. Unscrew the knurled ring which retains the tachometer cable to the cylinder head and lodge the cable clear of the engine. Disconnect the spark plug leads and lodge or tie them clear of the engine. Trace the alternator and ignition pickup leads back to their respective connectors. Separate the connectors and feed the wiring back until each part of the harness can be coiled neatly on top of the crankcase in a position which will not impede engine removal. Make sure any cable ties or clips holding the wiring to the frame have been released. In

particular, make sure that the clamp which holds the wiring to the frame upper bracing tube is released.

13 Disconnect the horn leads, then release the horn bracket mounting bolts, removing the horn (or where fitted, the twin horns) and the bracket as an assembly. On machines which have the indicator relay mounted below the frame tubes, this should be disconnected and removed to avoid damage (Note: On XJ750 models having the horn(s) housed in the headlamp nacelle, this stage can be ignored).

14 On UK market models it is now necessary to remove the oil cooler assembly. Start by unscrewing the large bolt which retains the oil cooler distributor block to the crankcase. This can be pulled clear together with the oil cooler hoses. Remove the Allen screw which retains the oil cooler hose clamp to the support bracket. The hoses can be removed with the oil cooler matrix as described below. Remove the two bolts which secure the bottom edge of the oil cooler to the frame. The oil cooler radiator is now held by its top mounting which consists of a rubber-mounted pin secured by an R-pin. Withdraw the pin and remove the oil cooler assembly complete with its hoses and spacer.

15 Unhook the ends of the coil spring retainer which secures the drive shaft gaiter to the crankcase. The gaiter can now be pulled back to expose the connecting flange and its mounting bolts. It will be necessary to prevent the drive shaft from turning while the flange bolts are removed, and this can best be accomplished by enlisting the aid of a nearby foot. Persuade the owner of the foot to apply it to the rear brake pedal whilst the bolt is slackened. The brake should then be released to allow the next bolt to be brought within reach by turning the rear wheel. Before separating the flanges, it is sound practice to scribe an alignment mark across adjacent edges, so that on reassembly the original position of the two items is restored.

16 Dismantle and remove the front engine plates, these being held by a total of six bolts and nuts. Remove the lower front mounting bolts and nuts (2 off). The weight of the engine may need to be taken during bolt removal. On all but the XJ750 J model the rear of the engine is retained by two large bolts which pass through the footrests before engaging the frame and crankcase lugs. Each bolt is held by a nut on its inner end. Remove the nuts and displace and remove the bolts together with the footrests. The engine fitted to the XJ750 J model is secured at the rear by a long through-bolt. To gain access to the bolt and securing nut the footrest mounting plates, together with the associated foot controls, must first be detached. Each mounting plate is retained by two bolts, the nuts of which are secured by R-pins. To remove the left-hand plate start by removing the plate mounting bolts after displacing the R-pins. Unscrew and remove the gearchange lever arm pinch bolt and

ease the arm off the splined shaft which projects from the gearbox. The complete plate together with the footrest and gearchange pedal can now be lifted away. The right-hand footrest mounting plate should be detached in a similar manner, but here the rear brake pedal arm must be detached from the splined shaft. Access can now be made to the engine bolt which should be removed.

17 The engine unit is removed from the right-hand side of the machine, and a suitable support should be placed next to the machine so that the engine can be rested on it during removal. A strong wooden box is ideal for this purpose. It should be stressed that the engine/transmission unit is both heavy and a very close fit in the frame, and this makes the removal operation rather awkward. A minimum of two people will be required to carry out the operation safely, and if possible a second assistant should be available to steady the machine and assist in manoeuvring the unit clear of the numerous frame obstructions.

18 With one person on each side of the machine lift the front of the engine slightly and move the engine forward to clear the drive shaft flange. Check that all wiring, cables and hoses are well clear, then manoeuvre the unit out to the right. Note that it will be necessary to manipulate the unit with a fair degree of skill to avoid the various projections from the frame. Once the unit is half out, rest it on the frame while both persons move to the right-hand side to complete removal. Once clear, place the unit on the workbench to await further dismantling.

4.4a Release clip or bolt to free rear of fuel tank

4.4b Squeeze clip ends together and prise off the fuel and vacuum pipes

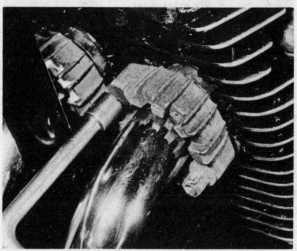

4.6a Release the exhaust retainer nuts, followed by ...

4.6b ...crankcase and silencer mountings to free system

4.8a Detach the battery negative (−) lead

4.8b Release A: Battery positive (+) link, and B: Starter motor cable, from the solenoid terminals

4.8c Remove the battery, followed by the battery tray

4.9a Release regulator/rectifier, TCI unit and support plate ...

4.9b ... noting that retaining screw also secures earth strap

4.10a Cleaner casing lower mounting bolt (arrowed)

4.10b Displace intake rubbers – note locating tab

4.11a Slacken clamp screw and release 'choke' cable

4.11b Unhook and release the throttle cable

4.11c Disengage and remove the carburettor assembly

4.12a Slacken and disengage clutch cable adjuster

4.12b Bend back security tang (arrowed) to free clutch cable

4.12c Unscrew knurled ring to free tachometer cable

4.13 Release horn bracket bolts

4.14a Distributor block (UK models) is retained by central bolt

4.14b Release distributor block to gain access to pipe unions

4.14c Single Allen screw secures hose clamp

4.14d Oil cooler lower mounting screws (arrowed)

4.14e Remove R-pin to free the oil cooler assembly

4.15a Unhook ends of spring retainer and displace ...

4.15b ... the rubber boot to expose flange bolts

4.16a Dismantle the engine front mountings (6 bolts) ...

4.16b ... and front lower mountings (2 bolts)

4.16c Note inboard nuts on rear mountings (except XJ750 J)

4.16d Withdraw footrest together with engine rear mounting bolt

4.17 The engine unit ready for removal from RH side

5   Dismantling the engine/gearbox unit: preliminaries

1   Before any dismantling work is undertaken the external surfaces of the unit should be thoroughly cleaned and degreased. This will prevent the contamination of the engine internals and will also make working a lot easier and cleaner. A high flash point solvent, such as paraffin (kerosene) can be used, or better still, a proprietary engine degreaser such as Gunk. Use old paintbrushes and toothbrushes to work the solvent into the various recesses of the engine castings. Take care to exclude solvent or water from the electrical components and inlet and exhaust ports. The use of petrol (gasoline) as a cleaning medium should be avoided, because the vapour is explosive and can be toxic if used in a confined space.
2   When clean and dry, arrange the unit on the workbench, leaving a suitable clear area for working. Gather a selection of small containers and plastic bags so that parts can be grouped together in an easily identifiable manner. Some paper and a pen should be on hand to permit notes to be made and labels attached where necessary. A supply of clean rag is also required.
3   Before commencing work read through the appropriate section so that some idea of the necessary procedure can be gained. When removing the various engine components it should be noted that great force is seldom required, unless specified. In many cases, a component's reluctance to be removed is indicative of an incorrect approach or removal method. If in any doubt, re-check with the text.
4   Note that a 'TORX' wrench will be needed prior to crankcase separation. Refer to Section 14 for details.

6   Dismantling the engine/gearbox unit: removing the cylinder head cover and camshafts

1   The camshafts may be removed with the engine removed from the frame and on a workbench, or with the unit installed in the frame. In the latter case it will be necessary to remove the fuel tank, horn and bracket assembly, and the indicator relay on models where this is mounted beneath the frame tubes. Note that if cam chain renewal is required, it is possible to cut the existing chain and thread a new chain into place rather than strip the complete engine; see paragraph 7 for details.
2   Slacken and remove the Allen bolts which secure the cylinder head cover. These should be removed in a diagonal sequence to prevent warpage of the cover. If the cover is stuck firmly to the cylinder head it can be dislodged by tapping around the joint with a hide mallet – do not attempt to lever it free. Once loose, lift the cover away.
3   Slacken and remove the two Allen bolts which secure the camshaft chain tensioner to the rear of the cylinder block. The tensioner assembly can now be lifted away. Release the four cross-head screws which retain the ignition pick-up inspection cover to the left-hand end of the crankcase. These may prove tight and an impact driver should be used to prevent damage to the screw heads. Lift the cover away to reveal the pickup and reluctor assembly. Using a spanner on the 19 mm flats provided, rotate the crankshaft anti-clockwise until the 'T' mark aligns with the fixed index mark, indicating that the engine is at TDC (Top Dead Centre).
4   Yamaha recommend that the camshaft sprocket bolts are removed in this position, but it was found in practice that one of the two bolts securing each sprocket will be masked by the cam chain tunnel. It is suggested that the engine is turned back and the hidden bolts removed first. Set the crankshaft at TDC once more, and remove the remaining bolts to free the sprockets.

5 Displace the sprockets until they drop clear of their locating shoulders, then lift out the chain guide which lies in the cylinder head recess between them. It should be noted that the crankshaft must not be rotated from this stage onwards since this may result in bent valves if the pistons contact those which are open. Before removing the camshaft caps make a note of the various identification marks on each. Viewed from the top, the exhaust camshaft caps are marked E1, E2 and E3 from left to right, whilst those of the inlet camshaft are marked I1, I2 and I3 in the same way. Each cap has a direction arrow which should face towards the right-hand (clutch) side. These marks should be checked, and any which are indistinct marked visibly prior to removal.

6 Slacken the camshaft cap retaining bolts in a diagonal sequence, unscrewing each one by about $\frac{1}{2}$ turn at a time. As the bolts are removed the camshafts will be pushed upwards by the valve springs. Remove the bolts and caps and place them in order on a clean surface. Pass a length of wire through the cam chain to retain it, noting that this should be done even if the engine unit is to be stripped completely, since it will allow the crankshaft to be turned with no risk of the chain becoming jammed around its sprocket. The camshafts and sprockets can now be disengaged from the chain and removed. **Note:** It is of great importance that the camshafts and related components are marked to avoid possible confusion during reassembly. This can be done using an indelible spirit-based marker once the area to be marked has been degreased. Alternatively, tie labels to the larger parts, such as the camshafts, and place smaller parts in clearly marked bags or boxes.

7 As described in paragraph 1 it is possible to renew the cam chain without stripping the engine. Care must be taken if using this method to rivet the chain ends securely on assembly; note that Yamaha can supply a cutting/riveting tool for this purpose. Remove the cylinder head cover and rotate the engine until the soft link in the chain appears between the sprockets, remove the tensioner and disconnect the chain at the soft link. Have ready some means of tying the chain's ends together to stop them falling free. Position the engine at TDC (cylinders 1 and 4) to prevent damage to the valves and remove the camshafts. Join the new chain to the old and rotate the crankshaft until the new chain is in position; keep the chain taut during the operation to prevent it bunching around the crankshaft sprocket. When in position join the chain ends at the soft link using the proper tool, refit the camshafts and set the valve timing – see Section 48.

6.3a Release camshaft chain tensioner from rear of block

6.3b Crankshaft should be turned until 'T' mark is aligned

6.4 Remove camshaft sprocket holding bolts

6.5a Disengage and remove centre guide

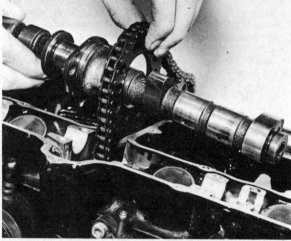

6.5b Remove camshaft caps and lift camshafts away

Fig. 1.1 Cam chain and camshaft – All 650 models except XJ650 J and XJ650 – 11N (UK)

1	Inlet side camshaft	12	Locknut
2	Exhaust side camshaft	13	Washer
3	Sprocket – 2 off	14	O-ring
4	Bolt – 4 off	15	Cam chain front guide
5	Cam chain	16	Cam chain tensioner
6	Cam chain joining link		blade
7	Cam chain guide	17	Bolt
8	Tensioner housing	18	Nut
9	Plunger	19	Washer
10	Spring	20	O-ring
11	Adjusting bolt	21	Bolt – 2 off

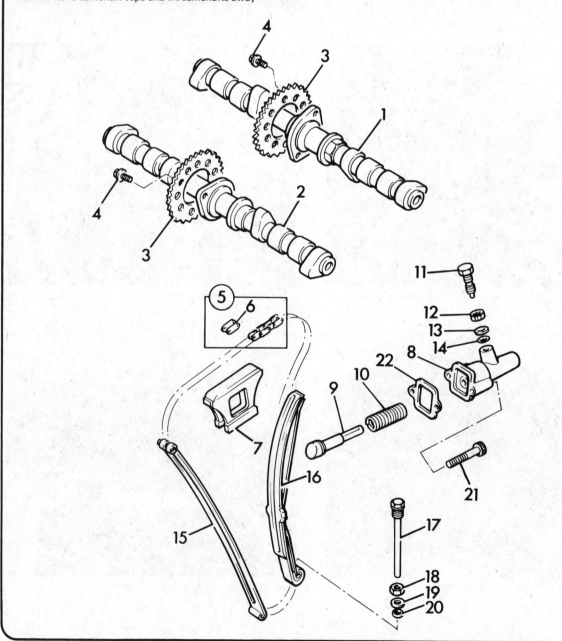

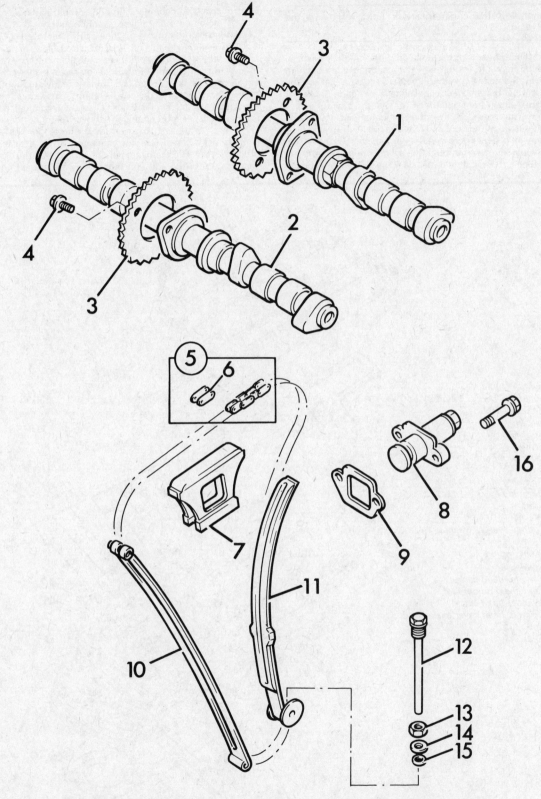

Fig. 1.2 Cam chain and camshafts – All 750 models, XJ650 J and XJ650 – 11N (UK)

1	Inlet side camshaft	6	Cam chain joining link	10	Cam chain front guide	13	Nut
2	Exhaust side camshaft	7	Cam chain guide	11	Cam chain tensioner	14	Washer
3	Sprocket – 2 off	8	Cam chain tensioner		blade	15	O-ring
4	Bolt – 4 off	9	Gasket	12	Bolt	16	Bolt – 2 off
5	Cam chain						

7 Dismantling the engine/gearbox unit: removing the cylinder head

1   The cylinder head can be removed with the engine unit in or out of the frame. In the former instance, refer to the opening remarks in Section 6 of this Chapter. In both cases, the camshafts must be removed as described in Section 6. Lift out the camshaft chain front guide and remove the spark plugs.

2   Remove the nuts from the studs which project down from the cylinder head at the front and rear of the unit. The remaining cylinder head fasteners are twelve domed nuts which screw onto studs passing down through the cylinder head and into the cylinder block. These should be slackened in the sequence indicated in the accompanying photograph, turning each nut in $\frac{1}{2}$ turn increments to release evenly pressure on the cylinder head. Once all nuts are free they can be removed.

3   The cylinder head can now be lifted clear of its holding studs and removed. Should this prove difficult it may be helpful to tap around the joint area with a soft faced mallet, taking care not to damage the cooling fins. In particularly stubborn cases a certain amount of judicious levering may prove necessary though this is undesirable unless absolutely essential. If levering is employed, use a broad tipped screwdriver, or better still a tyre lever, and take care to lever **only** where the fins are well supported by webs. Use the minimum of force necessary to dislodge the cylinder head, remembering that the light alloy castings are brittle and that broken fins are at best unsightly.

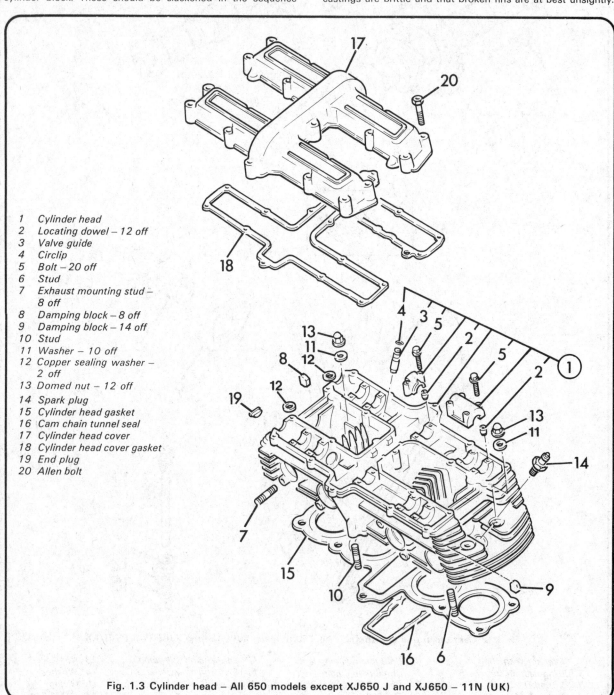

1   Cylinder head
2   Locating dowel – 12 off
3   Valve guide
4   Circlip
5   Bolt – 20 off
6   Stud
7   Exhaust mounting stud – 8 off
8   Damping block – 8 off
9   Damping block – 14 off
10 Stud
11 Washer – 10 off
12 Copper sealing washer – 2 off
13 Domed nut – 12 off
14 Spark plug
15 Cylinder head gasket
16 Cam chain tunnel seal
17 Cylinder head cover
18 Cylinder head cover gasket
19 End plug
20 Allen bolt

Fig. 1.3 Cylinder head – All 650 models except XJ650 J and XJ650 – 11N (UK)

7.2 Cylinder head nuts must be slackened in the sequence shown

7.3 Break block to head joint and remove cylinder head

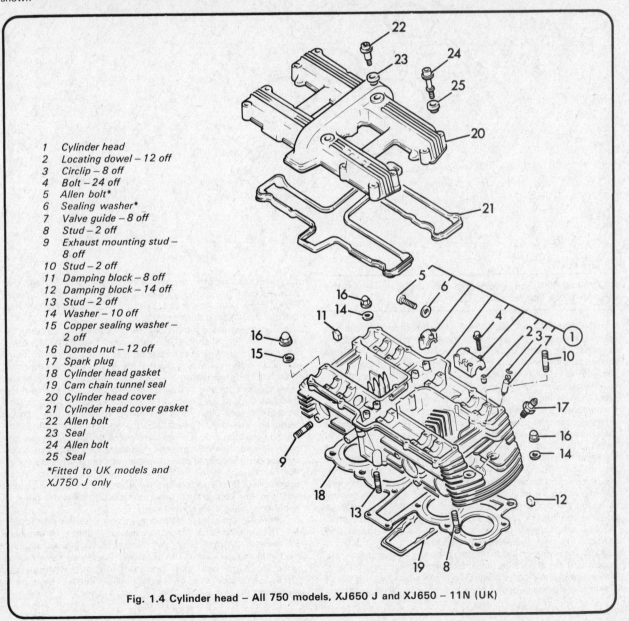

1 Cylinder head
2 Locating dowel – 12 off
3 Circlip – 8 off
4 Bolt – 24 off
5 Allen bolt*
6 Sealing washer*
7 Valve guide – 8 off
8 Stud – 2 off
9 Exhaust mounting stud – 8 off
10 Stud – 2 off
11 Damping block – 8 off
12 Damping block – 14 off
13 Stud – 2 off
14 Washer – 10 off
15 Copper sealing washer – 2 off
16 Domed nut – 12 off
17 Spark plug
18 Cylinder head gasket
19 Cam chain tunnel seal
20 Cylinder head cover
21 Cylinder head cover gasket
22 Allen bolt
23 Seal
24 Allen bolt
25 Seal

*Fitted to UK models and XJ750 J only

Fig. 1.4 Cylinder head – All 750 models, XJ650 J and XJ650 – 11N (UK)

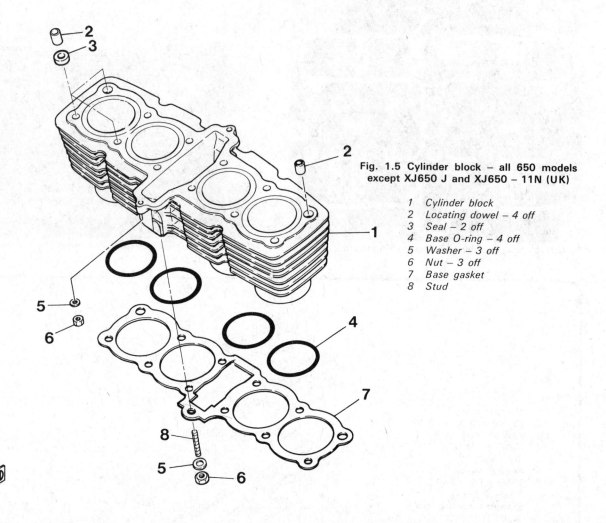

Fig. 1.5 Cylinder block – all 650 models
except XJ650 J and XJ650 – 11N (UK)

1   Cylinder block
2   Locating dowel – 4 off
3   Seal – 2 off
4   Base O-ring – 4 off
5   Washer – 3 off
6   Nut – 3 off
7   Base gasket
8   Stud

8   Dismantling the engine/gearbox unit: removing the cylinder block and pistons

1   The cylinder block can be removed after the camshafts and cylinder head have been released as described in the preceding sections. The design of the cylinder block casting means that the studs are exposed along part of their length, and this allows road dirt to accumulate around them. It follows that great care must be taken to prevent this debris from entering the crankcase during removal, particularly where it is not intended to separate the crankcase halves. To this end it is worth arranging the unit upside down so that the dirt drops clear of the crankcase. This does, however, pose problems if the work is being undertaken with the engine unit in the frame. The safest option is to remove the unit and proceed as described above. A less reliable option is to remove the block very carefully and to use a vacuum cleaner to catch the dirt before it can drop into the crankcase mouths. Commence the removal operation by releasing the single nut at the front of the cylinder block base and free the joint by tapping around it with a soft-faced mallet. DO NOT use screwdrivers or other levers between the mating surfaces; this will certainly lead to oil leaks.

2   Lift the cylinder block off the pistons. At this juncture a second person should be present to support each piston as it leaves the cylinder barrel spigot. To prevent broken particles of piston ring dropping in, and subsequently other foreign matter, the crankcase mouths should be padded with clean rag. This must be done before the rings leave the confines of the cylinder bores. Endeavour to lift the cylinder block squarely, so that pistons do not bind in the bores.

3   Before removing the pistons, mark each on the inside of the skirt to aid identification. It is important that the pistons are refitted to their original cylinders on reassembly. An arrow mark on each piston crown indicates the front, the pistons must therefore be fitted with the arrow facing forwards. Remove the outer circlip from one of the outermost pistons and push out the gudgeon pin. Lift the piston off the connecting rod. The gudgeon pins are a light push fit in the piston bosses so they can be removed with ease. If any difficulty is encountered, apply to the offending piston crown a rag over which boiling water has just been poured. This will give the necessary temporary expansion to the piston bosses to allow the gudgeon pin to be pushed out.

4   Each piston is ftted with two compression rings and an oil control ring. It is wise to leave the rings in place on the pistons until the time comes for their examination or renewal in order to avoid confusing their correct order.

5   Remove the remaining pistons, using a similar procedure.

6   The camshaft chain tensioner assembly should be removed to prevent it from becoming damaged during subsequent dismantling operations. The tensioner blade can be removed after the stop bolt and locknut, which project from the crankcase to the rear of the cam chain tunnel, have been slackened off.

8.2 Pistons must be supported as they emerge from bores

8.3 Pack crankcase with rag before releasing circlips

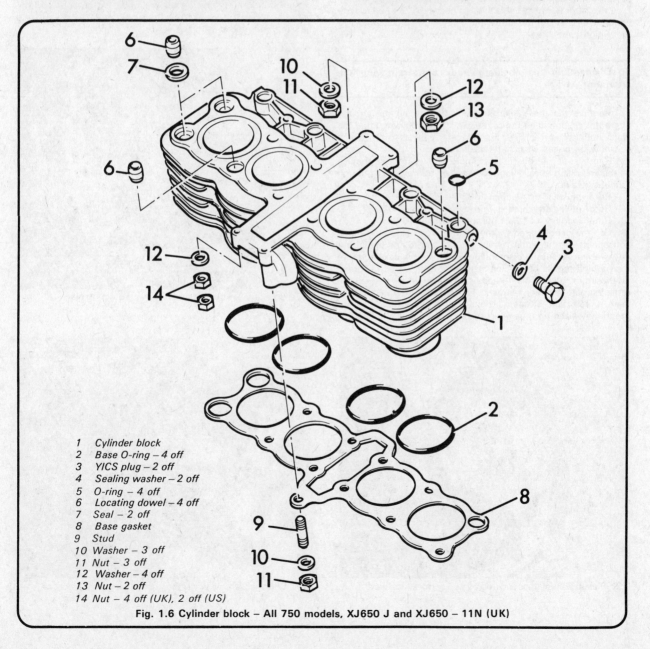

1   Cylinder block
2   Base O-ring – 4 off
3   YICS plug – 2 off
4   Sealing washer – 2 off
5   O-ring – 4 off
6   Locating dowel – 4 off
7   Seal – 2 off
8   Base gasket
9   Stud
10 Washer – 3 off
11 Nut – 3 off
12 Washer – 4 off
13 Nut – 2 off
14 Nut – 4 off (UK), 2 off (US)

Fig. 1.6 Cylinder block – All 750 models, XJ650 J and XJ650 – 11N (UK)

9 Dismantling the engine/gearbox unit: removing the ignition pickup assembly and crankshaft end covers

1   The crankshaft ends are housed behind light alloy inspection covers, the left-hand cover housing the ignition pickup assembly. The two covers are identical and can be removed after the four screws which retain each one have been released. Care should be taken not to damage the black finish of the screws and to this end an impact driver fitted with the correct size bit should be employed.

2   The ignition reluctor, or rotor, takes the form of a flat steel plate incorporating a trigger lobe and marks for timing purposes. It is held on the crankshaft end by a single Allen bolt. Remove the bolt and lift the reluctor away. Remove the two cross-head screws which secure the ignition pickup stator to the crankcase. These are at the one o'clock and seven o' clock positions when viewed from the end. Do not disturb any of the other screws, especially the one holding the timing index pointer. Remove the casing screw nearest to the gearbox pedal pivot and free the pickup wiring from behind the guide. The pickup stator can now be withdrawn together with its wiring. Free the latter from any cable clips as necessary, and where the engine unit is still in the frame, trace back to and disconnect at the wiring connector.

10 Dismantling the engine/gearbox unit: removing the alternator

1   Remove the three Allen screws which retain the circular alternator cover and lift the cover away. It will be noted that the cover doubles as a retainer for the stator coil assembly, which can now be pulled clear of its casing recess.

2   Removal of the rotor is less simple, and will require the use of the following special tools which can be obtained through Yamaha dealers:

Rotor puller, Part number 90890-01080
Rotor puller attachment, Part number 90890-04052
Rotor holding tool, Part number 90890-04043

Of the above tools, the first two (or home-made equivalents) are essential, whilst the holding tool is useful but not vital for successful dismantling. The rotor puller is no more than a high tensile bolt which screws into the extractor thread in the rotor boss. This thread is revealed after the central retaining bolt has been removed. For those fortunate enough to have a selection of metric bolts in the workshop, the appropriately sized item can be used to good effect.

3   The 'Rotor puller attachment' is no more than a plain steel pin having a diameter slightly smaller than that of the rotor retaining bolt. It is fitted into the shaft end to provide a bearing surface for the extractor bolt and thus to prevent damage to shaft material. Again, this part can be improvised quite easily if necessary, by cutting a scrap high tensile bolt down to length.

4   The 'Rotor holding tool' is worth using if it can be borrowed, otherwise it will be necessary to devise an alternative method to prevent the rotor and shaft from turning whilst the retaining bolt is removed. With the engine unit in the frame it is easiest to select top gear and apply the rear brake, thus locking the engine and transmission. If the unit has been removed from the frame, the transmission and crankshaft can be locked by selecting top gear and preventing the output flange from turning. This can be done by drilling two holes in one end of a piece of steel strip so that it can be bolted to two of the output flange bolt holes (see photograph). With the shaft locked by one of the above methods, slacken and removed the rotor holding bolt.

5   Fit the 'Rotor puller attachment' or its home-made e-quivalent into the end of the shaft, then screw the extractor bolt in until it bears on it. Gradually tighten the extractor bolt to draw the rotor off its taper, but do not overtighten it. If the rotor proves stubborn, try striking the end of the extractor bolt with a hammer to jar it free. The rotor can now be lifted clear of the housing and placed with the stator.

9.2a Unscrew centre bolt and remove the timing plate

9.2b Pickup backplate is secured by two screws (arrowed)

10.1 Stator assembly will often come away with the cover

10.4 Improvised holding tool (see text)

10.5 Use extractor bolt to draw rotor off its taper

11 Dismantling the engine/gearbox unit: removing the clutch assembly

1   The clutch may be removed with the engine unit in or out of the frame. Note that in the former instance it will be necessary to drain the engine oil, and that some provision should be made to catch the surplus which will be released as the cover is removed.

2   Slacken and remove the ten Allen bolts which secure the right-hand cover to the crankcase. Tap around the joint with a soft-faced mallet to free the cover, which can then be lifted away. The clutch pressure plate and springs are secured by five bolts. These should be removed evenly and progressively until spring tension has been released. The bolts can then be unscrewed fully and the pressure plate assembly lifted away. Slide the clutch plain and friction plates out of the clutch drum and place them to one side.

3   It will be necessary to lock the transmission while the clutch centre nut is removed. This can be done by employing one of the methods described in Section 10 paragraph 4, or by using the manufacturer's holding tool shown in Fig. 1.7. Straighten the tab washer which secures the clutch centre nut and remove the latter whilst the shaft is locked through the transmission. The tab washer, clutch centre and large plain spacer can now be slid off the input shaft end.

4   The clutch drum is carried on a large diameter centre sleeve with a caged needle roller bearing interposed between the two parts. The centre sleeve is extracted together with the bearing to provide clearance for the clutch drum to be removed. The sleeve is not a tight fit but some difficulty may be experienced in obtaining a good grip on it. To facilitate removal, two 6 mm threads are provided in the outer face of the sleeve, and one or two of the casing screws can be inserted to obtain purchase. Support the clutch drum with one hand and pull out the sleeve followed by the bearing. The clutch drum can now be moved towards the rear of the crankcase and lifted away.

11.2a Remove the clutch cover ...

11.2b ... and release clutch bolts and springs

11.2c Cover and clutch plates can now be removed

11.3 Remove clutch centre nut and clutch centre

11.4a Use 6mm screws to help extract clutch sleeve

11.4b Large clutch drum bearing can now be removed together with the clutch drum

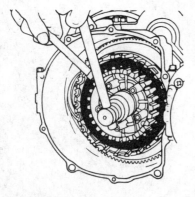

Fig. 1.7 Method of clutch retention using service tool

12 Dismantling the engine/gearbox unit: removing the oil pump and drive

1   It is possible to remove the oil pump with the engine unit in the frame, though this approach is far from convenient because much of the work will have to be done from the underside of the machine. If this method is unavoidable, start by draining the engine oil, then remove the clutch as described in Section 11 of this Chapter. To facilitate access it will be necessary to raise the machine leaving the sump area clear by placing suitable **secure** supports beneath the stand and wheels. Lash the machine to the roof or walls of the workshop using strong rope, or if possible webbing tie-downs, to prevent the machine from toppling over. Alternatively the machine can be leant over and propped securely to give better access. Both methods are somewhat precarious and great care must be taken to avoid damage or injury should the machine fall.

2   Slacken and remove the thirteen Allen screws which hold the sump to the crankcase underside. Lift the sump away, noting the wiring which runs from the oil level switch. Where appropriate, trace this back to the connector and separate it.

The pump is retained by three Allen screws to the crankcase underside; two of the screws also retain a pressed steel shroud around the pump pinion. Lift the shroud clear to expose the pinion and chain.

3   Working from the side of the unit, withdraw the pump drive pinion from the end of the gearbox input shaft, disengaging the chain as it is slid out of position.

13 Dismantling the engine/gearbox unit: removing the starter motor and gearchange mechanism

1   The above components can be removed with the engine unit in or out of the frame. In the case of the gearchange mechanism it will be necessary to drain the engine and middle gear case oil before the left-hand cover is removed, for obvious reasons.

2   The starter motor is retained by two bolts which pass through lugs at its left-hand end. Once these have been removed the motor can be pulled clear of the casing. Note that some resistance may be encountered due to the large O-ring which seals the motor boss and it may help if the motor is moved from side-to-side as it is pulled clear.

3   Where this has not been done before, remove the gearbox lever. Where this is of the conventional clamped type, remove the pinch bolt and slide it off its splines. In the case of machines fitted with a remote linkage pivoted on a pin projecting from the middle gearcase cover, prise off the E-clip which secures the lever to its pivot. Slide the lever clear to expose the rear link which can then be removed by releasing its clamp bolt and sliding it off its spline.

4   Release the ten Allen screws which secure the left-hand cover to the engine unit and lift it away. It will be noted that the cover serves to close two separate compartments. The rear section houses the middle gear assembly; the rubber blank which fits into the recess of the lower front division can be removed if required. The front part of the casing contains the gear selector mechanism. This consists of the gear selector arm and claw which engages the protruding end of the selector drum, and to the rear of this, a centralising arm. The latter is connected to the selector arm by gear teeth.

5   Remove the centralising arm by pulling it out of its casing hole, letting the spring ends slide off the locating pin. The selector arm is removed in a similar manner, having lifted the claw end to free it from the selector drum. The detent arm assembly is withdrawn together with the selector arm assembly.

12.2a Remove sump to expose oil pump

12.2b Release holding bolts and remove shroud

12.2c Disengage drive chain and remove the pump

13.2 Release bolts and pull starter motor clear of casing

13.4 Remove casing screws and lift cover away

13.5 Disengage and remove the selector mechanism

14 Dismantling the engine/gearbox unit: removing the middle gear assembly

1   The middle gear assembly comprises a third gearbox shaft consisting of a transmission shock absorber and terminating in a heavy bevel gear, and the middle driven gear running at 90° to the former and terminating in the output flange. To facilitate removal of the driving shaft and subsequent crankcase separation it will be necessary to remove two bearing retainers, each of which is retained by securely staked 'TORX' screws. These are similar in design to Allen screws, but have a star-shaped internal profile rather than a hexagon. It follows that a 'TORX' key will be required to remove them.

2   The best solution to this problem is to place an order with a Yamaha parts stockist for four new 'TORX' screws and the appropriate key, noting that new screws should be fitted during reassembly. An alternative is to obtain the key from a tool supplier, or, as was done during the workshop project, to modify an Allen key to fit the 'TORX' heads. This can be done by careful use of a three cornered file as shown in Fig. 1.8. It must be stressed that further dismantling must not be attempted until the 'TORX' screws have been removed. Whilst crankcase separation may just be possible with the retainers in position, they must be removed prior to reassembly of the crankcase halves.

3   Start by removing the four bolts which hold the driven middle gear housing to the rear of the crankcase. Note the position of the shims fitted between the housing and the crankcase. These control the gear meshing depth and must be refitted in the same positions as they were prior to removal. Grasp the output flange and pull the assembly out of the crankcase. If removal proves difficult, slacken the two casing bolts on either side of the housing and tap around the housing boss using a soft-faced mallet.

4   Using the appropriate key, remove the 'TORX' screws which secure the driving gear bearing retainers. It should be noted that these are staked into indentations in the retainers and that this will make removal difficult. Since the screws should be renewed it is advisable to drill out the staked areas to avoid any risk of destroying the internal profile of the screw heads. Take care during this operation to avoid drilling into or through the retainers. Removal of the driving gear and shaft cannot be completed until the crankcase halves have been separated.

14.3 Slacken casing bolts (arrowed) if housing is a tight fit

14.4a Drill out peened edge of screw to permit removal

14.4b Modified Allen key can be used in place of Torx key

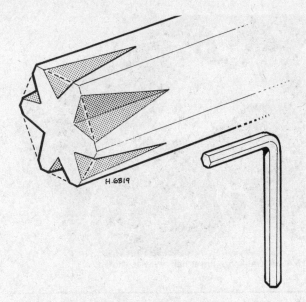

Fig. 1.8 Fabricated Torx bolt key

15 Dismantling the engine/geatbox unit: separating the crankcase halves

1 The crankcase halves are secured by a total of 38 bolts, the bolt numbers and locations being shown in Fig. 1.9. Starting with the upper crankcase half, slacken each bolt in sequence, starting at 38 and working backwards, turning each one by about a ½ turn at a time to prevent distortion. Repeat the sequence until every bolt is no more than finger-tight, then remove them completely.

2 Turn the unit over, taking care not to damage the timing pointer, and allow it to rest on the rear of the crankcase and the cylinder holding studs. Slacken the lower crankcase bolts in the same manner as described above, starting at bolt 23 and working back to 1. Note that the latter is located inside the oil filter housing aperture, and that bolts No 2, 19, 20, 21 and 23

are within the area normally covered by the sump. Note that it is inadvisable to subject the cylinder holding studs to excessive sideways pressure since this may bend them. It is worth placing blocks beneath the rear of the crankcase so that the unit rests squarely on the stud ends.

3 The crankcase halves are now ready for separation, this being accomplished by lifting the **lower** casing half off the upper half, the latter remaining supported on the bench. Separation may be impaired by the jointing compound and the locating dowels, both of which will resist separation until some initial movement has been made. It helps to tap around the joint with a soft-faced mallet, taking care not to strike the more fragile parts of the casing.

4 As the lower crankcase half is freed, check that the crankshaft, input shaft and the middle driving gear shaft remain in the inverted upper casing half. The gearbox output shaft, selector drum and the selector forks are contained in the crankcase lower half and will come away with it.

15.2a Note location of the bolt inside filter recess

15.2b Note also bolts within sump area

15.4 Lift crankcase lower half away as shown

16 Dismantling the engine/gearbox unit: removing the crankcase components

1   Lift the middle driving gear shaft from the inverted upper crankcase half. If necessary, tap the assembly a few times to free the bearings from their supporting bosses. Lay this assembly, and all parts which are removed subsequently, on a clean surface to await further inspection. The gearbox input shaft can be lifted from the casing as described above.

2   Using an impact driver, slacken and remove the three countersunk screws which hold the bearing housing to the inner face of the alternator housing. Hold the starter clutch and sprocket assembly in one hand and withdraw the bearing housing and shaft from the crankcase. The small oil spray nozzle should be displaced and placed to one side to avoid damage. Lift the starter clutch/sprocket assembly clear of the casing recess and disengage it from the chain. Mark one of the chain side plates at this stage to ensure that the chain is refitted facing in the same direction. If the direction of travel of a part worn Hy-vo chain is changed, vibration and noise can result.

3   Bend back the tab washer which locks the starter idler pinion shaft retaining bolt. Remove the bolt and the tab washer/retainer to allow the shaft to be displaced and the pinion removed. Grasp the ends of the crankshaft and lift it clear of the crankcase. If it proves to be firmly located try tapping it free using a soft faced mallet. Lift the crankshaft clear, guiding the camshaft chain through its recess in the crankcase.

4   Moving to the crankcase lower half, displace the gear selector fork shaft and withdraw it from the casing, removing each of the selector forks in turn. Note the position of each one and fit it back on the shaft in the correct order to act as a reminder during reassembly. The selector drum is retained by a locating pin which runs in a bore in the crankcase close to the neutral switch. The pin engages in a groove in the drum and is secured by a bolt and retaining plate at its upper end. Remove the bolt and retainer and displace the locating pin. A small magnet is useful here, or alternatively the casing can be inverted and the pin shaken free. Slacken and remove the neutral switch using a socket or box spanner to avoid damage to the terminal area, then displace and remove the selector drum.

5   Slacken and remove the three Allen screws which secure the bearing retainer plate to the crankcase at the left-hand end of the gearbox output shaft. The design of the crankcase casting is such that the left-hand bearing and the 5th gear pinion must be displaced and pulled off the shaft end before the rest of the cluster can be removed. The bearing is not a tight fit in the casing boss and can be pushed out by sliding the cluster

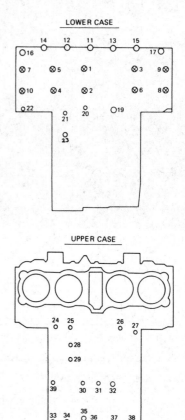

Fig. 1.9 Crankcase bolt tightening sequence

towards it. Once the bearing has been removed the 5th gear pinion can be slid off. Slide the rest of the gear cluster to the left to allow the right-hand end to be lifted out through the underside of the casing. The cluster can then be withdrawn completely.

17 Examination and renovation: general

1   Before examining the parts of the dismantled engine unit for wear it is essential that they should be cleaned thoroughly. Use a petrol/paraffin mix or a high flash-point solvent to remove all traces of old oil and sludge which may have accumulated within the engine. Where petrol is included in the cleaning agent normal fire precautions should be taken and cleaning should be carried out in a well ventilated place.

2   Examine the crankcase castings for cracks or other signs of damage. If a crack is discovered it will require a specialist repair.

3   Examine carefully each part to determine the extent of wear, checking with the tolerance figures listed in the Specifications section of this Chapter or in the main text. If there is any doubt about the condition of a particular component, play safe and renew.

4   Use a clean lint free rag for cleaning and drying the various components. This will obviate the risk of small particles obstructing the internal oilways and causing the lubrication system to fail.

5   Various instruments for measuring wear are required,

including a vernier gauge or external micrometer and a set of standard feeler gauges. The machine's manufacturer recommends the use of Plastigage for measuring radial clearance between working surfaces such as shell bearings and their journals. Plastigage consists of a fine strand of plastic material manufactured to an accurate diameter. A short length of Plastigage is placed between the two surfaces, the clearance of which is to be measured. The surfaces are assembled in their normal working positions and the securing nuts or bolts fastened to the correct torque loading; the surfaces are then separated. The amount of compression to which the gauge material is subjected and the resultant spreading indicates the clearance. This is measured directly, across the width of Plastigage, using a pre-marked indicator supplied with the Plastigage kit. If Plastigage is not available both an internal and external micrometer will be required to check wear limits. Additionally, although not absolutely necessary, a dial gauge and mounting bracket is invaluable for accurate measurement of end float, and play between components of very low diameter bores – where a micrometer cannot reach.

After some experience has been gained the state of wear of many components can be determined visually or by feel and thus a decision on their suitability for continued service can be made without resorting to direct measurement.

16.1 Input shaft can be lifted clear of crankcase

16.2a Remove retaining screws and withdraw shaft and bearing housing

16.2b Remove oil nozzle and put it somewhere safe

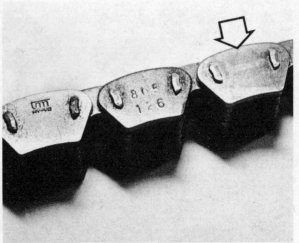

16.2c Starter clutch can be lifted clear and disengaged from chain

16.2d Mark one of the chain links to denote L or R

16.3 Crankshaft assembly can be lifted away from casing

16.4a Withdraw support shaft noting position of selector forks

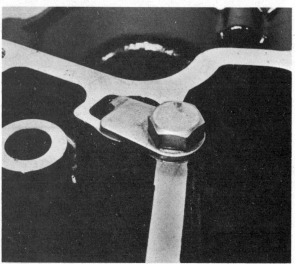

16.4b Release bolt and retainer plate ...

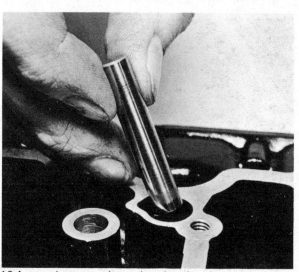

16.4c ... and remove selector drum locating pin

16.4d Neutral switch must be unscrewed ...

16.4e ... to allow removal of selector drum

16.5a Remove bearing retainer plate (Note oil feed drilling)

16.5b Displace the bearing and 5th gear pinion ...

16.5c ... and manoeuvre the output shaft clear of crankcase

18 Examination and renovation: main bearing and big-end bearing selection

1   With all modern motorcycle engines fitted with plain bearing crankshafts, careful measurement and checking is of vital importance during an overhaul. Such a requirement poses a considerable problem for the owner because of the fine tolerances used and the equipment necessary to check these with appropriate accuracy. It should be noted at the outset that wear is unlikely to be present unless an extremely high mileage has been covered or some major catastrophe has befallen the engine unit. In practice this means that in normal use, provided regular oil changes are made, the crankshaft will outlast the rest of the engine components by a considerable margin.

2   The crankshaft has finely ground main and big-end journals supported in renewable bearing shells or inserts. Both the crankshaft journals and the casing are coded to indicate their exact size within the manufacturing tolerances. To cater for variations within this range, the bearing inserts are supplied in a range of thicknesses so that the correct clearance can be achieved. Wear of the crankshaft journals is not normally to be expected, and no provision is made for reclaiming a worn crankshaft by re-grinding it and fitting undersize shells. If it proves to be worn or damaged it must be renewed.

3   Start by giving the journals a close visual examination. Each one should be clean and bright with no perceptible wear marks or scoring. Any evidence of this will suggest renewal of the crankshaft is required and it is best to seek specialist advice from a Yamaha dealer before proceeding further. Do bear in mind that an engine that has a badly worn or damaged crankshaft is very likely to require complete and extensive overhaul in all other areas. If this proves to be the case it will be a very expensive operation and consideration should be given to fitting a good secondhand unit as a more practicable alternative. If this course of action is chosen, check the local motorcycle breakers (wreckers) who may be able to supply a good low mileage engine unit at very reasonable cost. As with all secondhand purchases, however, check the prospective item very thoroughly before parting with any money. Should the crankshaft appear to be in good condition, the next items to be checked are the bearing inserts themselves.

4   Wear is usually evident in the form of scuffing or score marks in the bearing surface. It is not possible to polish these marks out in view of the very soft nature of the bearing surface and the increased clearance that will result. If wear of this nature is detected, the crankshaft must be checked for ovality as described in the following Section.

5   Failure of the big-end bearings is invariably accompanied by a pronounced knock within the crankcase. The knock will become progressively worse and vibration will also be experienced. It is essential that bearing failure is attended to without delay because if the engine is used in this condition there is a risk of breaking a connecting rod or even the crankshaft, causing more extensive damage.

6   Before the big-end bearings can be examined the bearing caps must be removed from each connecting rod. Each cap is retained by two high tensile bolts. Before removal, mark each cap in relation to its connecting rod so that it may be replaced correctly. As with the main bearings, wear will be evident as scuffing or scoring and the bearing shells must be replaced as complete sets.

7   Replacement bearing shells for either the big-end or main bearings are supplied on a selected fit basis (ie; bearings are selected for correct tolerance to fit the original journal diameter), and it is essential that the parts to be used for renewal are of identical size. Code numbers stamped on various components are used to identify the correct replacement bearings for both the crankshaft, main bearing and the big-end journals. The journal size numbers are stamped on the crankshaft left-hand outside web; the block of four numbers are for the big-end bearing journals and the block of five numbers

for the main bearing journals. The main bearing housing numbers are etched at the rear of the upper crankcase half as shown in the accompanying photographs.

8   A range of bearing shells (inserts) is available, selection being made by subtracting the crankshaft journal number from the appropriate crankcase number. In the example shown, bearing number 5 (the right-hand outer main bearing) would give a figure of 3. This figure can then be compared with the table below to find the correct insert colour:

Insert colour code

No 1	Blue
No 2	Black
No 3	Brown
No 4	Green
No 5	Yellow

In the example shown, a brown insert is required.

9   A similar method is used to select inserts for the big-end bearings. In this instance, subtract the number stamped on the crankshaft from the number marked in ink on the flat face of the connecting rod and cap. Again, compare the resulting figure with the table to obtain the correct insert colour for each bearing. If it is wished to avoid the above calculations and new inserts are to be ordered, be sure to quote **all** of the above numbers to the Yamaha dealer when placing the order.

10 Bearing shells are relatively inexpensive and it is prudent to renew the entire set of main and big-end bearing shells when the engine is dismantled completely, especially in view of the amount of work which will be necessary at a later date if any of the bearings fail. Always renew all sets of main and big-end bearings together.

19 Examination and renovation: crankshaft assembly

1   If wear has necessitated the renewal of the big-end and/or main bearing shells, the crankshaft should be checked with a micrometer to verify whether ovality has occurred. If the reading on any one journal varies by more than 0.04 mm (0.0015 in) the crankshaft should be renewed.

2   Mount the crankshaft by supporting both ends on V-blocks or between centres on a lathe and check the run-out at the centre main bearing surfaces by means of a dial gauge. The run-out will be half that of the gauge reading indicated. A measured run-out of more than 0.03 mm (0.001 in) indicates the need for crankshaft renewal. It is wise, however, before taking such drastic (and expensive) action, to consult a Yamaha specialist.

3   The clearance between any set of bearings and their respective journal may be checked by the use of Plastigage.

Plastigage is a graduated strip of plastic material that can be compressed between two mating surfaces. The resulting width of the material when measured against the graduated strip supplied will give the amount of clearance. For example if the clearance in the big-end bearing was to be measured, Plastigage should be used in the following manner.

4   Cut a strip of Plastigage to the width across the bearing to be measured. Place the Plastigage strip across the bearing journal so that it is parallel with the crankshaft. Place the connecting rod complete with its half shell on the journal and then carefully replace the bearing cap complete with half shell onto the connecting rod bolts. Replace and tighten the retaining nuts to the correct torque and then loosen and remove the nuts and the bearing cap. Do not rotate the crankshaft during this stage. Using the graduated markings on the Plastigage packet, compare the width of the markings with that of the compressed strip to find the bearing clearance (see photographs). Clearances may be checked also by direct measurement of each journal and bearing using external and internal micrometers.

5   The crankshaft has drilled oil passages which allow oil to be fed under pressure to the working surfaces. Blow the passages out with a high pressure air line to ensure they are absolutely free. Blanking plugs in the form of small steel balls are fitted in each web, to close off the outer ends of the passages. Check that these balls, which are peened into place, are not loose. A plug coming free in service will cause oil pressure loss and resultant bearing and journal damage.

6   When refitting the connecting rods and shell bearings, note that under no circumstances should the shells be adjusted with a shim, 'scraped in' or the fit 'corrected' by filing the connecting rod and bearing cap or by applying emery cloth to the bearing surface. Treatment such as this will end in disaster; if the bearing fit is not good, the parts concerned have not been assembled correctly. This advice also applies to main bearing shells. Use new big-end bolts too – the originals may have stretched and weakened.

7   Oil the bearing surfaces before reassembly takes place and make sure the tags of the bearing shells are located correctly. Lubricate the bolt threads with molybdenum disulphide grease and run the nuts into position. Yamaha advise that when tightening the big-end bolt nuts, the following should be noted; start by tightening the nuts evenly and then, when a torque setting of 2.0 kgf m (14.5 lbf ft) has been reached **do not** stop tightening until the final torque figure of 2.5 kgf m (18.1 lbf ft) has been achieved. If tightening is interrupted between these two figures, slacken the bolt to below the lower figure and start again. Once the big-end caps are secure, check each one to ensure smooth and free rotation. If there is any sign of tightness something is wrong. Dismantle the bearing and re-check the clearance with Plastigage.

18.7a Journal size coding is stamped on crankshaft web

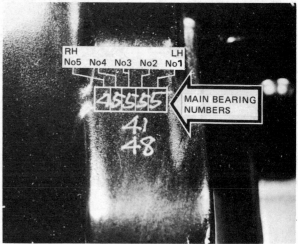

18.7b Main bearing numbers are etched on rear of crankcase

18.9 Connecting rod coding is etched as shown

19.4a Place Plastigage strip on bearing journals and assemble

19.4b Dismantle after tightening to normal torque values

19.4c Measure clearance against scale on edge of pack

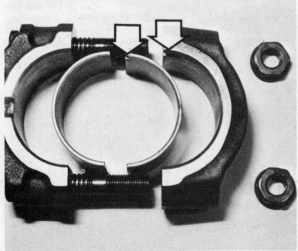

19.7a Note location tab on edge of bearing shells

19.7b Lubricate shells with oil and bolts with moly grease

19.7c Torque down bearing caps as described in text

20 Examination and renewal: gaskets, oil seals and O-rings

1   The engine/gearbox uses a combination of gaskets, oil seals and O-rings to contain engine compression and oil and to prevent the ingress of water or road dirt. By this stage of the overhaul it should be evident that it is not an operation to be taken lightly or to be repeated unnecessarily. To this end it is worth renewing all of the above using genuine Yamaha parts. The various parts are obtainable individually if required, but are normally purchased as a set. Yamaha supply two gaskets sets; a 'Top end' set covering the cylinder base gasket upwards and a second set covering the crankcase components.

2   In addition to the above, an oil seal is fitted to the left-hand end of the starter clutch shaft and another to the left-hand end of the crankshaft. The opposing crankshaft end is covered by a sealing cap. These components should be ordered in addition to the gasket sets.

3   It is worth noting that it is often difficult to judge the condition of an oil seal or gasket, and the type of mark on the sealing surface which would allow leakage after reassembly may not be obvious under visual examination. Some of the gaskets and seals may be in reusable condition, but given the inconvenience that would result if an error was made it should be considered better to renew all of them as a precautionary measure.

21 Examination and renovation: connecting rods

1   It is unlikely that any of the connecting rods will become damaged during normal usage unless an unusual occurrence such as a dropped valve causes the engine to lock. This may well bend the connecting rod in that cylinder. Carelessness when removing a tight gudgeon pin can also give rise to a similar problem. It is not advisable to straighten a bent connecting rod; renewal is the only satisfactory solution.

2   The bearing surface of each small-end eye is provided by a cold-metal-sprayed coating with a bronze base. If the small-end eye wears, the connecting rod in question must be renewed. If the clearance between a gudgeon pin and small-end is excessive, check first that the wear is in the eye and not the gudgeon pin. This will prevent the unnecessary renewal of a sound component. Always check that the oil hole in the small-end eye is not blocked since if the oil supply is cut off, the bearing surfaces will wear very rapidly.

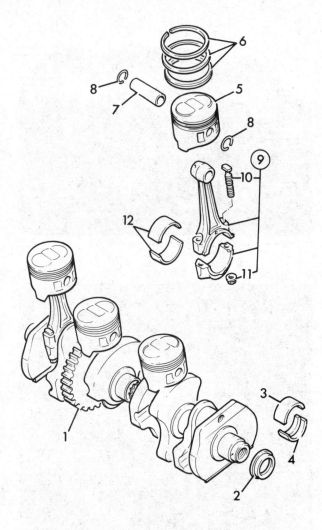

Fig. 1.10 Crankshaft and pistons

1	Crankshaft	6	Piston rings
2	Oil seal	7	Gudgeon pin
3	Main bearing upper half – 5 off	8	Circlip
4	Main bearing lower half – 5 off	9	Connecting rod
		10	Bolt
5	Piston – 4 off	11	Nut
		12	Big-end bearing

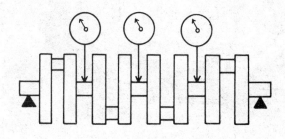

Fig. 1.11 Crankshaft run-out check using a dial gauge

22 Examination and renovation: cylinder bores

1   The usual indication of badly worn cylinder bores and pistons is excessive smoking from the exhausts, high crankcase compression which causes oil leaks, and piston slap, a metallic rattle that occurs when there is little or no load on the engine. If the top of the cylinder bore is examined carefully, it will be found that there is a ridge at the front and back the depth of which will indicate the amount of wear which has taken place. This ridge marks the limit of travel of the top piston ring.

2   Since there is a difference in cylinder wear in different directions, side to side and back to front measurements should be made. Take measurements at three different points down the length of the cylinder bore, starting just below the top piston ring ridge, then about 60 mm (2½ in) below the top of the bore and the last measurements about 25 mm (1 in) from the bottom of the cylinder bore. The cylinder measurement as standard and the service limit are as follows:

	XJ650	XJ750
Standard bore diameter	63.00 mm (2.480 in)	65.00 mm (2.559 in)
Service limit	63.10 mm (2.484 in)	65.10 mm (2.563 in)
Maximum allowable taper	0.05 mm (0.002 in)	
Maximum allowable ovality	0.01 mm (0.0004 in)	

Note that where the cylinder block has been rebored at a previous overhaul, the increased bore size must be taken into consideration when making bore wear measurements. The rebore sizes are in 0.25 mm (0.010 in) increments. If any of the cylinder bore inside diameter measurements exceed the service limit the cylinder must be bored out to take the next size of piston. If there is a difference of more than 0.05 mm (0.002 in) between any two measurements the cylinder should, in any case, be rebored.

3   Oversize pistons are available in four oversizes: 0.25 mm (0.010 in); 0.50 mm (0.020 in); 0.75 mm (0.030 in) and 1.0 mm (0.040 in).

4   Check that the surface of the cylinder bore is free from score marks or other damage that may have resulted from an earlier engine seizure or a displaced gudgeon pin. A rebore will be necessary to remove any deep scores, irrespective of the amount of bore wear that has taken place, otherwise a compression leak will occur.

5   Make sure the external cooling fins of the cylinder block are not clogged with oil or road dirt which will prevent the free flow of air and cause the engine to overheat.

6   If, for any reason, the cylinder block and cylinder head holding down studs are removed from the crankcase, they should be smeared with Loctite before they are reinserted.

23 Examination and renovation: pistons and piston rings

1   Attention to the pistons and piston rings can be overlooked if a rebore is necessary, since new components will be fitted.

2   If a rebore is not necessary, examine each piston carefully. Reject pistons that are scored or badly discoloured as the result of exhaust gases by-passing the rings.

3   Remove all carbon from the pistons crowns, using a blunt scraper, which will not damage the surface of the piston. Clean away carbon deposits from the valve cutaways and finish off with metal polish so that a smooth, shining surface is achieved. Carbon will not adhere so readily to a polished surface.

4   Small high spots on the back and front areas of the piston can be carefully eased back with a fine swiss file. Dipping the file in methylated spirits or rubbing its teeth with chalk will prevent the file clogging and eventually scoring the piston. Only very small quantities of material should be removed, and never enough to interfere with the correct tolerances. Never use emery paper or cloth to clean the piston skirt; the fine particles of emery are inclined to embed themselves in the soft aluminium and consequently accelerate the rate of wear between bore and piston.

5   Measure the outside diameter of the piston about 10 mm (0.4 in) up from the skirt at right angles to the line of the gudgeon pin. To determine the piston/cylinder barrel clearance, subtract the maximum piston measurement from the minimum bore measurement. If the clearance exceeds 0.1 mm (0.004 in), the piston should ideally be renewed (given that the cylinder bore is within limits). This however, is seeking perfection, and an additional clearance of perhaps 0.025 mm (0.001 in) will not reduce engine performance dramatically.

6   Check that the gudgeon pin bosses are not worn or the circlip grooves damaged. Check that the piston ring grooves are not enlarged. This can be done by measuring the clearance between the piston ring and groove. This should be as follows:

Piston ring to groove clearance

Top	0.03 – 0.07 mm (0.0012 – 0.0028 in)
2nd	0.02 – 0.06 mm (0.0008 – 0.0024 in)
Oil	0.00 mm (0.000 in)

7   Piston ring wear can be measured by inserting the rings in the bore from the top and pushing them down with the base of the piston so that they are square with the bore and close to the bottom of the bore where the cylinder wear is least. Place a feeler gauge between the ring ends. If the clearance exceeds the service limit the ring should be renewed. The expander band of the oil control ring cannot be measured. In practice, if wear of the two side rails exceeds the limit, the three components should be renewed.

Piston ring end gap (installed)

Top and 2nd	0.15 – 0.35 mm (0.0059 – 0.0318 in)
Service limit	1.00 mm (0.0394 in)
Oil (side rails only)	0.30 – 0.90 mm (0.012 – 0.035 in)
Service limit	1.50 mm (0.0591 in)

It is advised that, provided new rings have not been fitted recently, a complete set of rings be fitted as a matter of course whenever the engine is dismantled. This action will ensure maintenance of compression and performance. If new rings are to be fitted to the cylinder bores which are in good condition and do not require a rebore, it is essential to have the surface of the bores honed lightly. This operation is known as glazebusting and as the name suggests, it removes the mirror smooth surface which has been produced by the previous innumerable up and down strokes of the piston and rings. If the glaze is not removed, the new rings will glide over the surface, making the running-in process unnecessarily protracted. Note that new rings should not be fitted to a part-worn but un-honed cylinder bore, as the resulting wear ridge may break the top ring, particularly at high engine speeds. The resulting debris could have very expensive results. Ensure that the wear ridge is removed by honing.

8   Check that there is no build up of carbon either in the ring grooves or the inner surfaces of the rings. Any carbon deposits should be carefully scraped away. A short length of old piston ring fitted with a handle and sharpened at one end to a chisel point is ideal for scraping out encrusted piston ring grooves.

9   All pistons have their size stamped on the piston crown, original pistons being stamped standard (STD) and oversize

pistons having the amount of oversize indicated. Similarly oversize piston rings are stamped on the upper edge. The oil ring expander is marked with coloured paint to identify oversizes. These are as shown below:

1st oversize (+0.25 mm)	Brown
2nd oversize (+0.50 mm)	Blue
3rd oversize (+0.75 mm)	Black
4th oversize (+1.00 mm)	Yellow

23.2 Check condition of pistons and remove any carbon

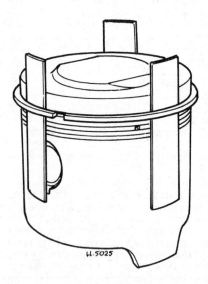

H.5025

Fig. 1.12 Method of removing and replacing piston rings

24 Examination and renovation: cylinder head and valves

1   Before dismantling the valve gear proper lift out each of the eight bucket type cam followers, together with the adjustment pads. Ensure that each follower is marked clearly, so that it may be replaced in the original recess in the cylinder head. It is good practice to obtain eight marked boxes or bags so that the appropriate valve components can be kept separate. It is best to remove all carbon deposits from the combustion chambers

before removing the valves for inspection and grinding-in. Use a blunt ended chisel or scraper so that the surfaces are not damaged. Finish off with a metal polish to achieve a smooth, shining surface. If a mirror finish is required, a high speed felt mop and polishing soap may be used. A chuck attached to a flexible drive will facilitate the polishing operation.

2   A valve spring compression tool must be used to compress each set of valve springs in turn, thereby allowing the split collets to be removed from the valve cap and the valve springs and caps to be freed. Keep each set of parts separate and mark each valve so that it can be replaced in the correct combustion chamber. There is no danger of inadvertently replacing an inlet valve in an exhaust position, or vice-versa, as the valve heads are of different sizes. The normal method of marking valves for later identification is by centre punching them on the valve head. This method is not recommended on valves, or any other highly stressed components, as it will produce high stress points and may lead to early failure. Tie-on labels, suitably inscribed, or a spirit-based marker, are ideal for the purpose. Because of the cylinder head design, modification of an existing valve spring compressor may be necessary so that it clears the high walls of the cam and valve spring compartments. Remove the oil seal cap from each valve guide. As each valve is removed, check that it will pass through the guide bore without resistance. After high mileages have been covered it is possible that the collet groove will have spread and if this increased diameter is pulled through the guide bore it will enlarge it. If any resistance is encountered, relieve the high spots with fine abrasive paper until the valve can be removed easily.

3   Before giving the valves and valve seats further attention, check the clearance between each valve stem and the guide in which it operates. Clearances are as follows:

Standard	Service limit
Inlet valve/guide clearance	
0.010 – 0.037 mm	0.10 mm (0.004 in)
(0.0004 – 0.0015 in)	
Exhaust valve/guide clearance	
0.025 – 0.052 mm	0.12 mm (0.005 in)
(0.0010 – 0.0020 in)	

The valve stem/guide clearance can be measured with the use of a dial gauge and a new valve. Place the new valve into the guide and measure the amount of shake with the dial gauge tip resting against the top of the stem. If the amount of wear is greater than the wear limit, the guide must be renewed.

4   To remove an old valve guide, place the cylinder head in an oven and heat it to about 100°C (212°F). The old guide can now be tapped out from the cylinder side. The correct drift should be shouldered with the smaller diameter the same size as the valve stem and the larger diameter slightly smaller than the OD of the valve guide. If a suitable drift is not available a plain brass drift may be utilised with great care. Even heating is essential, if warpage of the cylinder head is to be avoided. Before removing old guides scrape away any carbon deposits which have accumulated on the guide where it projects into the port. Removal of carbon will ease guide movement and help prevent broaching of the guide bore in the cylinder head. If in doubt, seek the advice of a Yamaha specialist. Each valve guide is fitted with an O-ring to ensure perfect sealing. The O-rings must be replaced with new components. New guides should be fitted with the head at the same heat as for removal, following which the seat must be recut to centre the seat with the guide axis.

5   Valve grinding is a simple task. Commence by smearing a trace of fine valve grinding compound (carborundum paste) on the valve seat and apply a suction tool to the head of the valve. Oil the valve stem and insert the valve in the guide so that the two surfaces to be ground in make contact with one another. With a semi-rotary motion, grind in the valve head to the seat, using a backward and forward action. Lift the valve occasionally so that the grinding compound is distributed evenly. Repeat the

application until an unbroken ring of light grey matt finish is obtained on both valve and seat. This denotes the grinding operation is now complete. Before passing to the next valve, make sure that all traces of the valve grinding compound have been removed from both the valve and its seat and that none has entered the valve guide. If this precaution is not observed, rapid wear will take place due to the highly abrasive nature of the carborundum base.

6   When deep pits are encountered, it will be necessary to use a valve refacing machine and a valve seat cutter, set to an angle of 45°. If, after recutting the seat, it is found that the seat width exceeds 2.0 mm (0.08 in) further cuts with 30° and 60° cutters must be made to reduce the seat width to the standard of 1.3 mm (0.05 in). In view of the high cost of the seat cutters and of the expertise involved it is suggested that this work be carried out by a Yamaha Service Agent or other specialist. Never resort to excessive grinding because this will only pocket the valves in the head and lead to reduced engine efficiency. If there is any doubt about the condition of a valve, fit a new one.

7   Examine the condition of the valve collets and the groove on the valve stem in which they seat. If there is any sign of damage, new parts should be fitted. Check that the valve spring collar is not cracked. If the collets work loose or the collar splits whilst the engine is running, a valve could drop into the cylinder and cause extensive damage.

8   Check the free length of each of the valve springs. The springs have reached their serviceable limit when they have compressed to the limit readings given in the Specifications Section of this Chapter.

9   Reassemble the valve and valve springs by reversing the dismantling procedure, referring to Section 47 for details.

10 Check the cylinder head for straightness, especially if it has shown a tendency to leak oil at the cylinder head joint. If there is any evidence of warpage, provided it is not too great, the gasket face may be lapped on a surface plate or a sheet of plate glass. Place a sheet of 400 or 600 grit abrasive paper on the surface plate. Lay the cylinder head, gasket-face down, on the paper and gently rub it with an oscillating motion to remove any high spots. Lift the head at frequent intervals to inspect the progress of the operation, and take care to remove the minimum amount of material necessary to restore a flat sealing surface.

11 In extreme cases, the above method may prove inadequate, necessitating the re-machining or renewal of the cylinder head casting. In this case, the advice of a Yamaha Service Agent or a competent machinist should be sought. It should be remembered that most cases of cylinder head warpage can be traced to unequal tensioning of the cylinder head nuts and bolts by tightening them in incorrect sequence or using incorrect or unmeasured torque settings.

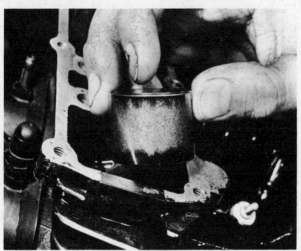

24.1 Remove cam followers and shims

24.2a Assemble valve spring compressor and release collets

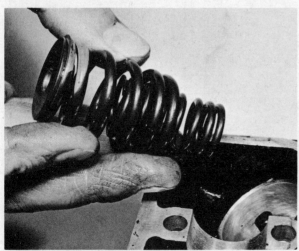

24.2b Remove valve springs and spring seat

24.2c Valve can now be displaced and removed

24.2d Valve stem oil seals should be renewed

25 Examination and renovation: camshafts, camshaft bearings and cam followers

1   The camshafts should be examined visually for wear, which will probably be most evident on the ramps of each cam and where the cam contour changes sharply. Also check the bearing surfaces for obvious wear and scoring. Cam lift can be checked by measuring the height of the cam from the bottom of the base circle to the top of the lobe. If the measurement is less than the service limit given in the Specifications the opening of that particular valve will be reduced resulting in poor performance. Measure the diameter of each bearing journal with a micrometer or vernier gauge. If the diameter is less than the service limit, renew the camshaft.

2   The camshaft bears directly on the cylinder head material and that of the bearing caps, there being no separate bearings. Check the bearing surfaces for wear and scoring. The clearance between the camshaft bearing journals and the aluminium bearing surfaces may be checked using Plastigage material in the same manner as described for crankshaft bearing clearance in Section 19.3 of this Chapter. If the clearance is greater than

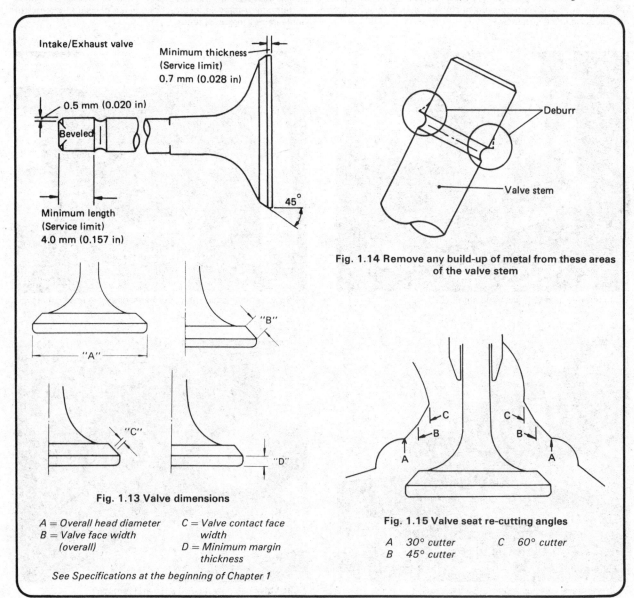

Fig. 1.13 Valve dimensions

A = Overall head diameter
B = Valve face width
    (overall)

C = Valve contact face
    width
D = Minimum margin
    thickness

See Specifications at the beginning of Chapter 1

Fig. 1.14 Remove any build-up of metal from these areas of the valve stem

Fig. 1.15 Valve seat re-cutting angles

A   30° cutter   C   60° cutter
B   45° cutter

given for the service limit the recommended course is to replace the camshaft. If bad scuffing is evident on the camshaft bearing surfaces, due to a lubrication failure, the only remedy is to renew the cylinder head and bearing caps and the camshaft if it transpires that it has been damaged also.

3   Inspect the outer surface of the cam followers for evidence of scoring or other damage. If a cam follower is in poor condition, it is probable that the bore in which it works is also damaged. In extreme cases this may necessitate renewal of both the follower and the cylinder head. Check for clearance between the followers and their bores. If excessive slack is evident, renew the follower.

26 Examination and renovation: camshaft chain, drive sprockets and tensioner blades

1   Check the camshaft drive chain for wear and chipped or broken rollers and links. The cam chain operates under almost ideal conditions and unless oil starvation or prolonged tension adjustment neglect has occurred, it will have a long life. If excessive wear is apparent, or cam chain adjustment has been difficult to maintain correctly, renew the chain.

2   The chain is tensioned via a steel-backed rubber blade by means of a spring loaded plunger. In addition, there is a second blade at the front of the cam chain tunnel which acts as a guide, and a cast bridge piece fitted between the camshaft sprockets which supports the chain upper run. If the rubber material of the guides or blades is worn, the damaged component should be renewed. Extreme wear may indicate a worn drive chain.

3   The cam chain drive sprockets are secured directly to the centre of each camshaft and in consequence are easily re-newable if the teeth become hooked, worn, chipped or broken. The lower sprocket is integral with the crankshaft and if any of these defects are evident, the complete crankshaft assembly

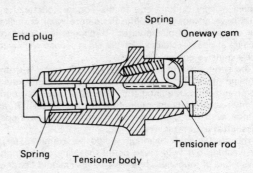

Fig. 1.16 Sectioned view of automatic cam chain tensioner assembly

must be renewed. Fortunately, this drastic course of action is rarely necessary since the parts concerned are fully enclosed and well lubricated, working under ideal conditions.

4   If the sprockets are renewed, the chain should be renewed at the same time. It is bad practice to run old and new parts together since the rate of wear will be accelerated.

27 Examination and renovation: starter shaft, chain and sprockets

1   The starter shaft runs immediately to the rear of the crankshaft and cylinder block and also carries the alternator assembly. It is driven by a Hy-Vo inverted tooth type chain from a sprocket which is integral with the crankshaft. Check the chain for damage or loose link plates and pivot pins. This type

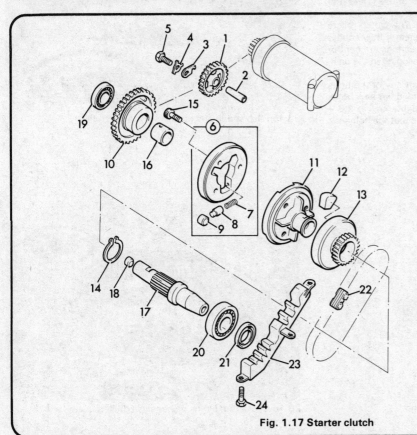

1   Starter idler pinion
2   Spindle
3   Retainer
4   Tab washer
5   Bolt
6   Clutch body
7   Spring – 3 off
8   Plunger – 3 off
9   Roller – 3 off
10 Driven pinion
11 Clutch hub
12 Rubber – 8 off
13 Drive sprocket
14 Circlip
15 Allen bolt – 3 off
16 Bush
17 Primary shaft
18 Plug
19 Bearing
20 Bearing
21 Oil seal
22 Hy-Vo chain
23 Guide
24 Bolt – 3 off

Fig. 1.17 Starter clutch

of chain is very durable and in the normal course of events should have a long service life. Premature wear is unlikely to occur except due to oil starvation. If damage is evident, renew the chain at once. A chain which breaks in service will invariably cause extensive engine damage.

2   The service life of the sprockets is in keeping with that of the chain itself. After considerable use the sprockets may become indented, requiring the renewal of both components. The drive sprocket is an integral part of the crankshaft, and in common with the cam drive sprocket, if wear develops, the crankshaft must be renewed.

3   The starter shaft bearings should be examined for wear after washing out with clean petrol (gasoline). A bearing that shows signs of roughness or excessive play should be renewed. Both bearings may be driven out of position in their housings.

4   The starter shaft itself is unlikely to suffer damage except, perhaps, after a very high mileage has been covered. Check the shaft for truth when rotated and inspect the shaft splines for damage.

28 Examination and renovation: starter clutch

1   The starter motor turns the crankshaft via a roller clutch mounted on the starter shaft. The clutch comprises a body containing three spring-loaded rollers which move along an inclined plane to the driven gear boss. When the starter is operated the rollers are pressed hard against the boss, thus locking the drive and conveying movement to the crankshaft. As soon as the engine fires, the primary shaft, and thus the clutch body, moves faster than the driven gear. The rollers move back against spring pressure and the unit freewheels.

2   This type of starter clutch is almost foolproof in operation and does not suffer from rapid wear. If necessary, the driven gear can be pulled clear of the body for examination, noting that the rollers will become displaced as this is done. Check the surface of the driven gear boss for signs of indentation, renewing the gear if it is worn or damaged. The clutch rollers should also be checked for wear, renewing them if they appear marked or flattened at any point. If the three springs are bent or broken, or if there has been any evidence of slipping in the past, they should be renewed.

3   Check the three Allen bolts for security. If any of the bolts are loose, they should be removed and discarded. Fit new bolts, having coated their threads with Locktite, tightening to 2.8 – 3.2 kgf m (20 – 23 lbf ft). Yamaha also advise that the bolt ends are staked over for additional security.

28.1a Remove hub and pinion unit to reveal rollers

28.1b Push spring and pin inwards to free roller

28.1c Rollers must be free from flats or scoring

28.1d Check that pin moves freely in drilling

29 Examination and renovation: clutch assembly

1 After an extended period of service the clutch linings will wear and promote clutch slip. The limit of wear measured across each inserted plate and the standard measurement is given in the Specifications. When the overall width reaches the limit, the inserted plates must be renewed, preferably as a complete set.

2 The plain plates should not show any excess heating (blueing). Check the warpage of each plate using plate glass or surface plate and a feeler gauge. The maximum allowable warpage is 0.05 mm (0.002 in).

3 Check the free length of each clutch spring with a vernier gauge. After considerable use the springs will take a permanent set thereby reducing the pressure applied to the clutch plates. The correct measurements are given in the Specifications.

4 Examine the clutch assembly for burrs or indentation on the edges of the protruding tongues of the inserted plates and slots worn in the edges of the outer drum with which they engage. Similar wear can occur between the inner tongues of the plain clutch plates and the slots in the clutch inner drum. Wear of this nature will cause clutch drag and slow disengagement during gear changes, since the plates will become trapped and will not free fully when the clutch is withdrawn. A small amount of wear can be corrected by dressing with a fine file; more extensive wear will necessitate renewal of the worn parts.

5 The clutch centre incorporates a spring shock absorber designed to damp out snatching or chattering as drive is taken up. The innermost plain plate is held in position by a large wire retaining clip and beneath it lies a thin diaphragm spring and a flat spring seat. It is unlikely that these components will need to be disturbed unless serious clutch chatter has been evident.

6 Examine the sleeve and caged needle roller bearing in conjunction with the clutch drum's internal bearing surface. If there are any signs of pitting or wear in any of the rubbing surfaces it will be necessary to renew the affected parts. It is unlikely that wear will be a significant problem during the life of the engine unit, but it should be noted that whilst the bearing and sleeve can be renewed separately, the outer race is integral with the clutch drum and thus these must be renewed as a unit.

7 The clutch release mechanism takes the form of a rack and pinion assembly contained in the outer cover and acting upon the clutch via a pull rod and needle roller release bearing. Again, wear is not likely to be a problem during normal use, but the assembly can be freed from the cover if necessary by prising off the E-clip which secures the pinion to the actuating shaft. The shaft can then be slid clear of the casing. Check the two needle roller bearings for wear or damage. If necessary, these can be tapped out of the casing for renewal, noting that it is best to heat the cover in very hot water to facilitate removal. The oil seal can be levered out of the casing bore. Before reassembly, lubricate the bearings and seal with grease.

29.7a Clutch release shaft is secured by an E-clip

29.7b Withdraw shaft to free pinion from spline

29.7c Shaft assembly can now be removed completely

29.7d Seal can be renewed if worn or damaged

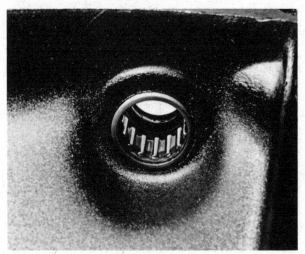

29.7e Lubricate bearings prior to reassembly

29.7f Do not forget to secure the shaft end

30 Examination and renovation: gearbox components

1   The gearbox comprises an input and output shaft carrying five pairs of gears. The input shaft protrudes from the right-hand side of the crankcase and carries the clutch assembly. Power from the crankshaft is transmitted through the clutch to the mainshaft, the selected gear pinion transferring motion to its counterpart on the output shaft. An additional pinion is fitted to the right-hand end of the output shaft forming the take-off point for drive to the middle gear assembly which runs in a compartment to the rear of the main gear train. The middle gear assembly is covered separately in Section 32 of this Chapter.

2   The gearbox is a substantial assembly and does not normally suffer much wear. Light general wear may be expected after extremely high mileages, but the usual causes of accelerated wear or damage can invariably be traced to misuse or poor lubrication. Initial examination can be carried out with the shafts intact, and should be directed at the pinion teeth. Damaged teeth will be self-evident and will of course demand renewal of the component(s) concerned.

3   Look for signs of general wear at the points of contact between the gear teeth. These should normally present a smooth, highly polished profile if in good condition. Pitting of the hardened faces will necessitate renewal of the affected pinions. This is only likely where the machine has been used for frequent short trips which may have prevented the engine unit reaching full operating temperature. Alternatively, neglected oil changes and the resulting thinned and dirty oil can have the same consequences. Pinions with chipped or pitted teeth must always be renewed, there being a real danger of breakage, if re-used. In view of the risk of extensive engine damage that this presents, do not be tempted to 'economise' at this point.

4   The gearbox ball bearings should be cleaned in a high flash-point solvent, or petrol (gasoline) if care is taken to avoid the obvious fire risk, and then checked for play or roughness when the bearing is rotated. Renew all bearings that show signs of damage or old age, noting that a worn bearing can cause accelerated wear of the gearbox components and may lead to poor gear selection.

5   If the general examination described above has indicated a likelihood of general wear, the gearbox shafts should be dismantled for further examination. Always deal with one shaft at a time to avoid confusion, and ensure that each component is kept scrupulously clean. The photographic sequence which accompanies this text shows the step-by-step reassembly order. The shafts are dismantled in the reverse order of this. See the accompanying figure for an exploded view of the gearbox assembly.

6   Check that each pinion moves freely on its shaft, but without undue free play. Check for blueing of the shaft or the gearbox pinions as this can indicate overheating due to inadequate lubrication. The dogs on each pinion should be checked for damage or rounded edges. Such damage can lead to poor gear engagement and will require renewal if extensively damaged. Look for signs of hairline cracks around the pinion bosses and dogs.

7   Set the shaft up on V-blocks and measure the runout at the shaft centre using a dial gauge. If this exceeds 0.08 mm (0.0031 in) the shaft should be renewed. Check the shaft surface for scuffing, scoring or cracks.

8   When rebuilding the gearbox shaft assemblies, check that all washers and circlips are fitted in the correct positions and renew any that appear weakened or bent. Follow the sequence illustrated in the accompanying photographic sequence, ensuring that the assembly is kept clean and is well lubricated.

30.5a Input shaft bearing may be removed with puller ...

30.5b ... or by judicious use of tyre levers, if it proves tight

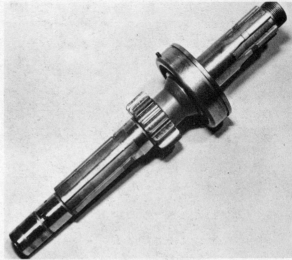

30.5c Input shaft and integral 1st gear pinion

30.5d Fit the 4th gear pinion as shown ...

30.5e ... and retain with plain washer and circlip

30.5f Combined 2nd/3rd gear pinion is fitted next ...

30.5g ... followed by the 5th gear pinion ...

30.5h ... which is secured by a circlip and washer

30.5i Fit bearing to shaft end to complete assembly

30.5j 3rd gear pinion is retained on output shaft as shown

30.5k Fit 4th gear pinion, noting direction of selector groove

30.5l Slide plain washer into position ...

30.5m ... then fit 1st gear pinion

30.5n Middle drive gear is fitted next ...

30.5o ... followed by the bearing

30.5p Fit 2nd gear pinion to LH end of output shaft ...

30.5q ... and secure with plain washer and circlip

30.5r Fit the 5th gear pnion ...

30.5s ... followed by LH bearing to complete

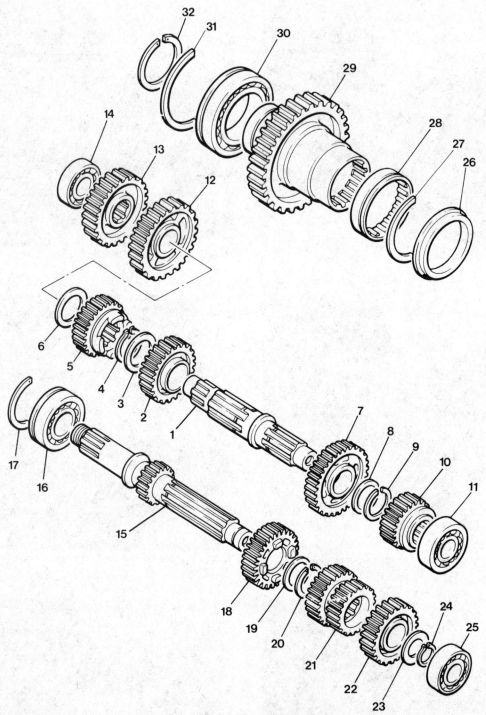

Fig. 1.18 Gearbox components

1	Output shaft	10	Output shaft 5th gear pinion	17	Bearing half-ring	25	Input shaft left-hand bearing
2	Output shaft 3rd gear pinion	11	Output shaft left-hand bearing	18	Input shaft 4th gear pinion	26	Oil seal*
3	Thrust washer	12	Output shaft 1st gear pinion	19	Thrust washer	27	Bearing half-ring
4	Circlip	13	Middle drive pinion	20	Circlip	28	Middle gear shaft left-hand bearing
5	Output shaft 4th gear pinion	14	Output shaft right-hand bearing	21	Input shaft 2nd and 3rd gear pinion	29	Middle gear shaft sleeve gear
6	Thrust washer	15	Input shaft	22	Input shaft 5th gear pinion	30	Middle gear shaft right-hand bearing
7	Output shaft 2nd gear pinion	16	Input shaft right-hand bearing	23	Thrust washer	31	Bearing half ring
8	Thrust washer			24	Circlip	32	Circlip
9	Circlip						

* Fitted to XJ650 – 4KO (UK), XJ650 G, H, LH, RJ, XJ750 RH, RJ

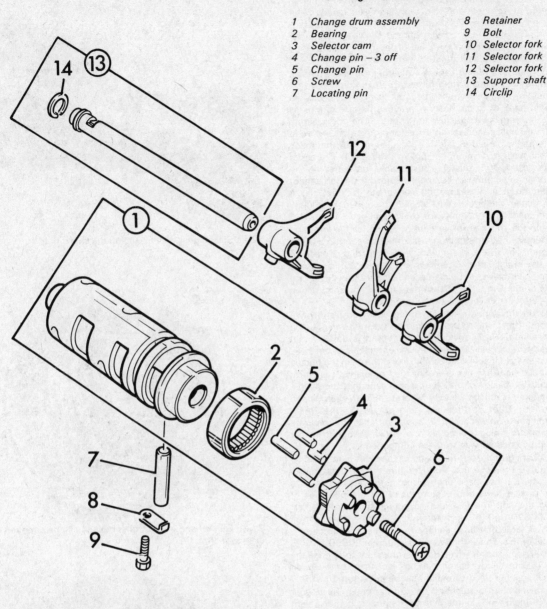

Fig. 1.19 Gear selector drum

1	Change drum assembly	8	Retainer
2	Bearing	9	Bolt
3	Selector cam	10	Selector fork
4	Change pin – 3 off	11	Selector fork
5	Change pin	12	Selector fork
6	Screw	13	Support shaft
7	Locating pin	14	Circlip

31 Examination and renovation: gear selector mechanism

1   Examine the selector forks carefully, noting any signs of wear on the fork ends where they engage with the groove in the gear pinion. It is important that the forked end is not bent in any way. Check the fit of the forks on the support shaft. These should be a light sliding fit with no appreciable free play. Any movement at the bore will be greatly magnified at the fork end, leading to imprecise gear selection.

2   Check the support shaft for wear or scoring, renewing it if obviously damaged. Check for straightness by rolling it across a surface plate or a sheet of plate glass. If the shaft is even slightly bent, it must be renewed.

3   Examine the tracks in the selector drum in conjunction with the selector fork pins which run in them. In practice, the tracks will not wear appreciably, even over high mileages, but the small pins on the selector forks may show signs of flattening, in which case they should be renewed to reduce play. Note that

the pins are integral with the fork, and thus the entire fork will have to be renewed if the pins are worn.

4   Examine the remainder of the selector components, looking for wear or damage, which should be self-evident. Renew any of the springs which appear weakened or are broken. Check the selector drum pins and claw for wear. The large locating pin is unlikely to warrant attention, as is the detent plunger assembly.

32 Examination and renovation: middle gear assembly

1   The middle gear assembly can be dealt with in two separate parts, namely the driving shaft and related components, and the driven shaft assembly. The driving shaft comprises a large sleeve gear which slides over the shaft and is connected by splines to a shock absorber cam. This in turn engages a similar spring-loaded cam which is splined to the shaft itself. The left-hand end of the shaft terminates in a heavy bevel gear.

2   Commence examination by sliding the sleeve gear off the shaft. The sleeve gear is supported by a large diameter journal ball bearing at its right-hand end, with a smaller needle roller bearing and, on some earlier models only, an oil seal on its left-hand end. The latter can be slid off the sleeve gear end for further inspection whilst in the case of the right-hand bearing the circlip which retains it must be removed first.

3   Give the shaft and shock absorber a close visual examination before proceeding further. It should be noted that to dismantle the shock absorber, to gain access to its component parts or to the bearing, requires the use of a press or similar arrangement to allow the spring to be compressed. The hydraulic puller and bearing extractor arranged as shown in the photographs proved successful, but it is not likely that many owners will have this type of equipment available. It is stressed that the assembly is under considerable pressure, so no makeshift arrangement can be applied safely. If the job cannot be tackled at home because of inadequate facilities, let a Yamaha dealer undertake this stage rather than risk injury should an inadequate compressor collapse in use.

4   Should the shock absorber cam or spring, or the left-hand bearing require further attention, assemble the shaft in a puller arrangement, or use a press with a U-shaped top piece to engage the spring seat. Compress the spring sufficiently to allow the collet halves to be displaced using a screwdriver or a magnetic probe. Gradually release pressure on the spring, then remove the puller or the assembly from the press.

5   Slide off the spring seat and spring, followed by the two cam pieces. These should be checked for wear or damage and renewed where this is obvious. In practice it is unlikely that these parts will wear significantly in normal use, but may become damaged if subjected to abuse or if the oil in the middle gear compartment has not been changed regularly. It is worth noting that the design of the crankcase means that the middle gear components run in their own separate oil bath even though a common filler orifice is shared with the main engine/transmission supply. It is important that the separate middle gear drain plug is removed at **every** oil change to ensure that fresh oil finds its way to these components.

6   The bearing at the left-hand end of the shaft is retained by a large nut. To remove this, first straighten the staked area of the nut where it has been peened into the shaft recess. Two Yamaha service tools are recommended for the purposes of removing the nut. The first is a hexagonal holding tool, Part number 90890-04046 which fits into the internal hexagon of the shaft. A large bolt head clamped in a vice will suffice as a substitute for this. The second tool, Part number 90890-04045, is a special deep socket which fits over the shaft to engage the nut. A 46 mm ring spanner, a box spanner, or even an adjustable spanner used carefully may be substituted.

7   The driven shaft assembly consists of a mating bevel gear and shaft running in a journal ball bearing contained in a light alloy housing. The housing and bearing are supplied as an assembly and thus must be renewed as such. The outer end of the shaft carries an output flange secured by a nut and washer and sealed with an O-ring. Dismantling is confined to removing the nut and flange to allow the shaft to be slid clear of the bearing housing. When reassembling the driven shaft note that Loctite should be applied to the flange securing nut threads and the nut staked into position.

8   The driving shaft assembly is rebuilt by reversing the dismantling sequence. When fitting the shock absorber driven cam piece it must be so positioned that the cam lobes are 90° from the row of oil holes in the shaft. A tolerance of ± 1 spline (15°) is allowed in positioning. Ensure when compressing the shock absorber spring that the same attention to safety is given as was exercised during dismantling. Note that the nut retaining the left-hand bearing must be staked to the shaft for security. One final component to be checked is the outrigger bearing which supports the front extension of the driven shaft. This is contained between the crankcase halves and should be examined for wear in the normal way.

32.2a Sleeve gear may be slid off the shaft

32.2b Oil seal and needle-roller bearing can be removed for examination

32.2c Note that RH bearing is retained by a circlip

32.4a Press or hydraulic puller is required to compress spring ...

32.4b ... to allow split collets to be removed

32.5a Slide off the heavy spring seat ...

32.5b ... and then remove the spring

32.5c Remove the two cam pieces ...

32.5d ... to permit examination for wear

32.6a Bearing is retained on shaft by large nut ...

32.6b ... which is staked for security (arrowed)

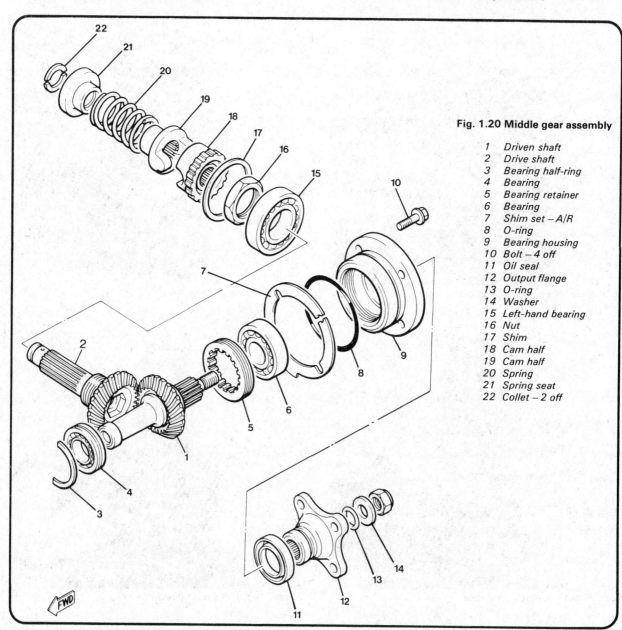

Fig. 1.20 Middle gear assembly

1   Driven shaft
2   Drive shaft
3   Bearing half-ring
4   Bearing
5   Bearing retainer
6   Bearing
7   Shim set – A/R
8   O-ring
9   Bearing housing
10 Bolt – 4 off
11 Oil seal
12 Output flange
13 O-ring
14 Washer
15 Left-hand bearing
16 Nut
17 Shim
18 Cam half
19 Cam half
20 Spring
21 Spring seat
22 Collet – 2 off

33 Engine reassembly: general

1   Before reassembly of the engine/gear unit is commenced, the various component parts should be cleaned thoroughly and placed on a sheet of clean paper, close to the working area.
2   Make sure all traces of old gaskets have been removed and that the mating surfaces are clean and undamaged. Great care should be taken when removing old gasket compound not to damage the mating surface. Most gasket compounds can be softened using a suitable solvent such as methylated spirits, acetone or cellulose thinner. The type of solvent required will depend on the type of compound used. Gasket compound of the non-hardening type can be removed using a soft brass-wire brush of the type used for cleaning suede shoes. A considerable amount of scrubbing can take place without fear of harming the mating surfaces. Some difficulty may be encountered when attempting to remove gaskets of the self-vulcanising type, the use of which is becoming widespread, particularly as cylinder head and base gaskets. The gasket should be pared from the mating surface using a scalpel or a small chisel with a finely honed edge. Do not, however, resort to scraping with a sharp instrument unless necessary.
3   Gather together all the necessary tools and have available an oil can filled with clean engine oil. Make sure that all new gaskets and oil seals are to hand, also all replacement parts required. Nothing is more frustrating than having to stop in the middle of a reassembly sequence because a vital gasket or replacement has been overlooked. As a general rule each moving engine component should be lubricated thoroughly as it is fitted into position.
4   Make sure that the reassembly area is clean and that there is adequate working space. Refer to the torque and clearance setting wherever they are given. Many of the smaller bolts are easily sheared if overtightened. Always use the correct size screwdriver bit for the cross-head screws and never an ordinary screwdriver or punch. If the existing screws show evidence of maltreatment in the past, it is advisable to renew them as a complete set.

34 Engine reassembly: refitting the upper crankcase components

1   Place the inverted upper crankcase on a clean workbench. Fit the union bearing shells to their journals ensuring that the shell tangs locate correctly in the housing cutouts. On all 750 models and later 650 models one bearing shell in each journal set is provided with an annular oil groove. The shell with this groove must be fitted in the lower casing in each instance. Sets of bearing shells of this type will probably be provided as replacement items for earlier models. If the starter shaft chain guide was removed, fit it back into the crankcase. It is a wise precaution to use Loctite on the threads of the retaining bolts. Fit the camshaft chain over the crankshaft sprocket, followed by the Hy-Vo starter shaft chain. If the latter is being re-used remember to fit it so that it runs in the original direction, using the reference mark made during removal. Fit the crankshaft oil seal and blanking plug to the left and right-hand ends of the crankshaft respectively, having greased the sealing lips of each. Lubricate the main bearing shells with engine oil and lower the crankshaft into position.
2   Install the starter idler pinion and fit the shaft and retainer plate. Fit the securing bolt using a new tab washer. Tighten the bolt to 1.0 kgf m (7.2 lbf ft) and bend up the locking tab to secure it. Engage the assembled starter clutch unit in the loop of the Hy-Vo chain and lay it in position in the crankcase. Fit a new O-ring to the oil spray nozzle. Slide the oil spray nozzle into position, ensuring that it locates correctly. Note that the nozzle is rather delicate and care must be taken to avoid damage when the starter shaft is fitted. Check that the bearing and oil seal are

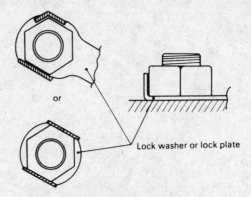

Fig. 1.21 Correct fitting of a tab washer

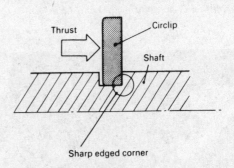

Fig. 1.22 Correct fitting of a circlip

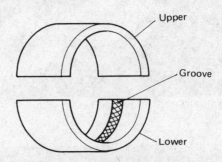

Fig. 1.23 Main bearing shell showing annular groove in lower shell

fitted correctly in the bearing housing, then lubricate the starter shaft and slide it into position through the bearing housing. Fit a new O-ring to the groove round the housing, then slide the assembly into place, guiding the shaft through the centre of the starter clutch. Fit the three retaining screws and tighten them evenly and firmly using an impact driver.
3   Install the gearbox input shaft in its casing recess, ensuring that the locating pin on the right-hand (clutch side) bearing faces towards the crankshaft and rests against the crankcase joint face. Make sure that the locating half-ring is engaged in the bearing and casing grooves. Fit the locating dowel in the hole just forward of the shaft right-hand bearing, using a new O-ring. Place the driving middle gear in its casing recess, noting that the shaft should not be fitted at this stage. Check that the bearing location half rings and the oil seal locating lip (where applicable) engage correctly in the casing grooves. Fit the driven gear outrigger bearing into the casing.

34.1a Fit starter chain guide using Loctite on bolts

34.1b Check that crankshaft seal and blanking plug locate correctly

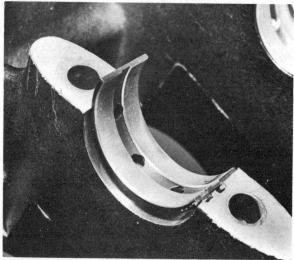

34.1c Locate and lubricate the main bearing shells

34.1d Fit cam and starter chains, then install crankshaft

34.2a Install idler gear in its casing recess

34.2b Engage assembled starter clutch in loop of chain

34.2c Fit oil nozzle noting locating pin and O-ring

34.2d Shaft and retainer may now be installed

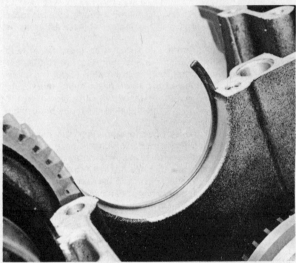

34.3a Make sure that half rings are in place ...

34.3b ... and then install the input shaft

34.3c Note pin (arrowed) must face forward

34.3d Middle gear assembly is located by half rings

34.3e Check that rings and seal are engaged

34.3f Do not forget to fit outrigger bearing

35 Engine reassembly: refitting the lower crankcase components

1   Place the lower crankcase half on the workbench leaving sufficient room around it to allow it to be inverted as necessary during assembly. Slide the gear selector drum into position in the casing. Working from the underside of the casing, drop the drum locating pin into its casing bore and secure it with the retainer plate and bolt, tightening the latter securely. Fit the neutral switch, ensuring that the sealing washer is in good condition or new, then tighten the switch using a socket or box spanner. Take care not to damage the switch terminal.
2   Manoeuvre the gearbox output shaft into the crankcase in the reverse of the removal sequence, feeding the partly disman-tled right-hand end of the assembly down through the sump aperture and out through the casing. Once the left-hand end has been located properly, fit the 5th gear pinion with the selector groove outermost, followed by the bearing, which should be fitted with the sealed side innermost. Fit a new O-ring to the recess in the bearing retainer, then offer it up and retain it with its Allen screws.
3   The selector forks can be fitted next, noting that each one is identified by a cast-in number, each of which must face towards the left-hand side of the crankcase. The forks are number 1, 2 and 3 from left to right. Position each fork in turn and slide the support shaft through it to retain it.

36 Engine reassembly: joining the crankcase halves

1   Check that the mating surfaces of each crankcase half are clean and dry, giving each one a final wipe with solvent to remove any residual grease or oil. Apply a silicone rubber jointing compound to the gasket faces, taking care to coat the surface evenly and completely. **Do not** allow the compound to come within 2-3 mm of the main bearing shells and do not apply it to the area around the central oil feed dowel. Note that it is especially important to ensure that the compound is applied around the main bearing cap bolt holes, bearing in mind the caution given above. If these areas are missed there is a risk of oil leakage between the bearing oil groove and bolt hole, allowing loss of oil pressure and the possibility of crankshaft seizure.
2   Using an oil can filled with clean engine oil, lubricate the crankshaft big-end and main bearings and the gearbox bearings and pinions. Bear in mind that when the rebuilt engine is first

started there will be a delay of several seconds before the oil begins to circulate, so careful oiling of vulnerable areas is essential.
3   At this stage it is invaluable to have an assistant to help during the joining operation. Check that the inverted upper crankcase is securely supported, then carefully lower the lower casing half into position. As the two halves meet, make sure that the central selector fork engages with the groove in the 2nd/3rd gear pinion. If this is not the case, it will be impossible to close the joint fully.
4   Having checked the selector fork position, the crankcase halves can be closed. Check around the unit to ensure that the various oil seals do not become displaced. It may prove helpful to tap around the crankcase to close the last few millimetres of gap. Use a soft-faced mallet, but avoid excessive force.
5   Once the crankcase halves have been closed, but before the retaining bolts are fitted, check that the gearbox selects properly by turning the end of the selector drum by hand. It will be found easier if the input shaft is turned to and fro to aid gear engagement.
6   Fit the crankcase retaining bolts, dropping each one into position, but do not tighten them at this stage. Note that the two bolts within the oil filter area do not have washers fitted beneath their heads. When all of the lower crankcase bolts are in place, screw them home finger tight only, then turn the unit over and repeat the procedure with the upper crankcase bolts. Note that bolt number 35 retains the engine earth cable (straight terminals) and that bolt number 37 retains the battery earth lead (right-angled terminal on battery end). When all 39 bolts are fitted, commence final tightening starting with the lower crankcase bolts.
7   The bolts are tightened in numerical sequence, each bolt being identified by a number cast into the crankcase near its head. Bolts 1 to 10 (main bearing bolts) and also bolts 33 and 34 on the upper crankcase are 8 mm, the remainder being 6 mm. Tighten each one evenly and smoothly to the torque figure shown below.

Crankcase bolt tightening torque

All 8 mm bolts	2.4 kgf m (17.5 lbf ft)
All 6 mm bolts	1.2 kgf m (8.7 lbf ft)

8   When all of the bolts are tightened, check that the crankshaft is free to rotate, remembering to support the camshaft chain to prevent it from becoming jammed around the crankshaft sprocket. Any undue stiffness or tight spots indicates that something is wrong, and the crankcase should be parted to rectify the problem before proceeding further.

35.2a Manoeuvre output shaft through sump aperture

35.2b Fit the 5th gear pinion as shown ...

35.2c ... followed by bearing (sealed face inwards)

35.2d Fit retainer, using new O-ring in oil feed hole

35.3a Selector forks and shaft may be fitted next ...

35.3b ... noting indentification marks (see text)

36.1 Fit dowel and a new O-ring to oil passage

36.3 Assemble crankcase halves, checking that the selector forks engage correctly

36.6a Note crankcase bolt inside filter recess has no copper washer

36.6b Note location of bolts inside sump area

36.6c Earth leads are attached to crankcase as shown

36.7 Tighten each bolt to the specified torque figure

37 Engine reassembly: refitting the oil pump, pump drive components and sump

1   If the engine unit has been stripped and overhauled it is a wise precaution to check the condition of the oil pump prior to its installation. Further details of this will be found in Chapter 2.

2   Fit a new O-ring to the outlet port of the pump to ensure a sound connection between it and the crankcase. Place the oil pump drive chain over the end of the gearbox input shaft, and feed the loop of chain through to the sump area. Offer up the pump, placing the chain around the pump sprocket. Fit the chain shroud around the underside of the sprocket, lining up the mounting holes with those of the pump mounting lugs. Fit the two shouldered Allen bolts through the shroud and pump and the remaining plain Allen bolt to the third pump mounting point. Tighten the bolts firmly, if possible using a torque wrench to secure them to 1.0 kgf m (7.2 lbf ft).

3   Slide the plain thrust washer fully home on the gearbox input shaft. Fit the pump drive sprocket over the shaft, without its centre sleeve, and loop the pump chain around it. Once the chain is in place slide the centre sleeve into position between the shaft and sprocket.

4   Make sure that the gasket faces of the sump and crankcase are clean and dry, then place a new gasket in position on the crankcase face. Offer up the sump and drop the retaining screws into place, noting that the rearmost screw on the left-hand side and the screw next to that each retain a wiring guide clip. Tighten the screws evenly and firmly. If possible, use a torque wrench to secure them to 1.0 kgf m (7.2 lbf ft). If the oil level switch was removed for any reason, check that its O-ring is in good condition and then refit it, tightening the retaining screws firmly. It is a sound precaution to use Loctite on the screw threads.

38 Engine reassembly: installing the middle gear components

1   Check that the driving shaft assembly is complete. On no account omit the shim which controls its position in the crankcase. Slide the assembly home through the centre of the sleeve gear, ensuring that the bearing and shim seat fully. Fit the bearing retainer plates followed by new Torx screws. The latter should be tightened firmly, using a torque wrench if possible to set them at 2.5 kgf m (18 lbf ft). Once secured, use a centre punch to stake the screws into the retainer indentations.

2   The driven shaft can be installed next, using the original semi-circular shims between the bearing housing and crankcase. Make sure that no dirt is trapped between the crankcase, shims or bearing housing. Clean the retaining bolt threads and apply Loctite before tightening them evenly to 2.5 kgf m (18 lbf ft). **IMPORTANT NOTE**: Refer to Section 39 before proceeding further with the rebuild. Also, make absolutely certain at this stage that the half-moon shaped rubber fillet is fitted to the cutout in the casing on the front lower wall of the bevel gear compartment. It must be noted that if this part is omitted the bevel gears will not receive full lubrication and may be destroyed.

37.1 Use a new O-ring at pump outlet

37.4a Fit sump (oil pan) screws and tighten evenly

37.4b Oil level switch is sealed by an O-ring

38.1a Slide driving shaft into position

38.1b Fit retainers and secure with new Torx screws

38.1c Tighten screws and stake them into retainer as shown

38.2a Assemble mesh depth shims on output shaft

38.2b Fit shaft, using Locktite on retaining screws

38.2c **Do not forget to fit this rubber fillet**

39 Engine reassembly: middle gear backlash – checking and adjustment

1   The middle gear assembly is set up by the manufacturer to give the correct amount of backlash between the two bevel gears. This setting is controlled by shimming the two shafts until the two gears run in correct alignment and their teeth mesh at the correct depth. During normal use this setting can be regarded as fixed, and routine checking and adjustment is not necessary. Engine overhaul, however, means that the various components have been disturbed and must be checked after installation.

2   If none of the middle gear components have been disturbed it will only be necessary to measure the amount of backlash in the gear train as described below. If, however, the crankcase, middle gears or the bearing housing have been renewed it will first be necessary to set up the position of the driving gear and shaft. If this is the case, refer to Section 40 before proceeding further.

3   To check the gear backlash a dial gauge and stand will be required. Yamaha also prescribe the use of a special holding fixture which locks the driving shaft in position, but with some degree of care the need for this can be avoided. The fixture can be obtained through Yamaha dealers as a 'Middle drive pinion holder', Part number 90890-04051. If required, a dial gauge can be obtained from the same source as Part number 90890-03097. It should be noted that this check can be made with the engine unit in the frame after removing the near left-hand engine cover and pulling back the output flange gaiter.

4   Assemble the holding tool, where available, so that the two tangs pass each side of the driven shaft. Secure it to the casing using the two fixing holes provided, then tighten the centre bolt to hold the driving gear. It should be noted that firm hand pressure on the bolt should prove adequate. In the absence of the holding tool ask an assistant to place a piece of steel strip between the casing and one pair of the driving gear teeth. This will allow the gear to be immobilised during the measuring operation.

5   Set up the dial gauge, using a stand to position the probe end against the edge of the flange in line with one of the bolt holes. The flange should be moved **gently** back and forth between the extremes of its free movement. Note the reading on the gauge which, all being well, will be within the limits for middle gear backlash which is 0.1 – 0.2 mm (0.004 – 0.008 in). Repeat the measurement on the three remaining flange lobes, noting the reading on each occasion. If the backlash exceeds the above figure it will be necessary to adjust the position of the drive shaft bearing housing by re-shimming. This requires clear access to the rear of the crankcase and thus should be done with the engine unit removed or with the swinging arm/drive shaft assembly detached.

6   Remove the four bolts which secure the driven shaft bearing housing to the crankcase. Tap around the housing until it can be pulled just clear of the crankcase and remove the shims. If necessary, slacken the crankcase holding bolt on each side of the housing to ease removal. The housing should be pulled out to leave a gap of about 2 mm (0.08 in) between the crankcase and the flanged face of the housing. Fit two of the bearing housing bolts spaced 180° apart and screw them home until they just touch the housing.

7   Using the dial gauge arrangement to check backlash, slowly turn both bolts inwards a fraction at a time until the gauge indicates 0.2 mm (0.008 in). Measure the clearance between the crankcase and housing using feeler gauges, the clearance indicated giving the shim thickness required. Shims are available in 0.05 mm increments from 0.1 mm to 0.5 mm.

8   Reassemble the bearing housing, having fitted the appropriate shims. Note that Loctite should be used on the retaining bolts. Re-check the backlash as described in paragraphs 3 to 5, making sure that the backlash is now within limits.

40 Engine reassembly: checking and shimming the middle drive shaft

1   If in the course of overhaul it has been necessary to renew the crankcase, the middle driving or driven gears or the driven gear bearing housing it will be necessary to set up the position of the driving shaft in the crankcase from scratch. The shaft is positioned by shims between the bevel gear and the bearing and serves to place the bevel teeth in the correct position in relation to the driven gear. To select the required shim thickness it is necessary to make the following calculation, based on the formula A = c - a - b.

2   The various letters shown in the equation represent the values shown below:

A is the thickness of shim required.

a represents a number which will be found etched on the end of the driving bevel gear. This is usually a decimal figure and is prefixed to indicate whether it is to be added to or subtracted from a nominal value of 43.0. Thus, if the marked number were '+03' or '−05' the resulting value of **a** would be 43.03 or 42.95 respectively.

b indicates the bearing thickness. This is taken to be a constant of 16.94 mm.

c is a value etched on the rear of the crankcase near the main bearing selection numbers. There are three sets of numerals grouped together, those required being the bottom line. Again, this will be a decimal figure to be added to the normal size of 60.00. As an example, if the numbers etched on the crankcase were 48, the total value of **c** would be 60.48.

3   To work through a hypothetical case, the following values are assumed:

A = shim size
a = 43.03
b = 16.94
c = 60.48

Using the equation given earlier, A = 60.48 - 43.03 - 16.94 = 0.51.

4   Shims are available in the following sizes:

0.15 mm
0.20 mm
0.30 mm
0.40 mm
0.50 mm

It will be noted that, given the shim sizes available, it is not possible to match the actual clearance precisely. To overcome this problem it will be necessary to round off the last digit as follows:

Last digit	Round off to
0, 1, 2	0
3, 4, 5, 6, 7	5
8, 9	10

In the example shown in paragraph 3, the shim size required will be 0.5 mm. To fit the shim, it will be necessary to remove the driving shaft and to release the bearing as described in Section 32.6. Once the shaft has been shimmed and fitted, set the bevel gear mesh depth (backlash) as described in Section 39.

41 Engine reassembly: refitting the clutch

1   Note the position of the driving tangs on the oil pump socket and the corresponding slots in the clutch outer drum. It is important to ensure that these align when the clutch drum is

fitted to avoid damage to the pump drive sprocket. Offer up the clutch drum **without** the caged needle roller bearing or inner sleeve. Manoeuvre it into position, holding it there while the bearing and then the sleeve are slid into place. Fit the large plain thrust plate with its grooved face outwards.

2   Slide the clutch centre into position and fit a new lock washer and the retaining nut. Lock the gearbox input shaft by the same method as was used during removal, and tighten the nut to 7.2 kgf m (52.0 lbf ft). Bend up the tab washer against one of the flats of the nut to secure it. Install the clutch plain and friction plates, starting and finishing with a friction plate and building them up alternately.

3   Fit the clutch release bearing over the end of the mushroom-headed pull rod, followed by the plain thrust washer. The assembly should then be fitted through the clutch pressure plate from the inside. Offer up the pressure plate, noting that the embossed dot on its outer edge must correspond with a similar dot on the clutch centre. Fit the clutch springs and

retaining bolts, tightening the latter evenly in a diagonal sequence to 1.0 kgf m (7.2 lbf ft).

4   Assemble the clutch release shaft on the inside of the outer cover, if this was removed during overhaul. The pull rod should be arranged so that it matches up with the release mechanism pinion when the cover is fitted. To this end, position the rack portion of the rod so that it faces towards the rear of the unit and downwards, approximately 45° from horizontal.

5   If the external clutch operating arm is already in place on the shaft, turn it until it lies parallel to the gasket face. Offer up the clutch cover, using a new gasket and allow the arm and shaft to turn back as it engages the pull rod. Fit and tighten the cover securing screws. If the external arm has yet to be fitted, note that it has an alignment dot which should coincide with the index mark on the cover when free play in the release mechanism has just been taken up. Check that the arm is correctly aligned and that the external spring (where fitted) is in place, then fit the retaining E-clip.

40.2 Lower numbers indicate driving shaft coding

41.1a Large plain washer is fitted against bearing

41.1b Offer up the oil pump drive sprocket and engage in chain loop ...

41.1c ... then slide bush into sprocket centre

41.1d Manoeuvre clutch drum into position, checking that it locates over sprocket tangs

41.1e Slide the needle roller bearing and sleeve into place

41.1f Fit the thrust plate with grooved face outwards

41.2a Slide the clutch centre into position ...

41.2b ... and tighten the securing nut to the specified torque setting

41.2c Do not omit to bend up the locking tab

41.2d Build up the clutch plain and ...

41.2e ... friction plates alternately

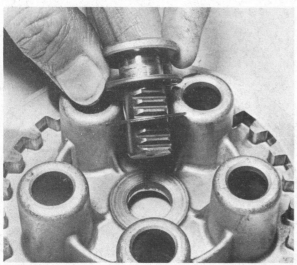

41.3a Fit bearing and washer to pull-rod and place through clutch pressure plate

41.3b Note index marks on clutch centre and pressure plate

41.3c Refit deflector plate if it was removed

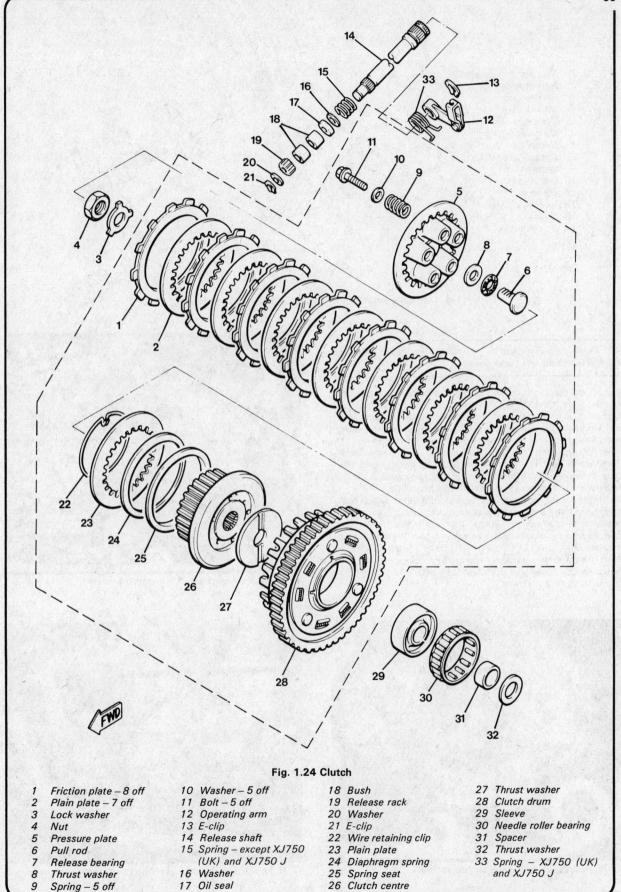

Fig. 1.24 Clutch

1 Friction plate – 8 off	10 Washer – 5 off	18 Bush	27 Thrust washer
2 Plain plate – 7 off	11 Bolt – 5 off	19 Release rack	28 Clutch drum
3 Lock washer	12 Operating arm	20 Washer	29 Sleeve
4 Nut	13 E-clip	21 E-clip	30 Needle roller bearing
5 Pressure plate	14 Release shaft	22 Wire retaining clip	31 Spacer
6 Pull rod	15 Spring – except XJ750	23 Plain plate	32 Thrust washer
7 Release bearing	(UK) and XJ750 J	24 Diaphragm spring	33 Spring – XJ750 (UK)
8 Thrust washer	16 Washer	25 Spring seat	and XJ750 J
9 Spring – 5 off	17 Oil seal	26 Clutch centre	

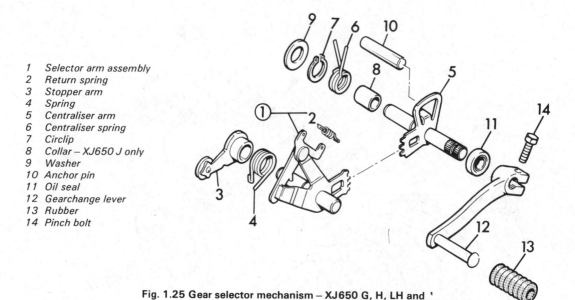

1   Selector arm assembly
2   Return spring
3   Stopper arm
4   Spring
5   Centraliser arm
6   Centraliser spring
7   Circlip
8   Collar – XJ650 J only
9   Washer
10 Anchor pin
11 Oil seal
12 Gearchange lever
13 Rubber
14 Pinch bolt

Fig. 1.25 Gear selector mechanism – XJ650 G, H, LH and '

42 Engine reassembly: refitting the gear selector mechanism and left-hand cover

1   Fit the selector arm and stopper assembly into the casing recess, lifting the selector arm claw and the stopper arm roller into position over the selector drum end. Fit the centraliser arm in its bore to the rear of the casing, ensuring that the springs ends fit each side of the locating pin. Note that gear teeth of the selector arm and centraliser arm must mesh symetrically. Temporarily refit the gearchange pedal or linkage and check gear selection before proceeding further. Engagement will be facilitated by rocking the output flange to and fro.

2   If they were removed for any reason, refit the two breather castings to the inside of the outer cover, using new gaskets. Make a final check to ensure that the rubber fillet which completes the middle gear casing is in position and properly located. Offer up and fit the left-hand cover using a new gasket. Note that a wiring guide for the ignition coil/pickup leads fits between the cover and the crankcase and is retained by the lower front casing screw.

42.1a Selector mechanism is installed as shown

42.1b Note return spring end and detent roller

42.1c Fit centering mechanism with teeth meshed as shown

42.2 Note wiring guide secured by cover screw

Fig. 1.26 Gear selector mechanism – XJ750 J

1	Selector arm assembly	12	Boot
2	Return spring	13	Rubber
3	Stopper arm	14	Pinch bolt
4	Spring	15	Change link
5	Centraliser arm assembly	16	Adjuster rod
6	Centraliser spring	17	Adjuster rod
7	E-clip	18	Adjuster barrel
8	Collar	19	Locknut – 2 off
9	Washer	20	Locknut – 2 off
10	Anchor pin	21	Boot
11	Gearchange arm	22	Washer
		23	E-clip

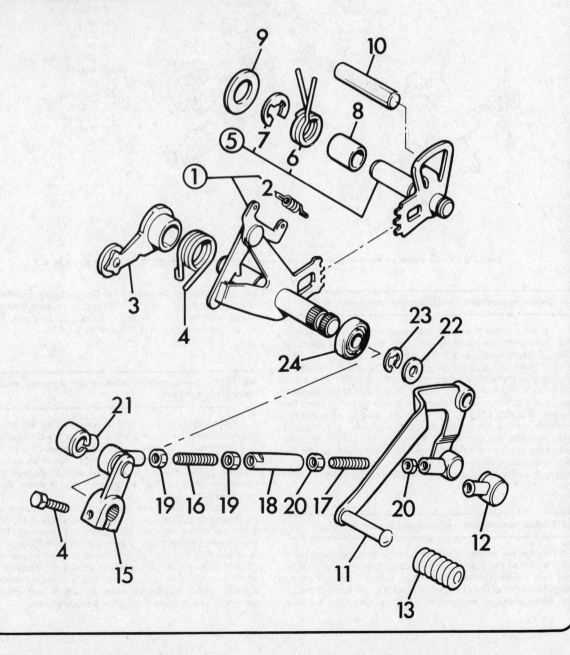

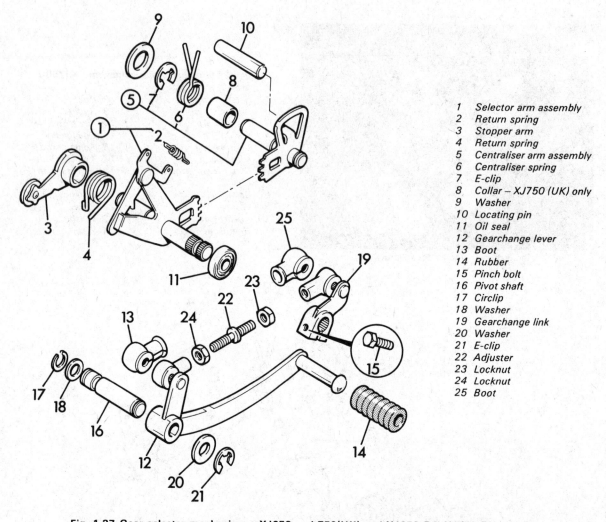

1   Selector arm assembly
2   Return spring
3   Stopper arm
4   Return spring
5   Centraliser arm assembly
6   Centraliser spring
7   E-clip
8   Collar – XJ750 (UK) only
9   Washer
10 Locating pin
11 Oil seal
12 Gearchange lever
13 Boot
14 Rubber
15 Pinch bolt
16 Pivot shaft
17 Circlip
18 Washer
19 Gearchange link
20 Washer
21 E-clip
22 Adjuster
23 Locknut
24 Locknut
25 Boot

Fig. 1.27 Gear selector mechanism – XJ650 and 750(UK) and XJ650 RJ, XJ750 RH and RJ

43 Engine reassembly: installing the alternator

1   Fit the alternator rotor to the projecting end of the starter shaft and insert the retaining bolt. Lock the shaft by whatever method was employed during removal, then tighten the retaining bolt to 5.5 kgf m (39.8 lbf ft). Place the stator assembly in the alternator cover, ensuring that the wiring and grommet fit into the slot in the cover. A thin film of RTV jointing compound at this point will ensure a good seal and prevent the ingress of water. Offer up the assembly, making sure that the retaining screws can pass through, the grooves in the stator edge and into the crankcase. A new gasket should be used unless the old item is in obviously serviceable condition. Fit the retaining screws and tighten them evenly and securely. If a square drive Allen key is available, torque to 1.0 kgf m (7.2 lbf ft).

44 Engine reassembly: refitting the starter motor

1   Arrange the alternator output leads so that they run behind and parallel to the starter shaft housing, then through 90° so that they will pass below the nose of the starter motor to exit near the centre of the crankcase. Check the condition of the O-ring which seals the starter motor to the crankcase. Care must

be taken to avoid damage to the O-ring during installation, and it is advisable to smear it with grease to ease fitting. Fit the motor and drop the retaining bolts into position. The bolts should be tightened to 1.2 kgf m (8.7 lbf ft).

45 Engine reassembly: refitting the ignition pickup and neutral switch lead

1   Fit the ignition stator into its recess at the left-hand end of the crankshaft, taking care not to damage the timing pointer. Fit and tighten the two retaining screws. The ignition reluctor/timing plate is installed next. It will be noted that a location pin in the crankshaft end should engage in a corresponding slot in the reluctor. Fit and tighten the central retaining bolt.
2   Check that the pickup wiring grommet is properly seated in its casing slot. To ensure a sound waterproof joint, smear the grommet edges with an RTV jointing compound. Route the output leads down, passing behind the wiring guide which is secured by the gear change cover forward screw then back parallel to the sump gasket face. Fit the neutral switch terminal to the switch unit near to the wiring clamp. Note that the ignition pickup inspection cover should be left off until the camshaft and ignition timing has been checked.

43.1a Place alternator rotor over tapered end of starter shaft ...

43.1b ... and tighten the retaining bolt to the specified torque

44.1 Reconnect starter cable and cover with rubber boot

45.2 Reconnect neutral switch lead

46 Engine reassembly: refitting the pistons and cylinder block

1　Before starting this stage of reassembly it should be noted that it will be necessary to arrange the cam chain so that it passes through its tunnel as the cylinder block is lowered into position. This is simple enough if an assistant is available, but if the job is to be tackled unaided it is best to arrange some stiff wire attached to the chain so that it can be pulled up through the chain tunnel. A magnetic probe can also be used to pull the chain into position. Another alternative is to wait until the rear chain guide is in place. The chain can be pulled straight and draped over the top of the guide where it can be secured with a rubber band. Whichever method is employed be sure that the chain is not bunched and jammed against the crankshaft and that it has not come off its sprocket.

2　Pad the crankcase mouths with clean rag to prevent the ingress of foreign matter during piston and cylinder barrel replacement. It is only too easy to drop a circlip while it is being inserted into the piston boss, which will necessitate a further strip down for its retrieval.

3　Fit the piston rings in their correct relative positions, taking great care to avoid breakage. With a little practice the ring ends can be eased apart with the thumbs and the ring lowered into position. Alternatively, a piston ring expander can be used. This is a plier-like tool that does much the same thing as described above. One of the safest (and cheapest) methods is to use three thin steel strips such as old feeler gauges to ease the ring into place. Arrange the various ring end gaps as shown in the accompanying figure.

4　Fit the pistons onto their original connecting rods, with the arrow embossed on each piston crown facing forwards. If the gudgeon pins are a tight fit in the piston bosses, warm each piston first to expand the metal. Do not forget to lubricate the gudgeon pin, small-end eye and the piston bosses before reassembly.

5　Use new circlips, **never** re-use old circlips. Check that each circlip has located correctly in its groove. A displaced circlip will cause severe engine damage. Note that the circlips should be fitted with the gap facing down towards the piston skirt.

6　Check that the mating surfaces of the cylinder block and crankcase are clean and free from oil or pieces of broken gasket, then fit a new cylinder base gasket. Fit the two large diameter dowels over the right-hand holding studs and push them firmly home into their recesses. Fit two new O-rings around the dowels. Place the smaller locating dowel in its recess around the rear left-hand stud or in the corresponding hole in the cylinder base. If the camshaft chain rear guide is not already in place it should be fitted at this stage. Place the lower end of the guide in its recess then fit the retaining bolt and locknut. Slacken the locknut fully and screw the bolt home until it seats against the end of the guide. Back off the bolt by $\frac{1}{4}$ of a turn and, holding it in this position, secure the locknut.

7   Rotate the engine so that cylinders No 2 and 3 are at TDC. Refitting of the cylinder block can be facilitated by the use of a piston ring clamp placed on each piston. This is by no means essential because the cylinder bore spigots have a good lead-in and the rings may therefore be hand-fed into the bores. Whichever method is adopted, an assistant should be available to guide the pistons.

8   Position a new sealing ring around each cylinder bore spigot and lubricate the bores thoroughly with clean engine oil. Carefully slide the cylinder block down the holding down studs until the pistons enter the cylinder bores; keeping the pistons square to the bores ease the block down until the piston clamps are displaced. Lower the cylinder block slightly further and remove the piston ring clamps (where used). Rotate the crankshaft slightly until the two outer pistons (1 and 4) approach their respective bores. These should be fitted as described above. Remove the piston ring clamps and the rag padding from the crankcase mouths and push the cylinder block down onto the base gasket.

9   Once all four pistons are securely in position, slide the cylinder block firmly onto the cylinder base gasket. Check that the cam chain is secured by passing a bar through the protruding loop, resting the ends across the gasket face. Fit the single cylinder retaining nut at the centre front of the cylinder block and tighten it to 2.0 kgf m (14.5 lb ft).

46.4a Arrow on piston crown must face forward

46.4b Push gudgeon (piston) pin into place ...

46.4c ... and retain using new circlips

46.6a Fit oil feed dowels and new O-rings

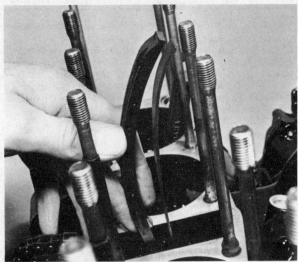

46.6b Place cam chain tensioner in crankcase recess

46.6c Fit adjuster bolt and locknut ...

46.6d ... and adjust (see text)

46.8a Remove old O-rings from cylinder base grooves ...

46.8b ... and work the new O-ring into position

46.8c Lower cylinder block, feeding pistons into each bore

47 Engine reassembly: refitting the cylinder head

1   Before the cylinder head can be refitted the valves, springs and cam followers must be replaced by reversing the dismantling procedure. Place a new valve guide seal onto each guide top. Lubricate the valve stems thoroughly. When fitting the springs do not omit the lower seating washer. Note that each spring has variable pitch coils. The springs must be fitted so that the more widely spaced coils are at the top. After fitting the valve collets and releasing the spring compressor strike smartly the top of each valve stem with a hammer. This will ensure that the collets are seated correctly. Lubricate the cam followers and install them, ensuring that each is returned to its original location. Fit also the adjuster pads in their original locations.

2   Fit a new cylinder head gasket to the top of the cylinder and check that the two locating dowels are in position around the right-hand outer holding studs and that **new** O-rings are fitted around them and around the YICS passages (where applicable). Fit the third locating dowel to the left-hand rear holding stud. On later XJ650 and all XJ750 models a fourth dowel is fitted to the front holding stud between cylinders 3 and 4. Fit the cam chain tunnel seal with its locating tabs facing downwards to engage with the head gasket.

3   Before the cylinder head is fitted it is advisable to check that the timing pointer is correctly positioned to indicate top dead centre (TDC) on cylinder No 1. Set up a dial gauge and stand so that the tip of the gauge will bear upon the piston crown at TDC. Using the flats provided on the timing plate, rock the crankshaft to and fro until top dead centre has been found. Check that the 'T' mark on the timing plate coincides with the fixed pointer, and if necessary slacken its retaining screw and set it correctly. It should be noted that this setting should not normally have moved, but that the pointer is easily knocked or bent during overhaul.

4   Lower the cylinder head into position over the holding down studs, ensuring that it locates over the dowels. Lightly oil the threads, then fit the cylinder head cap nuts and washers finger tight. The nuts should be tightened in the sequence shown in the accompanying illustration, first to half the torque value, and then to the full torque value. This will ensure that the head is pulled down evenly with no risk of distortion. The final torque figure is 3.2 kgf m (23.1 lbf ft). Fit the plain nuts to the downward-facing studs at the cylinder head/block joint at the front and rear of the cylinder block. The 8mm (thread size) nuts fitted to all models should be tightened to 2.0 kgfm (14.5 lbf ft); no setting is specified for the 6mm (thread size) nuts which are

also fitted on YICS models, but these should be tightened securely.

Note: Two out of the twelve washers fitted underneath the cylinder head cap nuts are copper sealing washers, and are placed on the extreme right-hand side of the head. These must be renewed and should never be omitted because their purpose is to seal the bores of the cylinder head stud holes, which also act as oilways.

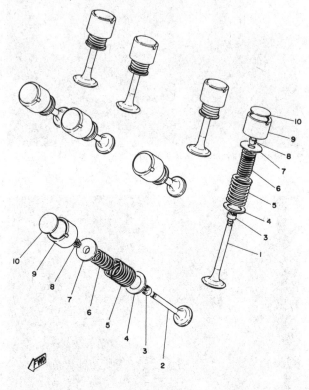

Fig. 1.28 Valve assembly

1	Inlet valve	6	Inner spring
2	Exhaust valve	7	Cap
3	Oil seal	8	Split collets
4	Spring seat	9	Cam followers
5	Outer spring	10	Adjustment pad

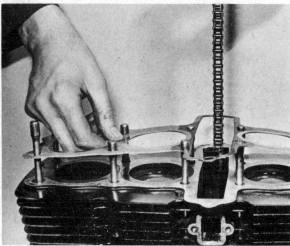

47.2a Place the cylinder head gasket over the studs

47.2b Fit dowels and **new** O-rings

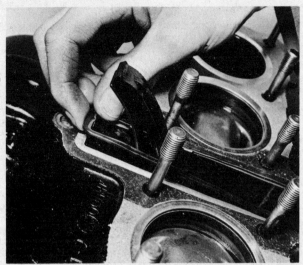
47.2c Do not forget camshaft chain tunnel seal

47.4a Lower cylinder head into place ...

47.4b ... and fit copper sealing washers (see text)

47.4c Fit centre nuts to projecting studs (2 off – non YICS models, 6 off YICS models)

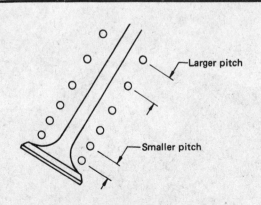
Fig. 1.29 Install valve springs with the closer wound coils towards the head of the valve

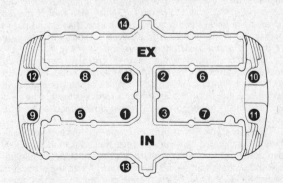
Fig. 1.30 Cylinder head tightening sequence – non YICS models

Note: Copper sealing washers should be fitted beneath nuts where oil ways are located.

48 Engine reassembly: refitting the camshafts and setting the valve timing

1   Fit each of the cam followers and shims in the order in which they were removed. It should be noted that it will be necessary to check and adjust the valve clearances after the camshafts have been installed because these will have been altered if the valves have been refaced, ground or renewed. It is, however, important that each follower is fitted in the bore from which it was removed. To this end, note the markings made during removal. If this precaution is not observed, slight discrepancies between the various part-worn components may result in rapid wear.

2   Fit the cam chain guide at the front of the tunnel, making sure that the lower end engages in its holding recess. The tensioner blade at the rear of the tunnel should already be in position as described in Section 46. Using the square head provided on the timing plate, rotate the crankshaft anti-clockwise until the 'T' mark aligns with the fixed pointer. Hold the cam chain reasonably taut while the crankshaft is being turned. This will prevent it from bunching and jamming around the crankshaft sprocket.

3   Lubricate the camshaft bearing surfaces in the cylinder head and caps with engine oil. Place the front (exhaust camshaft) sprocket inside the chain loop, then feed the exhaust cam through the sprocket and rest it loosely in position. Note the timing mark stamped on the camshaft; this should face upwards to set the camshaft roughly in the correct position. Repeat the above process to fit the inlet (rear) camshaft and sprocket.

4   Place the camshaft caps in their correct relative positions, noting that each one is coded 'I' (inlet) or 'E' (exhaust) and is numbered from 1 (left-hand side) to 3 (right-hand side). Do not omit to fit the locating dowels where necessary. Drop the bolts into position and locate them by screwing them in by one or two threads.

5   Tighten the cap bolts evenly and progressively, turning each one by about one turn. The object is to ensure that the camshaft and caps are drawn down evenly and squarely. It should be noted that as tightening progresses, one or more of the valves will be opened and this will tend to make the assembly pull down at an angle unless the above procedure is followed. Tighten all bolts to 0.5 kgf m (3.6 lbf ft) initially, then go over them again to bring them to the final torque figure of 1.0 kgf m (7.2 lbf ft).

6   Once the camshaft caps are fully tightened, make any final adjustment necessary to align the camshaft timing dot with the fixed index mark on cap No 2. Spanner flats are provided on the camshafts for this purpose, but note the following warning before moving them. **Important note:** When turning the camshafts, take great care to avoid levering against the cylinder head casting – it is easily damaged. Note also that if moved excessively, a valve head may contact a piston and can lead to a bent valve stem. If in any doubt **stop** rotation until you are sure that no undue resistance is evident.

7   Once the camshafts are aligned correctly, move each sprocket in relation to the chain until the bolt holes align with the camshaft flanges. This can be done by lifting the chain and working one link at a time around the sprocket teeth. The next stage is vital in ensuring correct valve timing, and should be followed meticulously.

8   Gently pull the front run of chain to remove all slack between the exhaust cam sprocket and the crankshaft sprocket. Grasp both sprockets and lift them onto the locating shoulders, keeping the chain taut as described above. Move the sprockets slightly to align the bolt holes, then fit one bolt to each to retain it. It should be noted that the bolts have specially hardened shoulders and should not be replaced with any other type of bolt. The bolts should be fitted finger tight only at this stage. Lift the top run of the chain slightly and insert the centre chain guide in its recess. The crankshaft should now be turned to align

the 'C' mark on the timing plate with the fixed pointer in preparation for cam chain tensioning.

Manual chain tensioner

9   Slacken the tensioner locknut and bolt and compress the tensioner spring by pushing the plunger into the body. Tighten the bolt to hold it in this position, then offer up the tensioner assembly, with a new gasket, and secure it with its two fixing bolts. Tighten the bolts to 1.0 kgf m (7.2 lbf ft). Unscrew the tensioner holding bolt until a click is heard, indicating that the plunger has extended and has taken up the chain slack. Tighten the holding bolt to 0.6 kgf m (4.3 lbf ft) and secure its locknut.

Automatic chain tensioner

10   Remove the end plug from the tensioner body to release the spring pressure on the tensioner plunger. Release the locking cam pawl by pushing it with a finger and push the tensioner into the body. Fit the assembly, with a new gasket, and tighten the fixing bolts to 1.0 kgf m (7.2 lbf ft). Refit the spring and end plug, tightening the latter to 1.5 kgf m (11 lbf ft). The spring will adjust the tensioner to the correct position automatically, its subsequent retraction being prevented by the cam arrangement.

11   Rotate the crankshaft by at least one full turn, bringing the 'T' mark back into alignment with the fixed pointer. Check that the dots on the camshafts align with the raised arrows on the adjacent bearing caps. If this is not the case it will be necessary to repeat the operation from paragraph 6 onwards. Once alignment of the timing marks has been checked, rotate the crankshaft to allow the remaining camshaft sprocket bolts to be fitted, tightening all four to 2.0 kgf m (14.5 lbf ft). Note that the valve clearances must now be checked and adjusted, following the sequence described in Routine Maintenance.

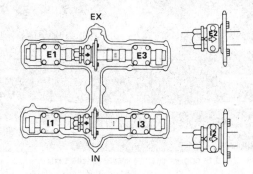

Fig. 1.31 Camshaft cap locations

48.1 Refit tachometer drive if this was removed

48.2a Place front guide into casing recess

48.2b Set crankshaft to align 'T' mark

48.3 Loop chain around sprockets and fit camshafts

48.4 Fit camshaft caps, noting locating dowels

48.6 Check that camshaft alignment mark is in the position shown

48.7 Fit centre guide between camshafts

48.8 Fit and tighten sprocket bolts

48.9 Installing the manual type tensioner

49 Engine reassembly: refitting the cylinder head cover

1   Check that the mating surfaces of the cylinder head and cover are clean and dry, then fit a new gasket in position. The type used is dependant on the model. Non YICS models employ a plain gasket in conjunction with separate half-moon shaped seals which blank off the cut-outs at the right-hand end of the cylinder head. Where these are supplied it is sound practice to locate them with the aid of a **fine** film of RTV sealant, after which the gasket can be fitted.
2   YICS models use a moulded synthetic rubber gasket in which the half-moon seals are integral. It follows that installation is simplified, the seals being positively located by the gasket. In either case, once the gasket is in place, lower the cylinder head cover into position and fit the 20 (early models) or 12 (làte models) retaining bolts. These should be tightened evenly and progressively to 1.0 kgf m (7.2 lbf ft).

50 Engine reassembly: refitting the engine/gearbox unit

1   Refitting the engine unit is a relatively straightforward operation, but the unit's considerable bulk and weight must be borne in mind. It is not feasible to install the unit unaided because it will be necessary to manoeuvre the crankcase mounting points into position, an operation which will require at least two pairs of hands. It was found in practice that two reasonably strong persons could just manage the refitting operation between them if the lifting phase was taken in stages.
2   Make sure that the area around the machine is clear, and that all cables and leads are lodged in a position which will not impede the engine fitting operation. Arrange the engine unit on a wooden crate or blocks on the right-hand side of the frame. Find a position where both persons can lift the unit safely, then place it halfway into the frame cradle. One person should steady the unit whilst the other moves round to the left-hand side. The engine can now be manoeuvred into its final position.
3   Assemble the front engine mounting plates and the six retaining bolts and nuts, fitting the latter finger tight at this stage to allow a degree of manoeuvring room where necessary. Check that the driveshaft coupling flange faces are in approx-imate alignment before proceeding further. Fit the lower front mounting bolts (2 off) and nuts, again finger tight. Assemble the large rear mounting bolts together with the footrest plates (where appropriate) or insert the single long through-bolt. Once

all are in position, the mounting bolts can be tightened to the torque figures shown below.

Engine mounting bolt torque settings
   8 mm (front mounting plate, 4 off) 2.0 kgf m (14.5 lbf ft)
   10 mm (front upper and lower) 4.2 kgf m (30.4 lbf ft)
   12 mm (rear mounting) 7.0 kgf m (50.6 lbf ft)

4   Turn the rear wheel until the marks on the edges of the driveshaft flange and middle gear case output flange align. Although this is not strictly essential it will ensure that there is no likelihood of driveshaft vibration due to any imbalance in the drive train. Position the flanges so that one of the four retaining bolts can be fitted, repeating the procedure with the remaining bolts. Tighten the bolts to 4.4 kgf m (31.8 lbf ft). Slide the gaiter back over the flanges and joint, and secure it with the spring retainer at each end.
5   On the machines fitted with an oil cooler, lift the radiator assembly into position and secure the upper locating pin with an R-pin after pushing it through its rubber mounting. Fit and tighten the two bolts which retain the lower edge. Assemble the oil cooler hose retainer plate which clamps the hoses to the extended cylinder head stud. Fit the hose unions to the distributor block, using new O-rings. Each union is secured by two Allen screws. Fit a new O-ring seal to the back of the block, then place it in position and secure it with the large internally-threaded bolt and plain washer; tighten to a torque setting of 5.0 kgf m (36 lbf ft). As an additional precaution, a thread locking compound may be used very sparingly.
6   On all models, fit the oil filter assembly using a new element and O-ring. Fit the central retaining bolt and tighten it to 1.5 kgf m (11.0 lbf ft).
7   Fit the tachometer drive cable (where applicable) to the cylinder head adaptor and secure its retaining ring. Refit the horn(s) and reconnect the horn wiring. Where appropriate, refit the indicator beneath the frame tubes.
8   Fit new exhaust port seals, retaining them with a smear of grease until the system is in place. Where appropriate, refit the control exhaust system mounting to the underside of the crankcase, noting that the rubber-mounted section is offset towards the front of the machine (see photograph). Reassemble and refit the exhaust system in the reverse of the removal sequence. Check that the system is correctly positioned before any of the fasteners are tightened, then secure them to the following torque figures, working from front to rear. Note that it

is important that no part of the system is forced into place, since this will invariably lead to stress fractures at a later date.

Exhaust system torque settings
Exhaust port nuts:
XJ650 models	1.0 kgf m (7.2 lbf ft)
XJ750 models	0.75 kgf m (5.4 lbf ft)
Silencer support to frame	4.3 kgf m (31.1 lbf ft)
Silencer to support bracket	2.5 kgf m (18.1 lbf ft)

49.1 Separate half-moon seals are used on early models

9   Reconnect the alternator and ignition wiring, taking care to route the wiring carefully so as to avoid chafing on frame fittings. Use the existing guides and cable clips, supplemented by additional cable ties where necessary. Reconnect the heavy duty lead to the starter motor terminal, remembering to slide the protective rubber boot into place. Refit the spark plug caps to their respective plugs noting that each is marked to identify to which cylinder it belongs. Reconnect the clutch cable and bend back the security tang to retain it. Adjust the cable to give 2-3 mm (0.08-0.12 in) free play measured between the lever and lever stock, then secure the adjuster locknut.

10   Reconnect the throttle and choke cables and manoeuvre the carburettor bank into position. This is a fairly awkward operation in view of the restricted space available and is much easier if an assistant is able to deal with one pair of instruments. Make sure that all four carburettors engage with the intake adaptors, then retain them by tightening the securing clamps. The air cleaner hoses are secured in a similar manner. Once the carburettors are in place the air filter casing can be secured. Refit the crankcase breather hose.

11   Refit and secure the injection moulded panel carrying the CDI and regulator units, and refit the battery case assembly. Check around the machine for any unconnected electrical leads, then refit the battery and connect the earth (–) lead and the positive (+) contact strip to the solenoid. Place the fuel tank in position and connect the fuel gauge sender lead (where fitted).

12   It is advisable at this juncture to check the electrical system, testing each circuit in turn to highlight any forgotten or badly made connections. On machines with Computer Monitor Systems, run throught the test sequence several times in addition to the visual checks mentioned above. If all appears to be well, switch off the ignition and complete reassembly.

13   Engage the front of the tank on its rubber buffers and connect the fuel feed and vacuum pipes. Lower the rear of the tank and secure it with its clip or mounting bolt as appropriate. Refit the side panels, making sure that they are correctly engaged. Refit, and where necessary adjust, the gearchange and brake pedals and linkages. Give the machine a final visual check and remember to check the work area for any forgotten clips or fittings.

49.2 Fit cover gasket and install cover

14   Fill the crankcase to the level specified below, using SAE 20W/40SE motor oil where the average ambient temperature does not fall below 5°C (40°F). Where the normal temperature rarely exceeds 15°C (60°F) use SAE 10W/30SE motor oil. Note that there is a considerable overlap, allowing either grade to be used in most areas.

Engine oil capacities – dry engine, new filter
XJ650 – 4K0 (UK) and XJ650 G, H, LH and J
   3.3 litre   (6.9/5.8 US/Imp pint)
All other models
   3.6 litre   (7.6/6.3 US/Imp pint)

It will probably appear that the engine has been over-filled at this stage, but much of the excess will be distributed around the engine once it has been run for a few minutes. Remember to re-check oil level after the initial start-up.

15   Once the machine has been started, the ignition timing should be checked using a stroboscopic timing lamp as described in Chapter 3, Section 8.

16   Refit the gearchange mechanism, and on XJ750 J models the footrests and brake pedal. Before tightening the control arm pinch bolts check that the controls are in the correct positions.

50.2 Lift the unit into position, allowing it to rest in the frame cradle until the mountings have been fitted

50.4 Reconnect the driveshaft flange and refit the rubber boot

50.5 Use new O-rings on oil cooler unions

50.8a Exhaust port seals can be held in place with grease

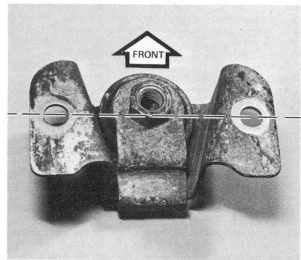
50.8b Note offset of exhaust centre mounting

50.8c Mounting bolts to crankcase underside as shown

50.8d Note special nut which secures silencer mounting

50.9a Cable routing along underside of unit

50.9b Bend over security tang to retain clutch cable

50.13 Align index marks on all control levers or pedals

51 Starting and running the rebuilt engine

1   Turn the fuel taps to the 'Prime' position and wait a few seconds whilst the carburettor float bowls fill. Return the fuel taps to the 'On' position once the engine is running. Check that the engine kill switch is set to the 'On' position. Start the engine using the electric starter. Raise the chokes as soon as the engine will run evenly and keep it running at a low speed for a few minutes to allow oil pressure to build up and the oil to circulate.

2   The engine may tend to smoke through the exhaust initially, due to the amount of oil used when assembling the various components. The excess of oil should gradually burn away as the engine settles down.

3   Check the exterior of the machine for oil leaks or blowing gaskets. Make sure that each gear engages correctly and that all the controls function effectively, particularly the brakes. This is an essential last check before taking the machine on the road.

52 Taking the rebuilt machine on the road

1   Any rebuilt machine will need time to settle down, even if parts have been replaced in their original order. For this reason it is highly advisable to treat the machine gently for the first few miles to ensure oil has circulated throughout the lubrication system and that any new parts fitted have begun to bed down.

2   Even greater care is necessary if the engine has been rebored or if a new crankshaft has been fitted. In the case of a rebore, the engine will have to be run-in again, as if the machine were new. This means greater use of the gearbox and a restraining hand on the throttle until at least 500 miles have been covered. There is no point in keeping to any set speed limit; the main requirement is to keep a light loading on the engine and to gradually work up performance until the 500 mile mark is reached. These recommendations can be lessened to an extent when only a new crankshaft is fitted. Experience is the best guide since it is easy to tell when an engine is running freely.

3   If at any time a lubrication failure is suspected, stop the engine immediately, and investigate the cause. If an engine is run without oil, even for a short period, irreparable engine damage is inevitable.

4   When the engine has cooled down completely after the initial run, recheck the various settings, especially the valve clearances. During the run most of the engine components will have settled into their normal working locations. Check the various oil levels, particularly that of the engine as it may have dropped slightly now that the various passages and recesses have filled.

Chapter 2 Fuel system and lubrication

Refer to Chapter 7 for information on the 1983 US models

Contents

Specifications

Fuel tank capacity

	XJ650 G,H, LH and J	XJ650 RJ and XJ650(UK)	XJ750 J	XJ750 RH, RJ and XJ750(UK)
Total	13.0 litre (3.4/2.9 US/Imp gal)	19.5 litre (5.2/4.3 US/Imp gal)	17.0 litre (4.5/3.7 US/Imp gal)	19.0 litre (5.0/4.2 US/Imp gal)
Reserve	3.4 litre (0.8/0.7 US/Imp gal)	3.8 litre (1.0/0.8 US/Imp gal)	4.1 litre (1.1/0.9 US/Imp gal)	4.1 litre (1.1/0.9 US/Imp gal)

Carburettors

	XJ650 models
Make	Hitachi
Model	HSC 32
Model ID number:	
XJ650 – 4KO (UK)	4K000
XJ650 – 11N (UK)	5N900
XJ650 J	5N800
All others	4H700
Main jet:	
XJ650 – 11N (UK)	112
All others	110
Needle jet	3.2 mm
Jet needle:	
XJ650 – 4KO (UK)	Y-11
XJ650 – 11N (UK)	Y-11
All others	Y-10
Pilot jet:	
XJ650 – 11N (UK)	43
All others	40
Starter jet	40
Main air jet	50
Pilot air jet:	
XJ650 J, XJ650 – 11N (UK)	205
All others	195

Pilot screw setting datum:

XJ650 RJ, XJ650 – 11N (UK)	$2\frac{1}{2}$ turns out
All others	Preset (no datum given)
Fuel level	3.0 ± 1 mm (0.118 ± 0.04 in)
Float height	17.5 ± 0.5 mm (0.7 ± 0.02 in)
Vacuum at idle	Above 180 mm Hg (7.09 in Hg)
Idle speed	1050 ± 50 rpm

Carburettors

XJ750 models

Make	Hitachi
Model	HSC32
Model ID number:	
XJ750(UK)	5N100
XJ750 RH, RJ	5G200
XJ750 J	15R00
Main jet	120
Needle jet	3.2 mm
Jet needle:	
XJ750(UK)	Y-14
All others	Y-13
Pilot jet:	
XJ750(UK)	43
All others	40
Starter jet:	
XJ750 J	43
All others	40
Main air jet	80
Pilot air jet:	
XJ750(UK)	195
All others	225
Pilot screw setting datum	Preset (no datum given)
Fuel level	3.0 ± 1 mm (0.118 ± 0.04 in)
Float height	17.5 ± 0.5 mm (0.7 ± 0.02 in)
Vacuum at idle	185 ± 10 mm Hg (7.3 ± 0.4 in Hg)
Idle speed	1050 ± 50 rpm

Engine oil capacity

	XJ650 – 11N (UK) and XJ650 RJ	All others XJ650	XJ750 models
Dry	3.60 litre (7.6/6.3 US/Imp pint)	3.30 litre (6.9/5.8 US/Imp pint)	3.60 litre (7.6/6.3 US/Imp pint)
Oil and filter change	2.95 litre (6.2/5.1 US/Imp pint)	2.65 litre (5.6/4.7 US/Imp pint)	2.80 litre (5.9/4.9 US/Imp pint)
Oil change	2.65 litre (5.6/4.7 US/Imp pint)	2.35 litre (5.0/4.1 US/Imp pint)	2.50 litre (5.2/4.4 US/Imp pint)

Engine oil grade

Above 5°C (40°F)	SAE 20W/40 SE motor oil
Below 15°C (60°F)	SAE 10W/30 SE motor oil

Oil pump

Type	Trochoid
Outer rotor/housing clearance	0.09 – 0.15 mm (0.0035 – 0.0059 in)
Inner rotor/housing clearance	0.03 – 0.09 mm (0.0012 – 0.0035 in)
End float	0.03 – 0.08 mm (0.0012 – 0.0031 in)
Oil pressure relief valve:	
Opening pressure	71 psi (5.0 kg/cm^2)
By-pass valve:	
Opening pressure	14 psi (1.0 kg/cm^2)

1 General description

The fuel system comprises a petrol tank from which petrol is fed by gravity to the four CD (constant depression) carburettors, via a vacuum-controlled fuel tap. In the normal 'On' position the fuel flow from the tap is regulated by a diaphragm and plunger. The diaphragm chamber is connected by a small rubber pipe to the inlet tract, and thus will only open the plunger valve when the engine is running. Thus for normal running the fuel cock lever is set to the 'On' position at all times, although the fuel supply is turned off as soon as the engine stops. The 'Reserve' lever position provides a small emergency supply of fuel in the event that the fuel gauge (where fitted) is ignored, whilst the 'Prime' setting is provided to fill the carburettor float bowls should these have been dismantled for any reason or if the machine has run completely dry.

The throttle twistgrip is connected by cable to the four throttle butterfly valves. These can be opened or closed to control the overall air flow through the instruments, and thus the engine speed. Each carburettor contains a diaphragm-type throttle valve which moves in response to changes in manifold depression and in this manner automatically controls the column and strength of the mixture entering the combustion chamber. Because the carburettors react automatically the engine runs at the optimum setting at any given throttle twistgrip setting and engine load condition, and with compensation for variations in atmospheric pressure due to changes in altitude.

The CD carburettor is ideally suited to provide an accurately controlled mixture which conforms with the increasingly stringent emission laws in the US and Europe. This allows the overall mixture to be proportionally weaker than that of a conventional slide-type instrument, and in turn should give better fuel economy.

Engine lubrication is by a wet sump arrangement which is shared by the primary transmission and gearbox components. Oil from the sump is picked up by an engine-driven trochoid oil pump and delivered under pressure to the working surfaces of the engine and gearbox components. The accompanying line drawing shows the layout of the lubrication system.

2 Fuel tank: removal and replacement

1   The fuel tank is retained at the forward end by two rubber buffers fitted either side of the under side of the tank which fit into cups on the frame top tube. The rear of the tank sits on a small rubber saddle placed across the frame top tube and is retained by a single bolt passing through the projecting rear of the tank. On those models fitted with a spring clip fitted over the bolt, merely withdraw the clip to release the rear of the tank. On other models, the bolt must be removed.

2   Turn the fuel tap to the 'On' position and pull off the fuel line and the vacuum line from the tap unions. The 'ears' of the securing spring clips should be pinched during removal to release the tension on the pipes. On machines equipped with a low fuel warning lamp, trace and disconnect the leads from the sender unit in the tank base. The fuel tank can now be lifted at the rear and eased back to free it from the front mounting rubbers.

3 Fuel tap: removal, overhaul and reassembly

1   The fuel tap is unlikely to require removal unless leakage has developed or contaminated fuel has necessitated removal for cleaning. Before removal can take place it will be necessary to ensure that any fuel in the tank is drained off, and this is obviously less of a problem if the tank is nearly empty before work commences. Alternatively, the tank can be removed from the machine and, provided that it is not too full, placed on its side on some soft rag. Any remaining fuel will thus run to the lower half of the tank and clear of the tap orifice. If draining the tank, use a long piece of fuel pipe and turn the tap to the prime (PRI) position and transfer the fuel to a suitable container. Note that this operation must be carried out in a well ventilated place, preferably outdoors, and with due regard for fire risks. Be absolutely certain that there is no smoking or naked lights in the vicinity because a large amount of fuel vapour will be released.

2   The top assembly is retained by two screws which pass through its flanged base into the underside of the tank. Once these have been removed the tap can be lifted away. Care should be taken not to damage the O-ring which seals the tap

flange. This can be re-used if in good condition, but must be discarded and renewed if it is marked or split.

3   Examine and clean the filter gauze which projects from the flange. Look for signs of water or debris from the tank and if necessary flush the tank with clean fuel to prevent subsequent blockages.

4   It is seldom necessary to remove the lever which operates the petrol tap, although occasions may occur when a leakage develops at the joint. Although the tank must be drained before the lever assembly can be removed, there is no need to disturb the body of the tap.

5   To dismantle the lever assembly, remove the two crosshead screws passing through the plate on which the operating positions are inscribed. The plate can then be lifted away, followed by a spring, the lever itself and the seal behind the lever. The seal will have to be renewed if leakage has occurred, as will the O-ring which sits in the annular groove in the tap lever boss. Reassemble the tap in the reverse order. Gasket cement or any other sealing medium is **not** necessary to secure a petrol tight seal. It is important to note that the popular silicone rubber or RTV instant gasket compounds must never be used on any part of the fuel system. The compound is attacked by fuel and will break-up allowing small rubber-linked particles to obstruct the fuel filter or carburettor jets.

6   The fuel flow control diaphragm and plunger assembly is housed behind a square cover on the inboard face of the tap body. In the event of the diaphragm becoming holed or split the plunger valve will close, blocking the fuel supply. As a temporary expedient, the machine can be ridden by selecting the prime position. Repair presents something of a problem because Yamaha do not list the diaphragm as a separate part, indicating that the entire fuel cock assembly must be renewed.

7   Reassembly and installation are a straightforward reversal of the removal sequence, bearing in mind the above mentioned caution against the use of jointing compounds. Petrol has a remarkable ability to locate and exploit the smallest leaks, so it is best to renew all O-rings and seals as a precaution. Apart from environmental considerations, the resultant fuel vapour would create a very dangerous fire hazard. Note that there is an O-ring seal between the petrol tap body and the petrol tank, which must be renewed if it is damaged or if petrol leakage has occurred.

8   In the event of tap failure, check the condition of the vacuum pipe between the inlet manifold and the diaphragm stub. If this is damaged it will not operate the tap manually and the resulting air leakage into the inlet port will cause misfiring and, possibly, overheating. The remarks in Section 4 can be applied to the vacuum pipe.

2.1 The tank fixing clip (shown displaced for clarity)

2.2 Squeeze 'ears' of clip together, then prise off fuel pipe

3.2 Fuel tap is retained to tank by two screws

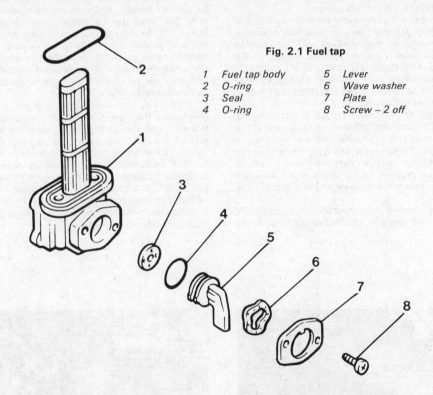

Fig. 2.1 Fuel tap

1	Fuel tap body	5	Lever
2	O-ring	6	Wave washer
3	Seal	7	Plate
4	O-ring	8	Screw – 2 off

4 Fuel feed pipe: examination

1 A synthetic rubber pipe is fitted between the fuel tap and the left-hand carburettor float bowl, the remaining float bowls being linked via passages in the carburettor bodies by short connecting stubs. The pipe is pushed over a projecting stub at each end and is secured by wire clips. Check periodically to ensure that the pipe has not begun to split or crack and that the wire clips have not begun to bite into the pipe wall. Damage at the pipe ends can often be corrected by removing the pipe and cutting off the damaged section, provided that sufficient length remains.

2 Do not replace a broken pipe with one of natural rubber, even temporarily. Petrol causes natural rubber to swell very rapidly and disintegrate, with the result that minute particles of rubber would easily pass into the carburettors and cause blockages of the internal passageways. Plastic pipe of the correct bore size can be used as a temporary substitute but it should be replaced with the correct type of tubing as soon as possible since it will not have the same degree of flexibility.

5 Carburettor adjustment and exhaust emissions: general note

In some countries legal provision is made for describing and controlling the types and levels of toxic emissions from motor vehicles.

In the USA exhaust emission legislation is administered by the Environmental Protection Agency (EPA) which has introduced stringent regulations relating to motor vehicles. The Federal law entitled the Clean Air Act, specifically prohibits the removal (other than temporary) or modification of any component incorporated by the vehicle manufacturer to comply with the requirements of the law. The law extends the prohibition to any tampering which includes the addition of components, use of unsuitable replacement parts or maladjustment of components which allows the exhaust emissions to exceed the prescribed levels. Violations of the provisions of this law may result in penalties of up to $10 000 for each violation. It is strongly recommended that appropriate requirements are determined and understood prior to making any change to or adjustments of components in the fuel, ignition, crankcase breather or exhaust systems.

To help ensure compliance with the emission standards some manufacturers have fitted to the relevant systems fixed or pre-set adjustment screws as anti-tamper devices. In most cases this is restricted to plastic or metal limiter caps fitted to the carburettor pilot adjustment screws, which allow normal adjustment only within narrow limits. Occasionally the pilot screw may be recessed and sealed behind a small metal blanking plug, or locked in position with a thread-locking compound, which prevents normal adjustment.

It should be understood that none of the various methods of discouraging tampering actually prevents adjustment, nor, in itself, is re-adjustment an infringement of the current regulations. Maladjustment, however, which results in the emission levels exceeding those laid down, is a violation. It follows that no adjustments should be made unless the owner feels confident that he can make those adjustments in such a way that the resulting emissions comply with the limits. For all practical purposes a gas analyser will be required to monitor the exhaust gases during adjustment, together with EPA data of the permissible Hydrocarbon and CO levels. Obviously, the home mechanic is unlikely to have access to this type of equipment or the expertise required for its use, and, therefore, it will be necessary to place the machine in the hands of a competent motorcycle dealer who has the equipment and skill to check the exhaust gas content.

For those owners who feel competent to carry out correctly the various adjustments, specific information relating to the anti-tamper components fitted to the machines covered in this manual is given in the relevant Sections of this Chapter.

6 Carburettors: removal

1   Carburettor removal will be necessary if any serious dismantling or overhaul work is necessary. Although it is possible to effect adjustment and a small amount of dismantling with the assembly in situ, the cramped location makes this an awkward proposition. It is almost invariably best to start by removing the entire bank of instruments from the machine.

2   Start by lifting the seat, then remove the fuel tank and side panels. The air cleaner casing should be freed to allow it to be pushed clear of the instruments during removal. This is accomplished by removing its two mounting bolts, one of which is located between the frame top tubes and the other at the lower left-hand side of the casing.

3   Slacken the clips which retain the carburettors to the inlet adaptor and air cleaner rubbers. Disengage the rubbers from the air cleaner casing and pull them away from the carburettors. Slacken the screw which secures the cold start (choke) cable to its support bracket and unhook the nipple from its operating arm. The cable can be pulled clear and lodged against the frame. Lift the throttle cable outer clear of its seating and turn the inner cable until the nipple can be disengaged. Note that if access proves awkward leave the throttle cable removal until the assembly has been manoeuvred clear of the inlet adaptor rubbers. Free the clutch cable from its guide on the rear of the carburettor assembly.

4   Grasp the carburettor bank and pull it rearwards until it disengages the inlet rubbers. Where appropriate, release the throttle cable, then withdraw the assembly from the right-hand side of the machine. As access to the carburettor becomes easier, pull off the overflow drain tubes to allow removal.

5   The carburettors are refitted by reversing the removal sequence, remembering to fit the drain tubes and throttle cable before the instruments are fully home. If an assistant is available, installation will prove much easier. Once installation is complete, check the carburettor settings and adjustments as described later in this Chapter.

6.3a Adaptor hoses locate as shown – note tab and projections

6.3b Slacken screw, then disengage the 'choke' cable

6.3c Unhook and disengage throttle cable

6.4 The carburettor assembly can now be removed

7 Carburettors: dismantling, overhaul and reassembly

1   Most of the dismantling work that is likely to be required can be undertaken without the need for separating the instruments. This is a rather laborious task which is best avoided if at all possible. Should the need arise, however, details are given later in this section. It should be noted that Yamaha advise that separation of the individual instruments should be avoided since it can lead to misalignment of the operating linkages. Whilst separation may prove inevitable in some instances it should, therefore, be avoided where possible.

2   To gain access to the float chamber components, namely the float assembly and jets, the float bowl concerned is removed after its four retaining screws have been released. It is recommended that one instrument at a time should be dealt with to preclude any possibility of parts becoming interchanged. Note that it is not necessary to remove the connecting bracket.

3   Using a small piece of wire or a pair of pointed-nose pliers displace the headed pivot pin which locates the twin float assembly and lift the float from position. This will expose the float needle. The needle is very small and should be put away in a safe place so that it is not misplaced. Make sure that the float chamber gasket is in good condition. Do not disturb the gasket unless leakage has occurred or it appears damaged.

4   Check that the twin floats are in good condition and not punctured. Any leakage will allow fuel to find its way into the float making it less bouyant than normal and upsetting the fuel level in the float chamber concerned. A quick check can be made by shaking the float and listening for signs of fuel inside it. A more reliable method is to hold the suspect float under hot water. This will cause the air inside it to expand, and if a leak exists, to be expelled as small bubbles. Repairing a damaged float is not a practicable proposition and renewal is the best course of action.

5   Examine the float needle, checking for a wear ridge at the point where it contacts the seat face. Once this has worn badly leakage can occur causing flooding of the carburettor float bowl. The resulting over-rich mixture will make normal running impossible and will be especially noticeable at idle speed. Less severe wear may not be so obvious but may cause excessive fuel consumption. If the needle is worn it must be renewed, together with the seat if this is also worn or scored. It should be noted that if these components are not functioning correctly it will not be possible to adjust the carburettors since the flooding will mask the effects of such adjustment. The brass seat has a hexagon head and may be unscrewed if required. It incorporates a fine gauze filter, which should be checked and cleaned.

6   The cold start mechanism, which is generally and incorrectly known as a 'choke', consists of a fuel enriching circuit built into each carburettor and is controlled by a spring-loaded plunger valve in each body. These are opened or closed in unison by rocker arms secured by grub screws to the operating shaft which runs across the bank of instruments. To remove the shaft, slacken the screws which secure each rocker arm and the operating lever. Slide the shaft clear of the carburettors lifting away each component as it is freed. Lay out the arms, bushes and lever in the exact order of removal to ensure that they are refitted in the same sequence. If required, each plunger valve can now be unscrewed for inspection. Wear or damage is not likely, but if discovered will necessitate the renewal of the valve. The vulnerable area is the valve tip and the corresponding seating face in the carburettor body.

7   Each diaphragm chamber cover is retained by four screws and can be lifted clear once these have been removed. The diaphragm will normally remain in the carburettor body but take care when lifting the cover in case part of it sticks to the underside. Lift out the return spring, then carefully ease the edge of the diaphragm away from the carburettor body taking care not to damage it. The diaphragm can now be removed together with the valve and needle.

8   The jet needle is retained by a spring and an Allen-headed plastic plug. The latter should be unscrewed to allow the needle and spring to be displaced. The valve body and diaphragm are not available separately and should be examined as a unit. Check the diaphragm carefully for any signs of splitting or tearing. The valve surface should be smooth and free from scoring. If either part is defective the assembly must be renewed. Check that the jet needle is straight and undamaged, noting that close scrutiny will be necessary. Even light scoring or wear will upset fuel metering, and examination should be carried out in conjunction with the needle jet. If any wear is found renew **both** components, since needle jet wear is difficult to assess visually.

9   The main air jet and pilot air jet are located in bores below the diaphragm, and are covered by a retainer plate. This can be removed by unscrewing its single retaining screw. When unscrewing the jets note their position in the carburettor as a guide to reassembly. The main and pilot (fuel) jets are located on the underside of the instrument, in casting extensions which

project into the float bowl. The longer, pilot, jet screws directly into the carburettor casting, whilst the shorter main jet screws into the bottom of the needle jet. Once the main jet has been unscrewed the instrument can be inverted and the needle jet tipped out.

10 Examine the main and pilot jet bores, looking for debris or water droplets which may have obstructed them. Such obstructions can be removed by blowing the jets through with compressed air, or as a last resort, by cleaning them with a fine **nylon** bristle. On no account should wire be used to clear a jet because it can enlarge or score the orifice, either of which will upset its fuel metering rate. The air jets should be checked in a similar manner, although their comparatively large size and the fact that the air is unlikely to contain any large particles means that obstructions should not be common. The starter jet, where fitted, meters fuel to the cold start circuit, and is screwed into the top edge of the float bowl. The machine featured in this manual also had a pair of air compensator jets which are screwed into drillings on the air cleaner side of the main choke. These will not normally require attention and can be left undisturbed. Note that in the official service literature Yamaha do not make any mention of these jets and in consequence no sizes are given.

11 As mentioned earlier in this Section, there are few occasions where it is advantageous to separate the individual instruments, and unless it proves essential to do so this should be avoided. Before commencing separation, note that a surface plate or a sheet of plate glass will be required during reassembly, to ensure alignment.

12 Start by removing the cold start operating shaft as described in paragraph 6. Slacken the screws which secure the carburettors to the two support brackets. These will invariably be tight due to the use of Loctite on their threads, and it is advisable to use an impact driver to loosen them. Once the support brackets have been freed the instruments will be held together only by fuel connections and the throttle linkages, and may be eased apart. The throttles are connected by spring-loaded links which incorporate the synchronising adjustment screws and these will pull free as each instrument is displaced.

13 When reassembling the bank of instruments, check that the fuel stub O-rings are in good condition and renew any which appear suspect. As each instrument is joined, make sure that the fuel stubs seat correctly in their bores, and that the tang of each throttle link engages between the corresponding spring-loaded pin and adjustment screw. Fit the mounting brackets

and screw the retaining screws **loosely** home, having coated the threads with Loctite. Place the assembly on the surface plate or plate glass sheet with the air cleaner side of the main chokes downward. Holding the assembly flat, tighten the retaining screws. This procedure will ensure accurate alignment.

14 Assemble the remaining carburettor parts by reversing the dismantling sequence. Note that each diaphragm has a small locating tab which should be aligned to fit in the cutout in the carburettor casting. When installing the carburettor jets, remember that brass is soft and will strip or shear easily if excessive force is employed. It is best to renew the float bowl gasket as a matter of course, though it is permissible to re-use the old one if necessary, assuming that it is in serviceable condition. Before the carburettors are refitted, check that the cold start and throttle linkages work smoothly. If the carburettor synchronising screws were disturbed for any reason, check that all four throttle butterflies move in unison and are as accurately synchronised as possible. It will, of course, be necessary to check this more accurately when assembly is complete (see Section 10 or 11).

7.2 The float bowl is retained by four screws

7.3a Displace pivot pin (arrowed) with a piece of wire

7.3b Lift away the float and needle assembly

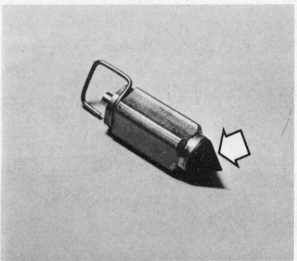

7.5a Examine the float needle for wear on the sealing tip

7.5b Float needle seat can be unscrewed for renewal ...

7.5c ... or for access to the gauze screen

7.6a Release the screws (arrowed) which retain links to the cold start shaft

7.6b Withdraw the shaft and lift each link clear

7.6c Plunger assembly can be unscrewed from carburettor

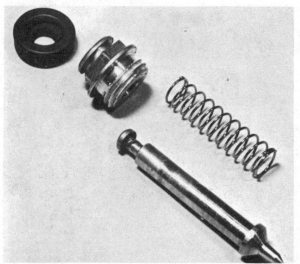

7.6d Cold start plunger components

7.7a Remove diaphragm cover and spring

7.7b Diaphragm, valve and needle are removed together

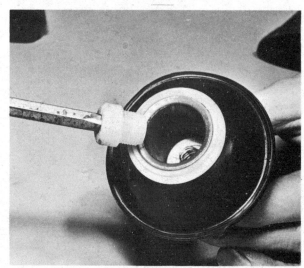

7.8a Unscrew the white plastic plug ...

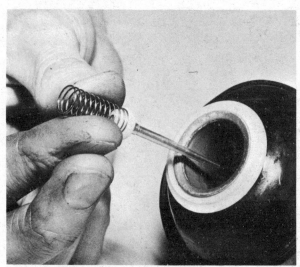

7.8b ... to release the spring and needle

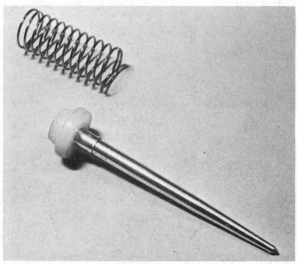

7.8c Check the needle for straightness and scoring

7.9a Air jets are covered by plate

7.9b A: Main air jet B: Pilot air jet

7.9c Pilot jet location

7.9d Remove the main jet and washer ...

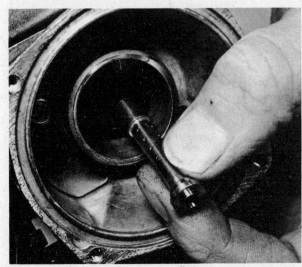

7.9e ... and displace and remove needle jet

7.10a Check needle jet bleed holes for obstruction

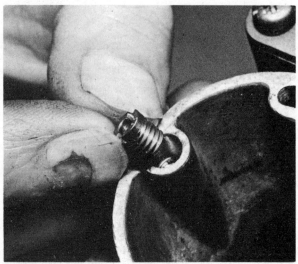

7.10b Air compensator jet location

7.14a Refit needle jet from top as shown ...

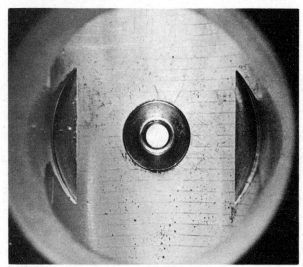

7.14b ... checking that it seats correctly in recess

Fig. 2.2 Carburettor

1 Diaphragm cover – 4 off
2 Throttle cable bracket
3 Screw – 5 off
4 Spring washer – 5 off
5 Diaphragm spring – 4 off
6 Screw – 10 off
7 Spring washer – 10 off
8 Needle retaining plug – 4 off
9 Spring – 4 off
10 Jet needle – 4 off
11 Diaphragm/piston – 4 off
12 Needle jet (main nozzle) – 4 off
13 Cover – 4 off
14 Special screw – 4 off
15 Main air jet – 4 off
16 Pilot air jet – 4 off
17 Cold start plunger – 4 off
18 Spring – 4 off
19 Housing nut – 4 off
20 Boot – 4 off
21 Bracket
22 Cable guide
23 Screw – 3 off
24 Washer – 3 off
25 Choke arm – 4 off
26 Adjuster screw – 4 off
27 Transfer pipe
28 Bush
29 Screw
30 Spring washer
31 Pilot jet – 4 off
32 Sealing washer – 4 off
33 Main jet – 4 off
34 Sealing washer – 4 off
35 Float needle valve – 4 off
36 Starter jet – 4 off
37 Float assembly – 4 off
38 Float pivot – 4 off
39 Gasket – 4 off
40 Float bowl – 4 off
41 Washer – 16 off
42 Spring washer – 16 off
43 Screw – 16 off
44 Drain plug – 4 off
45 Bracket
46 Screw – 4 off
47 Spring washer – 4 off
48 Throttle stop screw
49 Spring
50 Cold start shaft
51 Throttle adjuster – 3 off
52 Spring – 3 off
53 Buffer – 3 off
54 Spring – 3 off
55 Cable guide
56 Cable guide
57 Bush
58 Cold start cable bracket

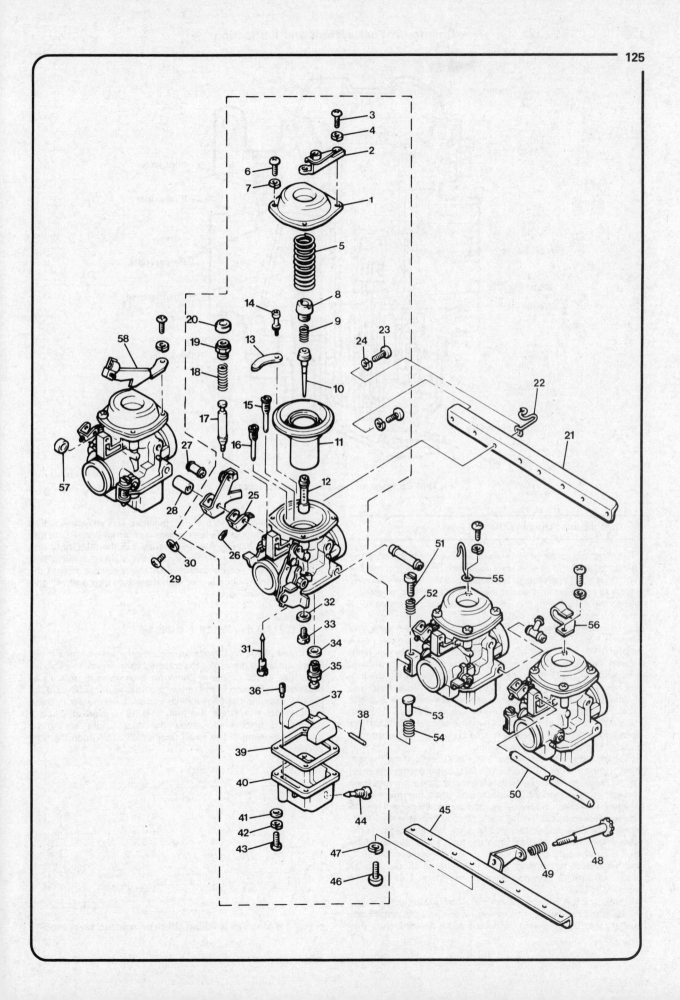

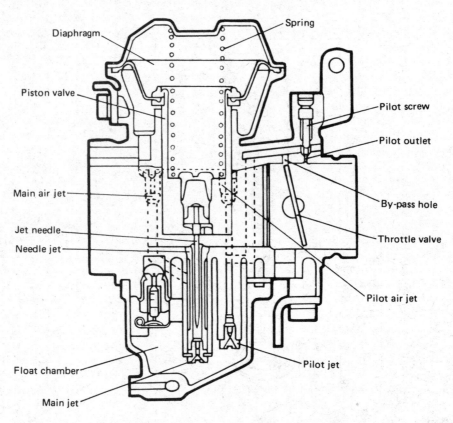

Diaphragm

Spring

Piston valve

Pilot screw

Pilot outlet

By-pass hole

Main air jet

Jet needle

Needle jet

Throttle valve

Pilot air jet

Float chamber

Pilot jet

Main jet

Fig. 2.3 Carburettor main components location

8   Carburettors: checking the fuel level

1   Before any attempt is made to adjust or synchronise the four carburettors, it is necessary to ensure that the fuel level in each float bowl conforms to the prescribed limits. The fuel level affects to some extent the mixture strength at all engine speeds, and so the accuracy of this setting is fundamental to that of all other carburettor adjustments.

2   To measure the fuel level some form of level gauge is needed. Yamaha can supply a gauge for this purpose, or a length of 6 mm diameter clear plastic tubing can be used instead. One end of the pipe can then be pushed over the fuel drain stub on the bottom of the carburettor float bowl, the free end being held along the side of the caburettor body to form a U-tube. When the drain screw is opened fuel will flow into the tube indicating the fuel level inside the float bowl. A useful sophistication would be to fit a small crocodile clip near the open end of the tube to enable it to be clipped to the carburettor body during the checking operation.

3   Start and run the engine for a few minutes to allow the fuel level to find its normal position before commencing the level check. Place the machine on its centre stand on a smooth level surface. It is important to make sure that the machine is absolutely vertical otherwise inaccurate readings will result. This is checked by connecting the tube to carburettor No 1 (left-hand). Open the drain screw and note the fuel level against the side of the float bowl. Now pass the tube across the machine until it can be held against the float bowl of carburettor No 4, and again note the reading. If the two differ at all, place small pieces of card or plywood beneath the stand feet until the machine is level.

4   Next, connect the tube to each carburettor in turn and measure the fuel level in millimeters below the carburettor body gasket face. The following points should be noted. Ensure that

the fuel tap is turned to the 'PRI' position. This will allow fuel to flow into the float bowls to compensate for that drained out into the level tube. When checking the level in the tube, note that the height to be measured is that of the main reservoir of fuel in the tube. Do not be confused by the meniscus around the tube walls; this will be higher than the actual level and will give an incorrect reading.

Fuel level
   3.0 ± 1.0 mm (0.118 ± 0.039 in)

5   If any of the fuel levels are outside the above limits it will be necessary to dismantle the affected carburettors to establish the cause of the problem. Check for float damage or leaks as described in Section 6 and renew the float as necessary. Do not omit to examine the float needle and seat since wear in these components will affect the fuel level. If no sign of wear or damage is discovered, adjust the float height setting by judicious bending of the small tang which bears upon the float needle.

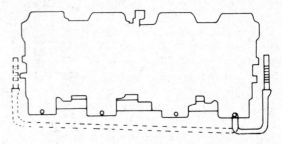

Fig. 2.4 Machine levelling check prior to fuel level check

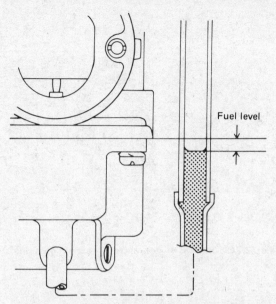

Fig. 2.5 Measuring the carburettor fuel level

Fig. 2.6 YICS blanking tool to permit carburettor synchronisation

9 Yamaha YICS system: general description – All XJ750, XJ650 J and XJ650 – 11N (UK) models

1   The Yamaha Induction Control System (YICS) consists of a cross drilling of the cylinder head, immediately below the inlet ports, and smaller connecting bores from it to the inlet tract just before the valve, and angled so that they face the cylinder walls. Yamaha call the resulting balance pipe arrangement 'sub-intake ports'.

2   When any one cylinder is on its induction stroke, the end of its 'sub-intake port' is subjected to fairly high depression. This draws mixture through it and from the rest of the YICS drillings, and to some extent through the remaining three carburettors. The mixture emerges near the opened valve at relatively high velocity and enters the cylinder where it is deflected in a swirling motion by the cylinder wall. This small extra charge of mixture adds slightly to the overall mixture content of the cylinder, but more importantly, the swirling action is said to improve combustion efficiency, and thus extracts more power from a given amout of fuel.

10 Carburettors: synchronisation – YICS models

1   The procedure for synchronizing the carburettors on the 750 cc and later 650 cc models is slightly complicated by the YICS passages. Since the inlet ports are interconnected it is impossible to obtain individual vacuum gauge readings using the usual approach. To overcome this, Yamaha produce a special tool in the form of a long rod with a locking handle and seals along its length. Before synchronisation takes place, one of the two end plugs in the main YICS passage is removed and the tool (part number 90890-04068) is inserted and locked in place. The tool effectively blanks off the YICS system to allow normal synchronisation adjustment, following the procedure described in Section 10 of this Chapter. It should be noted that the idle speed may fall slightly after the blanking tool is fitted, and if necessary should be increased to 950–1000 rpm using the large central throttle stop control. Synchronise the carburettors as described in the following Section, then remove the tool and refit the blanking plug. Check the idle speed, with the engine warm, and make any necessary adjustment to restore it to the specified 1050 rpm.

11 Carburettors: synchronisation – non YICS models

1   On any multi-carburettor engine the accurate synchronisation of the carburettors is essential if smooth running and good performance and fuel economy are to be obtained. This is especially true of four cylinder engines, and in cases of extreme mal-adjustment the engine will refuse to idle reliably and may even produce expensive-sounding noises due to backlash in the primary transmission. It will be appreciated that if one or more of the carburettors is out of adjustment, the related cylinder will have to be 'carried' by the remaining cylinders, thus the engine will be attempting to run at two different speeds for any given throttle setting.

2   Synchronisation poses something of a problem in that there is no way in which it can be carried out accurately without the use of a vacuum gauge set. This may be of the manometer (mercury column) type or may have one or more clock-type gauges. Yamaha can supply a single clock type gauge with a four way selector switch; its part number is 90890-03094. Alternatively a number of mail order suppliers can supply vacuum gauge sets of various types. They all have something in common in that they are rather expensive, so a decision must be made as to whether it is worth spending money on the equipment or if it is best to entrust this operation to a Yamaha dealer. If gauges are available, proceed as described below.

3   Remove the seat, and prop the rear of the fuel tank so that access to the vacuum take-off points and synchronising screws is possible. Remove the vacuum take-off caps from the carburettor intake adaptors of cylinders 1, 2 and 4, and pull off the vacuum pipe from No 3. Turn the fuel tap to 'PRI' so that a supply of fuel is available during the test sequence. Connect the vacuum gauge adaptors and hoses to the four take-off points.

4   Start the engine and set up the vacuum gauge(s) to the manufacturer's recommendations. This normally entails setting a small damping valve or valves so that the gauge(s) respond quickly as the throttle is opened, but do not oscillate too wildly with each engine revolution. Allow the engine to idle until it reaches its normal operating temperature, then set the large throttle stop control to give an idle speed of about 1000 rpm.

5   Start by synchronising carburettors No 1 and 2 by turning the synchronising screw located between the two instruments. No specific vacuum reading is given by the manufacturer, but it is essential that the two carburettors give the same reading. Now synchronise carburettors No 3 and 4 in the same way, using the right-hand screw.

6   The two pairs of carburettors must now be synchronised to each other. This is done by turning the centre synchronising screw until No 1 and 2 carburettors give the same reading as No 3. It is likely that the engine speed will have increased by now, and this should be checked and reset at 1000 rpm. If the adjustment sequence was carried out accurately, the four carburettors should show a similar reading. If this is not the case, repeat the procedure until the correct balance is obtained. After synchronisation has been carried out, set the idle speed to within the prescribed limits of 950 – 1050 rpm using the central throttle stop control.

   Important note: On no account should the pilot mixture screws be moved. See Section 12 for details.

11.5 Synchronisation is achieved by connecting link adjuster screws between throttles

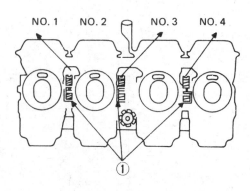

Fig. 2.7 Carburettor synchronisation screw locations

12 Carburettor settings

1   Some of the carburettor settings, such as the sizes of the needle jets, main jets and needle positions, etc are pre-determined by the manufacturer. Under normal circumstances it is unlikely that these settings will require modification, even though there is provision made. If a change appears necessary, it can often be attributed to a developing engine fault.

2   If a particular fault has been traced to the carburettors, check first that the jets are unobstructed and of the specified size, that the fuel level is correct and that the carburettors are synchronised. The only other permissible adjustment is setting the idle speed. This is controlled by the large central knob and should be set to give an idle speed of 1050 ± 50 rpm.

3   It should be noted that no mention has been made concerning pilot mixture adjustment. This is because the screw settings are 'pre-set' by the factory and no adjustment is recommended. Even dealers are advised not to attempt adjustment, so it is impossible to give advice on this point. Yamaha have chosen to take this precaution to avoid possible infringement of the exhaust emission regulations in certain US states. If incorrect mixture settings are suspected, it will be necessary to solicit the aid of a competent motorcycle dealer with full exhaust analysis facilities. Using this equipment it is possible to set the pilot mixture screws accurately and within the required legal limits. Those living in areas where stringent anti-pollution laws are in operation should check whether **any** proposed carburettor adjustment or modification is permitted.

13 Exhaust system

1   Unlike a two-stroke, the exhaust system does not require frequent attention because the exhaust gases are usually of a less oily nature.

2   Do not run the machine with the exhaust baffles removed, or with a quite different type of silencer fitted. The standard production silencers have been designed to give the best possible performance, whilst subduing the exhaust note to an acceptable level.

3   Whilst there are a number of good quality after-market exhaust systems available, there are others which may be of poor construction and fit, and which may reduce performance rather than improve it. When purchasing such systems it is helpful to obtain recommendations from other owners who have had time to evaluate the system under consideration. Do not forget that there are noise limits which will be met by the more reputable manufacturers. It is not advised that a non-standard system is fitted during the warranty period, because this could result in subsequent claims being refuted.

4   It is advisable to pay close attention to the condition of the exhaust system in the hope of prolonging its useful life. Any accumulated road dirt will encourage surface rusting, and this is of particular relevance to the collector box/balance pipe area. Unfortunately, the system will eventually rust through from the inside due to the acidic nature of the exhaust gases, and little can be done to prevent this. The problem could be avoided if a stainless steel system were employed, but with all standard mild steel types it will remain a regular expense. Routine examination will at least give warning of its impending demise and allow the owner to start saving.

12.3 The mixture screw. **Do not attempt adjustment**

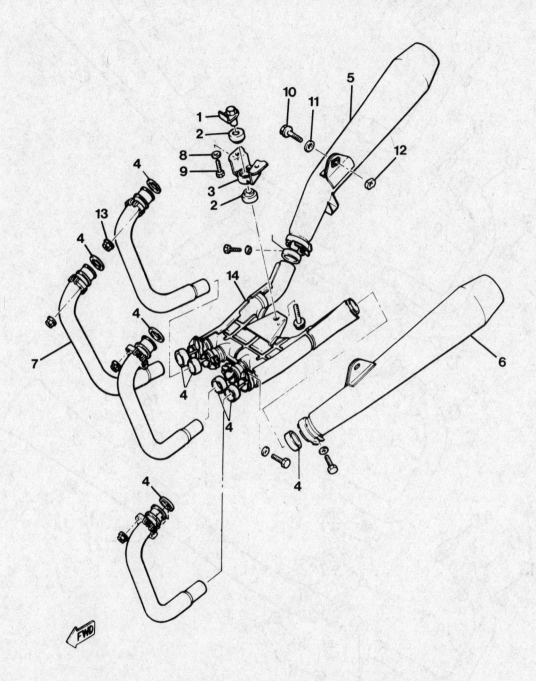

Fig. 2.8 Exhaust system – XJ650 and 750(UK) and XJ650 RJ, XJ750 J, RJ and RH

1	Special bolt	6	LH silencer	10	Bolt – 2 off
2	Rubber – 2 off	7	Exhaust pipe	11	Washer – 2 off
3	Mounting bracket	8	Washer – 2 off	12	Special nut – 2 off
4	Seal set – 4 off	9	Bolt – 2 off	13	Nut – 8 off
5	RH silencer				

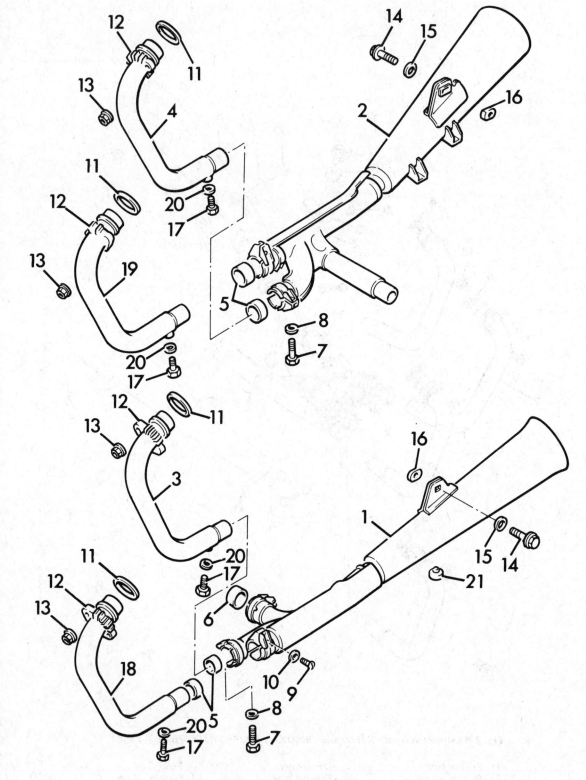

Fig. 2.9 Exhaust system – XJ650 G, J, H and LH

1	LH silencer assembly	6	Sealing sleeve
2	RH silencer assembly	7	Pinch bolt – 2 off
3	Exhaust pipe	8	Washer – 2 off
4	Exhaust pipe	9	Screw – 2 off
5	Sealing sleeve – 4 off	10	Washer – 2 off

11	Exhaust gasket – 4 off	16	Special nut – 2 off
12	Flange – 4 off	17	Screw – 4 off
13	Nut – 8 off	18	Exhaust pipe
14	Bolt – 2 off	19	Exhaust pipe
15	Washer – 2 off	20	Washer – 4 off
		21	Rubber buffer

14 Air cleaner: location and maintenance

1 The air cleaner consists of a pleated paper element housed in a plastic casing beneath the seat. The casing serves to silence induction noise and acts as a small plenum chamber. To remove the element, lift the seat and remove the tool tray. Remove the three screws which secure the lid of the filter compartment and lift it clear. Note that the fuse box is mounted on the lid, but need not be disturbed. The element can now be lifted out for cleaning.

2 Cleaning is carried out by tapping the element sharply to dislodge any loose dust after which it can be cleaned further by blowing compressed air through from the inner surface. Cleaning should be carried out at the recommended routine maintenance interval or more frequently where the machine is used in particularly dusty conditions. If any damage is discovered the element must be renewed promptly.

3 If the element is damp or oily it must be renewed. A damp or oily element will have a restrictive effect on the breathing of the carburettor and will almost certainly affect the engine performance.

4 On no account run the engine without the air cleaner attached, or with the element missing. The jetting of the carburettors takes into account the presence of the air cleaners and engine performance will be seriously affected if this balance is upset.

5 To replace the element, reverse the dismantling procedure.

Give a visual check to ensure that the inlet hoses are correctly located and not kinked, split or otherwise damaged. Check that the air cleaner casing is free from splits or cracks.

14.1 Removing the air filter element

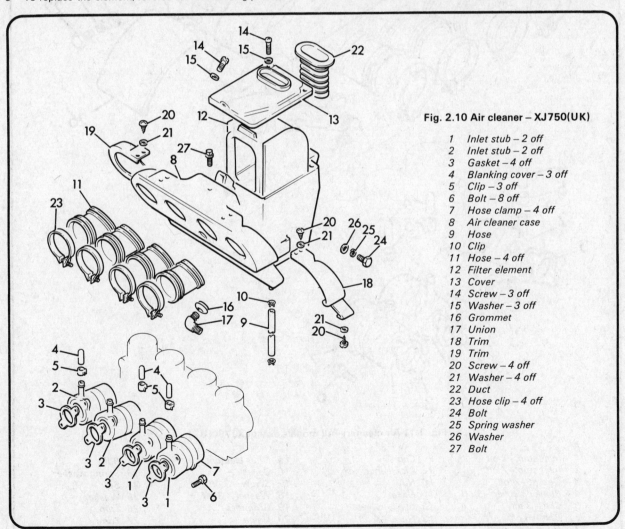

Fig. 2.10 Air cleaner – XJ750(UK)

1 Inlet stub – 2 off
2 Inlet stub – 2 off
3 Gasket – 4 off
4 Blanking cover – 3 off
5 Clip – 3 off
6 Bolt – 8 off
7 Hose clamp – 4 off
8 Air cleaner case
9 Hose
10 Clip
11 Hose – 4 off
12 Filter element
13 Cover
14 Screw – 3 off
15 Washer – 3 off
16 Grommet
17 Union
18 Trim
19 Trim
20 Screw – 4 off
21 Washer – 4 off
22 Duct
23 Hose clip – 4 off
24 Bolt
25 Spring washer
26 Washer
27 Bolt

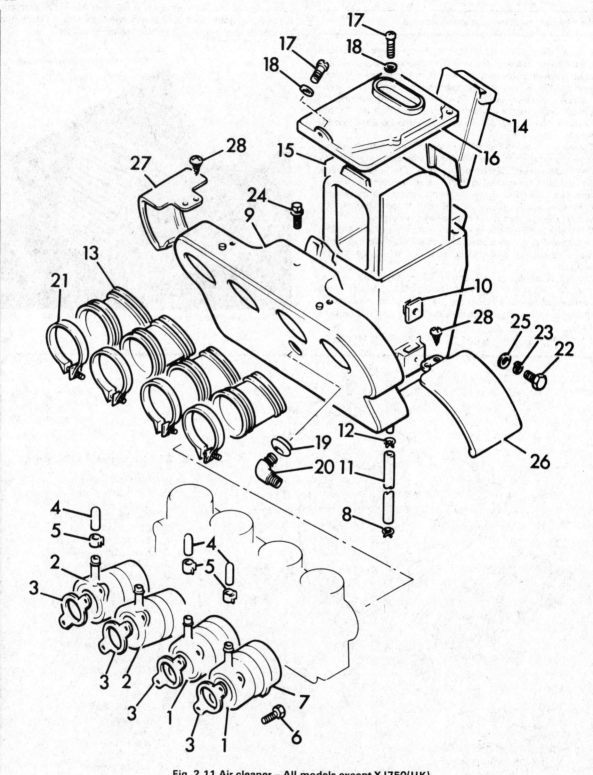

Fig. 2.11 Air cleaner – All models except XJ750(UK)

1	Inlet stub – 2 off	8 Clip	15 Filter element	22 Bolt
2	Inlet stub – 2 off	9 Air cleaner case	16 Cover	23 Spring washer
3	Gasket – 4 off	10 Nut	17 Screw – 3 off	24 Bolt
4	Blanking cover – 3 off	11 Hose	18 Washer – 3 off	25 Washer
5	Clip – 3 off	12 Clip	19 Grommet	26 Trim
6	Bolt – 8 off	13 Hose – 4 off	20 Union	27 Trim
7	Hose clip – 4 off	14 Duct*	21 Hose clip – 4 off	28 Screw – 4 off

* Not fitted to XJ750 RH, J and RJ

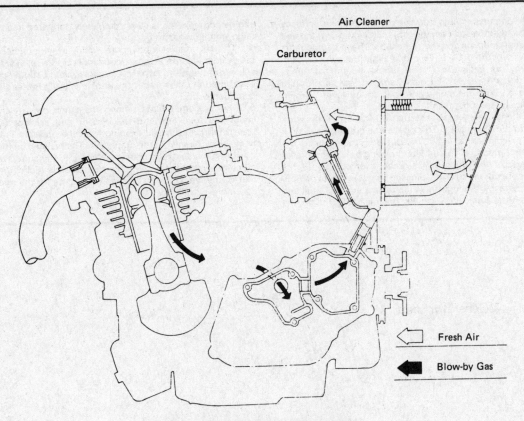

Fig. 2.12 Crankcase ventilation system

Air Cleaner

Carburetor

⇦ Fresh Air

⬅ Blow-by Gas

15 Crankcase ventilation system: description and maintenance

1   In common with most modern engines, the Yamaha XJ650 and 750 models employ a system to prevent oil mist or vapour from being expelled into the atmosphere. A hose is connected to the air cleaner casing, just below the line of intake adaptors. The lower end of the hose is connected to a stub on the engine.

2   It will be appreciated that, during normal running, a small amount of air is displaced beneath the descending pistons, and it follows that this air, and the oil mist it carries must be vented to avoid pressurising the crankcase. The vent provides exits via a labyrinth-like oil separator mounted inside the left-hand engine casing, and thence to the hose and air cleaner casing. The displaced air exits as a series of pulses, and most of the oil droplets will be trapped in the oil separator and drain back into the sump. Any remaining fine oil mist will be drawn into the engine and burnt before being expelled with the exhaust gases.

3   The crankcase ventilation system is fully automatic in operation and will require no attention during normal use. Only in unusual circumstances, where a machine is used for very short journeys in a cold, damp climate is there a risk of the oil separator becoming choked with emulsified oil (oil sludge). Should this be noted during oil changes, in the form of a thick creamy white scum, it is advisable to remove the left hand cover to expose the separator casting which can then be detached and cleaned out.

16 Engine lubrication

1   The engine and transmission assembly share a common crankcase and thus the same oil supply. A reservoir of oil is contained in the light alloy sump, from which the trochoid oil

15.3 The oil separator casting and gaskets

pump draws oil through a gauze strainer and is fed, via the oil cooler on the UK market models, to the oil filter. Once any impurities have been removed, the oil is forced into a transverse main oil gallery for distribution to the engine components. The accompanying illustration shows oil distribution throughout the engine unit, and it is worth noting that few moving parts are not directly pressure-fed. This arrangement indicates that the unit is, in general, well protected and can be expected to cover very high mileages before wear begins to show. It also underlines just how important regular oil changes and filter changes are. In an engine of this type, neglected oil and filter changes will account for almost all catastrophic failures.

2   The unit employs a large number of plain bearing surfaces to support highly-stressed components such as the big-end and main crankshaft journals and the camshafts. These components are actually supported by a very thin high-pressure oil film, which is why the engine will seize almost instantly if this is interrupted for any reason. To guard against loss of pressure through the oil level becoming too low, an oil level warning system is fitted (see Chapter 6).

3   An additional safeguard takes the form of an oil pressure relief valve in the pump body. This opens at a pressure of 5.0 kg cm² (71.0 psi) and thus maintains but does not exceed the correct system pressure. The oil filter housing bolt incorporates a spring-loaded filter bypass valve. This is forced open if the filter element becomes so badly blocked that it restricts oil flow through its surface. Although this maintains a supply of oil, and thus avoids total lubrication failure, the oil being fed to the engine components is now completely unfiltered and will allow rapid engine wear to occur.

4   The UK market models are equipped with oil coolers, and the system incorporates a secondary pressure relief valve which will allow the oil to bypass the cooler circuit should it become obstructed. It follows that the engine will run hotter should this condition occur.

5   Certain components, namely the clutch, primary chain, cam chain and followers and the pistons and cylinder bores, are splash lubricated by oil exuding from the pressure-fed components or by oil mist in the crankcase. The residual oil then drains down the inside of the crankcase and into the sump, where the process is repeated. A small pocket of oil is maintained at the rear of the crankcase to provide an oil bath for the middle gear assembly.

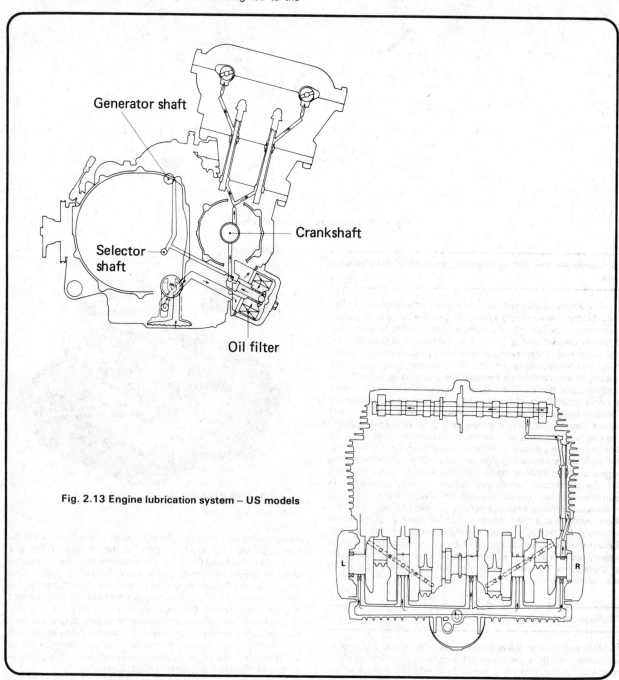

Generator shaft

Selector shaft

Crankshaft

Oil filter

Fig. 2.13 Engine lubrication system – US models

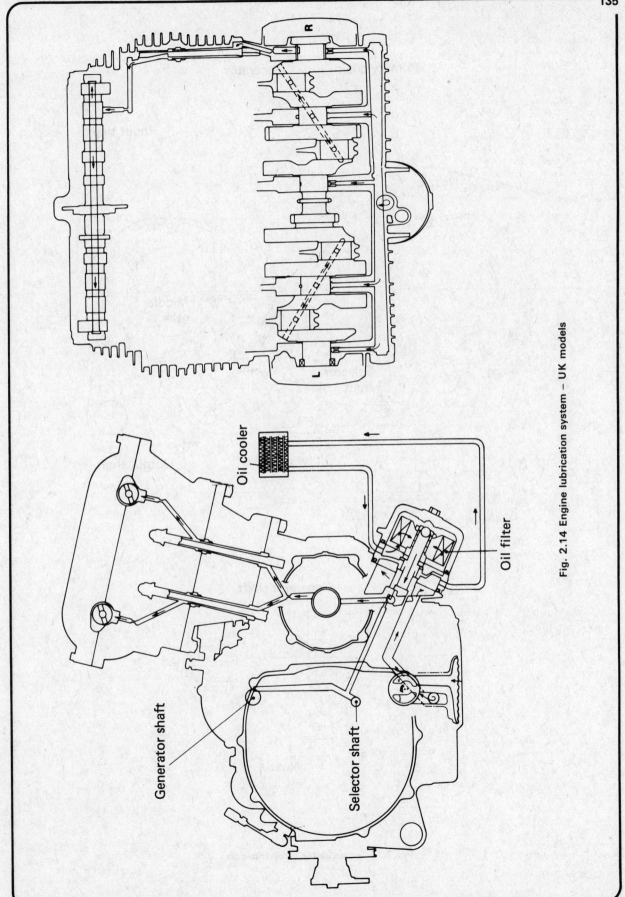

Fig. 2.14 Engine lubrication system – UK models

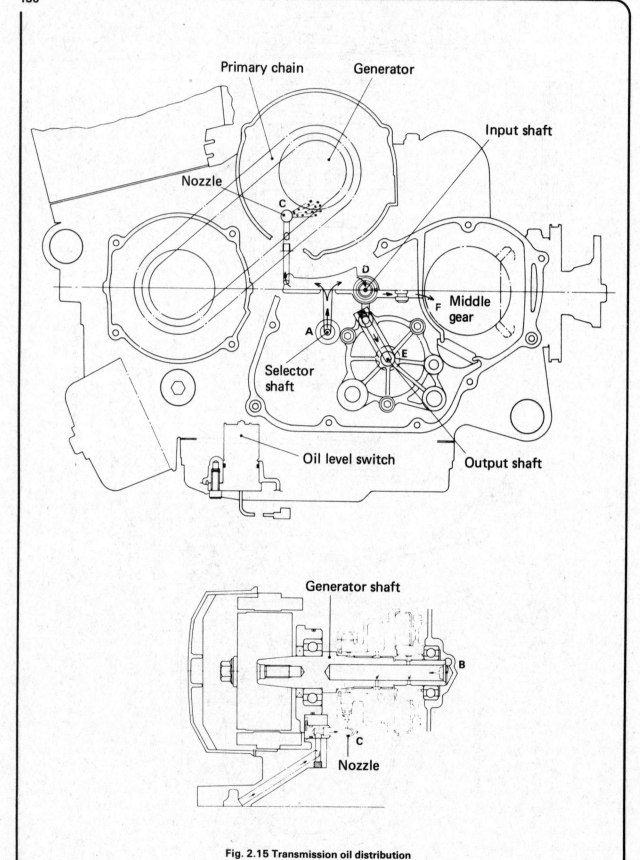

Fig. 2.15 Transmission oil distribution

2

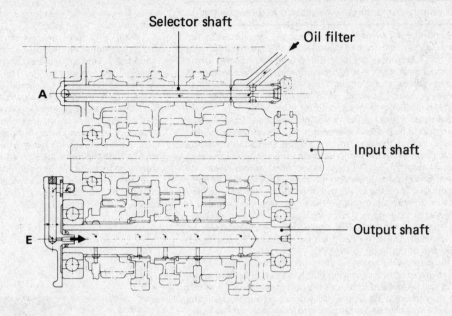

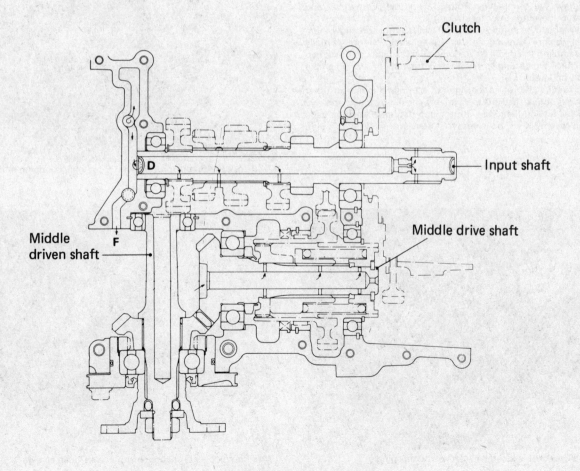

Fig. 2.16 Gearbox lubrication system

17 Oil pump: dismantling, examination and reassembly

1   The oil pump can be removed from the engine while the engine is still in the frame. However, it will be necessary to detach the sump after draining the engine oil. Access to the sump area is not easy, but can be improved if the machine is supported **safely** at an angle. Note that the pump removal sequence is also covered in Chapter 1, and that the photographs relating to this should be referred to.

2   Slacken and remove the two shouldered, and single plain, Allen screws which retain the pump to the underside of the crankcase, noting that the pump sprocket shroud will also be released. Removal of the pump normally requires the removal of the clutch assembly so that the pump drive sprocket can be removed. This in turn allows enough chain slack for the pump to be disengaged from the chain and lifted away. Before resorting to the above approach try tilting the pump to check whether it is possible to disengage the pump sprocket from the chain. Failing this, try removing the single bolt which secures the pump sprocket. The latter can then be pulled off its shaft end, allowing the pump body to be freed.

3   If the pump was removed with the sprocket attached, this can be released after its securing bolt has been unscrewed. The sprocket will be inclined to turn as the bolt is slackened, and this can be prevented by passing a screwdriver through one of the holes in the sprocket and arranging it to lodge against the pump body.

4   Slacken and remove the four screws which secure the pump cover, noting that the cover will be pushed clear by spring pressure from the pressure relief valve. Remove the cover and spring, followed by the valve body. Grasp the pump spindle and withdraw the pump inner rotor. Invert the pump body and displace the outer rotor by shaking it.

5   Wash all the pump components with petrol and allow them to dry before carrying out an examination. Before partially reassembling the pump for various measurements to be carried out, check the casting for breakage or fracture, or scoring on the inside perimeter.

6   Examine the rotors for signs of scoring, chipping or other damage, noting that this is invariably caused by small abrasive particles finding their way into the pump. Should such damage be discovered it will be necessary to renew the pump as a unit, noting that component parts are not supplied separately. The same remarks and checking methods can be applied to the pressure relief valve plunger.

7   Wear, if present, will be evident in the form of increased clearances between the rotors and between the outer rotor and pump body. This is checked using feeler gauges as shown in the accompanying photographs. The clearances should be within the limits given in the Specifications, or low oil pressure will result. Again, the only remedy is renewal of the pump unit.

8   If all is well, reassemble and refit the pump by reversing the above sequence, having first checked and cleaned the gauze strainer. Use a **new** O-ring between the pump and crankcase to ensure proper sealing. Note when fitting the pump mounting screws that the two shouldered types also retain the pressed steel sprocket shroud. Though not essential, it is good practice to prime the pump before installation. This can be done by pouring oil into the pump outlet whilst rotating the drive sprocket to distribute the oil.

17.3a Lock the pump sprocket with a bar whilst undoing bolt

17.3b Sprocket is located by flats on pump spindle

17.4a Pump cover is secured by four cross-head screws

17.4b Cover will be displaced by relief valve spring

17.4c Remove valve body from pump

17.4d Spindle and inner rotor can be withdrawn ...

17.4e ... followed by the outer rotor

17.7a Check clearances between the two rotors ...

17.7b ... and between outer rotor and pump body

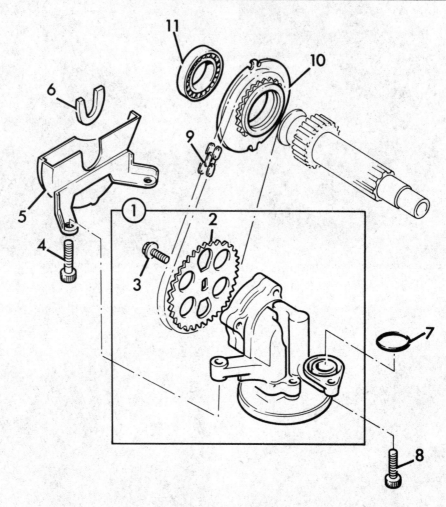

Fig. 2.17 Oil pump

1 Oil pump assembly	4 Bolt – 2 off	7 O-ring	10 Drive sprocket
2 Driven sprocket	5 Guard	8 Screw	11 Bearing
3 Screw	6 Grommet	9 Drive chain	

18 Oil filter: function and renewal

1   The oil filter consists of a renewable filter element housed in a light alloy bowl mounted at the front of the crankcase. Oil from the pump is passed to the outside of the filter surface, which is pleated to present a large area and thus offer low resistance to the oil. As the oil is forced through the filter pores any small particles carried in suspension are trapped leaving clean oil to be circulated around the engine. Periodically, the element must be removed and renewed before the surface becomes blocked by contaminants. As mentioned earlier in this Chapter, failure to do so will cause the opening of a bypass valve, maintaining the oil supply but leaving it unfiltered.

2   The filter element is removed after the contents of the sump has been drained. The filter bowl is secured by a single central bolt which incorporates the bypass valve mentioned above. As the bowl and element are detached, a small tray or bowl will be required to catch the residual oil that will be released.

3   Discard the used element and clean the inside of the bowl, taking care to remove all traces of sediment. It is worthwhile examining this during cleaning, since the sudden appearance of an unusual amount of metal particles may indicate that a bush or bearing surface has failed, particularly if excessive mechan-

ical noise has suddenly appeared. Should there be cause for suspicion it is wise to drop the sump and examine it and the oil pickup strainer for further clues. This may seem to involve a lot of extra work, but could save the engine unit if a failing component is spotted in time. Check also the bypass valve is clean and that the ball can be pushed off its seat against spring pressure.

4   It is recommended that the sealing grommets, one each side of the element, are renewed along with the large O-ring which seals the bowl and the smaller O-ring around the centre bolt head. Push the centre bolt/bypass valve into position in the bowl, then fit the spring, plain washer and element. Offer up the assembly to the crankcase, holding the bowl square whilst the bolt is tightened to 1.5 kgf m (11.0 lbf ft). **Do not** overtighten this bolt because it may shear or make subsequent removal very difficult.

19 Oil cooler: general description – UK models

1   An oil cooler is fitted to limit the oil temperature during hard riding. The oil cooler matrix is mounted on the frame front

downtubes to gain the best possible benefit from the cooling airflow. The feed and return pipes are interconnected at their lower ends with a distributor block interposed between the oil filter chamber and crankcase.

2   To maintain peak efficiency the matrix should be kept clear of any debris, preferably using an air jet directed from behind to blow out the air channels. Avoid using sharp instruments to dislodge any foreign matter; this may easily lead to damaged vanes. Should leakage of the matrix occur renewal of the complete component will probably be the only satisfactory solution. Repair is unlikely to be successful.

3   Leakage at the hose unions will result from deteriorating O-rings. These can be renewed without difficulty. To release the hose unions, remove the filter assembly as described in Section 17, then unscrew the large adaptor bolt which holds the spacer block to the crankcase. Once this has been swung clear of the crankcase and exhaust system the union Allen screws can be released.

4   Details on removing the oil cooler and hoses from the frame will be found in Chapter 1, Section 4, together with the relevant photographs.

19.3 Oil cooler hose adaptors can be released after pulling distributor block clear

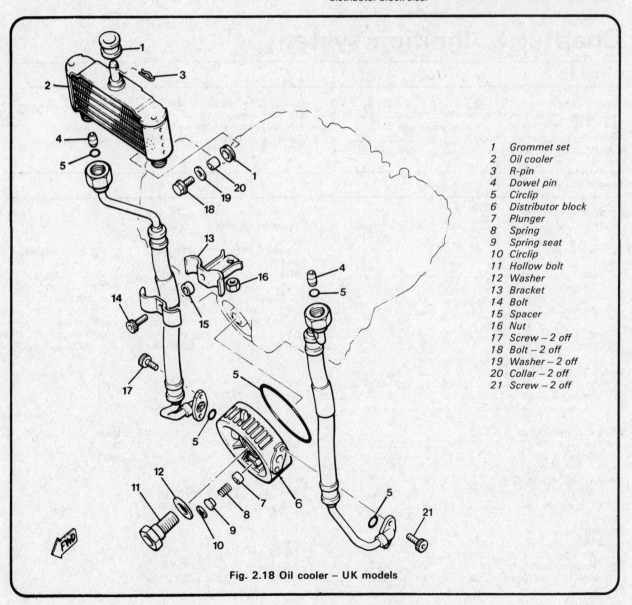

1	Grommet set
2	Oil cooler
3	R-pin
4	Dowel pin
5	Circlip
6	Distributor block
7	Plunger
8	Spring
9	Spring seat
10	Circlip
11	Hollow bolt
12	Washer
13	Bracket
14	Bolt
15	Spacer
16	Nut
17	Screw – 2 off
18	Bolt – 2 off
19	Washer – 2 off
20	Collar – 2 off
21	Screw – 2 off

Fig. 2.18 Oil cooler – UK models

Chapter 3 Ignition system

Contents

Specifications

Ignition system
Type .... TCI, fully electronic

Ignition timing

	XJ650 G,H,LH and RJ	XJ750 J and XJ750(UK)	All other models
Retarded	10° @ 1050rpm	7° @ 1050rpm	7° @ 1050rpm
Intermediate	12° @ 1800 ± 200rpm	9° @ 1600 ± 200rpm	12° @ 1800 ± 200rpm
Intermediate	35.5° @ 4100 ± 400rpm	35.5° @ 3300 ± 300rpm	35.5° @ 4100 ± 400rpm
Full advance	37.5° ± 2° @ 5000rpm	37.5° ± 2° @ 4000rpm	37.5° ± 2° @ 5000rpm

Spark plugs
Make .... NGK or ND
Type .... BP7ES or W22EP
Gap .... 0.7 - 0.8 mm (0.023 - 0.032 in)

Plug cap resistance

	UK models	US models
Cylinders 1 and 4	5.5 K ohm	5 K ohm
Cylinders 2 and 3	5.5 K ohm	10 K ohm

Pickup coil resistance
XJ650 G, H, LH and RJ .... 700 ohm ± 20% @ 20°C (68°F)
All other models .... 650 ohm ± 10% @ 20°C (68°F)

Ignition coil
Make .... Hitachi
Type CM12-09
Primary resistance .... 2.5 ohm ± 10% @ 20°C (68°F)
Secondary resistance .... 11.0 K ohm ± 20% @ 20°C (68°F)

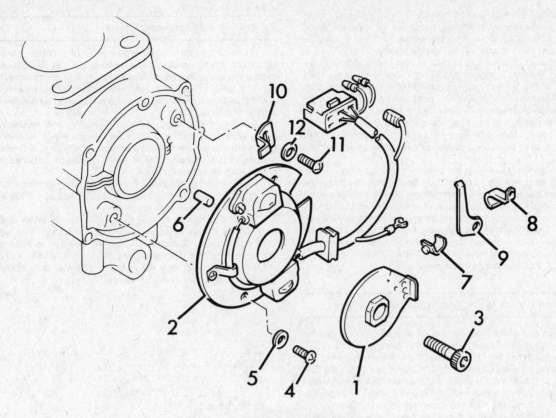

Fig. 3.1 Pick-up coil assembly

1	Reluctor	4	Screw – 2 off	7	Clamp	10	Timing index
2	Pick-up stator	5	Washer – 2 off	8	Clamp	11	Screw
3	Bolt	6	Drive pin	9	Clamp	12	Washer

1 General description

The Yamaha XJ650 and XJ750 models employ a fully transistor-electronic ignition system, described by the manufacturer as TCI. The system is controlled by the TCI unit, a large rectangular 'black box' mounted at the rear of the machine beneath the dual seat. The system is triggered, electronically rather than mechanically, by a pickup assembly mounted on the crankshaft end. Two pickup coils are fitted, one triggering the spark for cylinders 1 and 4 and the other controlling cylinders 2 and 3. It should be noted that each pickup provides a spark simultaneously at **both** of the related spark plugs. However, only one of the cylinders concerned will be near TDC on the compression stroke, and thus ignition will only take place in this cylinder, the remaining spark being wasted. This principle is normal on most four cylinder motorcycle engines and is known as the spare spark system.

The pickups consist of small electromagnetic coils mounted on opposite sides of a moving reluctor. As the raised tang of the reluctor passes the coil poles, a small signal current is induced and passed to the TCI unit. The unit senses the signal from the pickup and switches off the current to the primary windings of the relevant pair of coils. This induces a high tension pulse in the secondary winding which is fed via the HT leads to the appropriate spark plugs.

At the moment that the unit initiates the spark in one set of coils, it switches the current on to the remaining pair, thus setting them up for the same cycle once the reluctor reaches the next pickup coil. This cycle takes place once in each crankshaft revolution.

The TCI unit incorporates a bypass circuit which works in conjunction with a ballast resistor to provide a heavy spark during starting. Once the engine is running the ballast resistor reduces the power to the coils, lowering their power consumption and extending coil life. The TCI unit also incorporates a coil protection circuit which switches off the supply to the coils should the ignition be left switched on without the engine running for more than a few seconds.

When the starter button is next operated, the supply to the coils is restored after the crankshaft has turned through 180°.

As engine speed rises it is necessary to advance the ignition so that the spark occurs at an earlier point in the engine cycle. This is to ensure that full combustion pressure is available at higher speeds. The TCI unit is able to 'read' the engine speed from the reluctor and can then advance the ignition as required. This arrangement obviates the need for a centrifugal timing unit. The system, once set correctly, does not suffer from mechanical wear, and thus does not require regular adjustment or maintenance.

2 Testing the ignition system: general information

1 An electronic ignition system does not require maintenance in the generally accepted sense. There are no mechanical parts, thus wear does not take place and the need for compensation by adjustment does not arise. Ignition problems in this type of system can be broken down as follows.

a Loose, broken or corroded connections
b Damaged or broken wiring
c Faulty or inoperative electronic components

2   The above are arranged in the order in which they are most likely to be found, and with the exception of item C should provide no undue problem in the event of fault finding or rectification. Where part of the electronic side of the system fails, however, diagnosis becomes rather more difficult. The following sections provide details of the necessary test procedures, but it must be remembered that basic test equipment will be required. Most of the tests can be carried out with an inexpensive pocket multimeter. Many home mechanics will already have one and be conversant with its use. Failing this, they are easily obtainable from mail order companies or from electronics specialists, or can be ordered through a Yamaha dealer.

3   When carrying out tests on the electronic ignition system, bear in mind that wrong connections could easily damage the component being tested. Adhere strictly to the test sequence described and be particularly careful to avoid reversed battery connections. Note that the system must **not** operate with one or more HT leads isolated, as the very high secondary winding voltage may destroy the ignition coil. Beware of shocks from the high tension leads or connections. Although not inherently dangerous, they can be rather unpleasant.

3   Ignition system faults: tracing the defective area

1   Ignition problems must be tackled in a methodical fashion if time is not to be wasted in jumping to incorrect conclusions. The flow chart in Fig. 3.4 shows the correct sequence for investigating a failure or partial failure of the system. Note that apart from the TCI unit there are two distinct sub-systems controlling cylinders 1 and 4 and cylinders 2 and 3 respectively. It follows that the appropriate sub-system should be investigated in the event of a malfunction of one pair of spark plugs.

4   Ignition coils: spark gap testing

1   The individual ignition coils may be tested under load by connecting them to a device known as a spark gap tester. This allows the spark to be observed under load as the gap is progressively increased until failure occurs. A sound coil will produce a reliable spark across a minimum gap of 6 mm (0.24 in). As the equipment used is unlikely to be available to the owner it is suggested that a suspect coil is entrusted to a Yamaha dealer for testing or the following test is carried out to gain some indication of the coil's condition.

4.1 Ignition coils are mounted beneath fuel tank

5   Ignition coils: resistance tests

1   If one of the igniton coils is suspected of partial or complete failure, its internal resistance and insulation can be checked by measuring the primary and secondary winding resistances. Note that it is very unlikely that both coils would fail simultaneously, and if this appears to be the case, be prepared to look elsewhere for the problem.

2   Set the multimeter to the ohms scale, and connect one probe lead to each of the thin low tension wires. Note that it does not matter which probe is connected to which lead. A reading of 2.5 ohms ± 10% at 20°C should be obtained if the primary windings are in good order.

3   Repeat the test for the high tension leads, this time with the meter set on the kilo ohms scale. A resistance of 11 Kohms ± 20% at 20°C should be indicated.

4   If the coil has failed it is likely to have either an open or short circuit in the primary or secondary windings. This type of fault would be immediately obvious and would of course require the renewal of the coil concerned. Where the fault is less clear cut it is advisable to have the suspect coil tested on a spark gap tester by a Yamaha Service Agent.

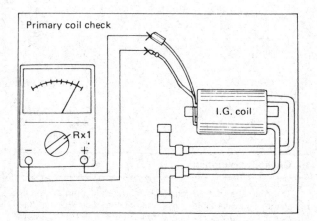

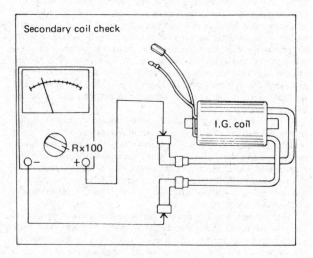

Fig. 3.2 Ignition coils resistance tests

Electrode gap check - use a wire type gauge for best results

Electrode gap adjustment - bend the side electrode using the correct tool

Normal condition - A brown, tan or grey firing end indicates that the engine is in good condition and that the plug type is correct

Ash deposits - Light brown deposits encrusted on the electrodes and insulator, leading to misfire and hesitation. Caused by excessive amounts of oil in the combustion chamber or poor quality fuel/oil

Carbon fouling - Dry, black sooty deposits leading to misfire and weak spark. Caused by an over-rich fuel/air mixture, faulty choke operation or blocked air filter

Oil fouling - Wet oily deposits leading to misfire and weak spark. Caused by oil leakage past piston rings or valve guides (4-stroke engine), or excess lubricant (2-stroke engine)

Overheating - A blistered white insulator and glazed electrodes. Caused by ignition system fault, incorrect fuel, or cooling system fault

Worn plug - Worn electrodes will cause poor starting in damp or cold weather and will also waste fuel

6 Pickup coils: resistance test

1   The pickup coils are unlikely to cause problems unless they
have suffered mechanical damage or have been electrically
overloaded. They can be checked using a multimeter set on the
resistance scale. Trace the pickup wiring back to the four pin
connector at the TCI unit and unplug it. Each coil has two
connecting leads, one of each is blue in each case, and the other
orange or grey. It will be necessary to determine by experiment
which of the blue leads belongs to which coil.
2   Measure the resistance between the orange lead and the
appropriate blue lead and note the reading obtained. Repeat the
test, this time using the grey lead and the remaining blue lead.
In each case a resistance of 700 ohms ± 20% at 20°C (68°F)
should be indicated. If the coil has broken it will normally give
a reading indicating infinite resistance (open circuit) or zero
resistance (short circuit). In both cases it will be necessary to
renew the pickup baseplate assembly as a unit. The coils are not
available separately.

6.1 The pickup assembly

7 TCI unit: testing

1   Yamaha do not produce test figures for the TCI unit, so a
faulty unit should be checked by eliminating all other possible
causes. If a sound TCI unit can be borrowed, it would be
possible to make sure of any diagnosis by substitution. Failing
this, it is worthwhile seeking confirmation from a Yamaha
dealer before purchasing a new component.

8 Ignition timing: checking and resetting

1   The ignition timing should be checked at the intervals
specified in Routine Maintenance (where appropriate), and
whenever the ignition pickup assembly has been dismantled for
repair or renewal. It should be noted that adjustment is unlikely
to be necessary unless the pickup has been disturbed. Before
starting the timing check, it should be noted that the fixed
timing pointer is adjustable, and if it has been removed or bent,
or if there is some doubt about its accuracy, check its position
as described at the end of this Section. For normal timing
checks, this stage can be ignored.
2   A stroboscopic timing lamp, or 'strobe', will be required for
the timing check, because the test must be carried out with the
engine running. Two basic types of strobe are available, namely
the neon and xenon tube types. Of the two, the neon type is

much cheaper and will usually suffice if used in a shaded
position, its light output being rather limited. The brighter but
more expensive xenon types are preferable if funds permit,
because they produce a much clearer image.
3   Connect the strobe to the left-hand (No 1 cylinder) high
tension lead, following the maker's instructions. If an external
12 volt power source is required it is best **not** to use the
machine's battery as spurious impulses can be picked up from
the electrical system. A separate 12 volt car or motorcycle
battery is preferable. Remove the timing inspection plate, then
start the engine. Check that it is within the prescribed idle speed
limits of 1050 ± 50 rpm, making any necessary adjustments.
4   Direct the timing lamp towards the timing disc on the
crankshaft end. The flashing light pulses will make the disc
appear to freeze in relation to the pointer, and the inverted U-
shaped timing mark should align precisely with this.
5   If the timing mark does not freeze in a position which puts
the fixed pointer mark within the range indicated by the arms of
the timing mark, check that the pickup baseplate assembly is
fitted correctly and that the two securing screws are tight. Note
that the timing is not adjustable, and if the marks still fail to
align correctly, it will be necessary to check that the fixed
pointer is accurately positioned, as described below.
6   The check will require the use of either a dial gauge and
spark plug thread adaptor, or a degree disc and some form of
piston stop. Of the two methods, the degree disc arrangement
is more accurate, but either should suffice.
7   If using a dial gauge arrangemennt, remove the spark plug
from cylinder No 1 and fit the adaptor and gauge using
extension probes as required. Turn the crankshaft to a few
degrees before TDC, then continue turning until the gauge
needle reaches the dead point at actual TDC. Note that the dead
point covers a few degrees of crankshaft movement and the
correct crankshaft position is at the centre of this dead point.
Check the fixed pointer, which should align precisely with the 'T'
mark on the timing plate. If this is not the case, slacken the
screw which secures the fixed pointer and move it as required.
8   If a degree disc is to be used it will be necessary to contrive
a piston stop which will screw into the spark plug hole. A
suitable device can be made up using an old spark plug. Clamp
the plug in a vice and hacksaw around the spun lip at the top
of the metal body. Once sawn around its circumference, the
porcelain insulator and centre electrode can be displaced and
discarded. File off the earth electrode, then pass a bolt down
through the body so that it projects by about $\frac{1}{2}$ inch or so.
Secure the bolt with a nut, noting that the overall width of the
nut must be smaller than the plug thread. Screw the assembled
stop into the plug hole, setting it so that the piston is stopped
a few degrees from TDC on each side.
9   Slacken the recessed Allen bolt which secures the timing
plate to the crankshaft. Remove the bolt and then refit it
together with the degree disc. The bolt need only be fitted finger
tight. Make up a pointer from a length of wire and fix it to the
crankcase, using one of the casing screws. Position the end of
the pointer so that it runs close to the edge of the timing disc.
Turn the crankshaft until it contacts the piston stop on one side
of TDC, and set the degree disc at zero against the wire pointer.
Turn the crankshaft back until the piston meets the stop at the
other side of TDC, and note the reading in degrees on the timing
disc. If, for example, a difference of 64° is shown, this should be
halved to set the disc at true TDC. Adjust the disc to read 32°
against the pointer. The stop can now be removed and the
crankshaft rotated until the disc shows zero (0° or 360°)
against the pointer. Check and adjust the fixed timing pointer as
described at the end of paragraph 7.
10   Once the pointer has been adjusted it should require no
further attention unless it is removed or has become bent.
Recheck the ignition timing as described earlier in this Section
and note whether the timing mark is now within limits. If not, a
fault in the TCI system is indicated, and further checks of the
component parts will be necessary. Note that it may prove
easier and quicker to have this done by a dealer, who will have
the facility to check by substitution where necessary.

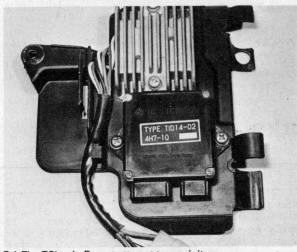

7.1 The TCI unit. Do not attempt to repair it

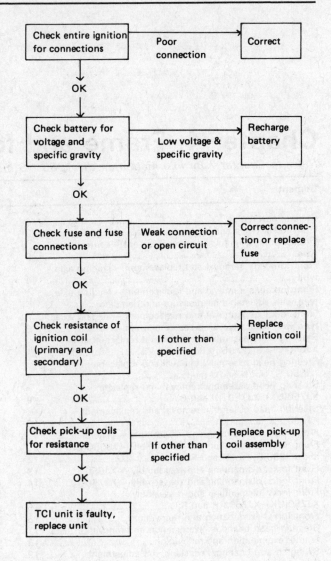

Fig. 3.4 Ignition system fault diagnosis flow chart

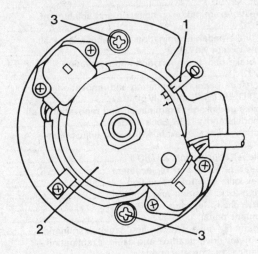

Fig. 3.3 Ignition timing alignment marks

1 Fixed pointer   3 Securing screws
2 Reluctor

---

9 Spark plugs: checking and resetting the gaps

1   The standard spark plug types recommended for the Yamaha models are NGK BP7ES or ND W22EP. Certain operating conditions may require a change in spark plug grade, but generally the type recommended by the manufacturer gives the best all round service.

2   Check the gap of the plug points at the recommended Routine Maintenance interval. To reset the gap, bend the outer electrode to bring it closer to, or further away from the central electrode until a 0.7 mm (0.028 in) feeler gauge can be inserted. Never bend the centre electrode or the insulator will crack, causing engine damage if the particles fall into the cylinder whilst the engine is running.

3   With some experience, the condition of the spark plug electrodes and insulator can be used as a reliable guide to engine operating conditions. See the accompanying photographs.

4   Always carry a spare spark plug of the recommended grade. In the rare event of plug failure, it will enable the engine to be restarted.

5   Beware of over-tightening the spark plugs, otherwise there is risk of stripping the threads from the aluminium alloy cylinder head. The plugs should be sufficiently tight to seat firmly on their copper sealing washers, and no more. Use a spanner which is a good fit to prevent the spanner from slipping and breaking the insulator.

6   If the threads in the cylinder head strip as a result of overtightening the spark plugs, it is possible to reclaim the head by the use of a Helicoil thread insert. This is a cheap and convenient method of replacing the threads; most motorcycle dealers operate a service of this nature at an economic price.

7   Make sure the plug insulating caps are a good fit and have their rubber seals. They should also be kept clean to prevent tracking. These caps contain the suppressors that eliminate both radio and TV interference.

Chapter 4 Frame and forks

Refer to Chapter 7 for information on the 1983 US models

Contents

Specifications

Front forks

	XJ650 G, H and LH	XJ650 J	XJ650 RJ and XJ650(UK)
Type	Oil damped, telescopic	Oil damped, air assisted, telescopic	Oil damped, telescopic
Oil capacity (per leg)	262 cc (9.24 US fl oz)	278 cc (9.40 US fl oz)	236 cc (7.18/8.31 US/Imp fl oz)
Oil grade	SAE 10W/30 motor oil or SAE 10W fork oil		
Air pressure:			
Standard	N/A	5.7 psi (0.4 kg/cm^2)	N/A
Range	N/A	0.17 psi (0-1.2 kg/cm^2)	N/A
Maximum	N/A	17 psi (1.2 kg/cm^2)	N/A
Travel	150 mm (5.91 in)	150 mm (5.91 in)	150 mm (5.91 in)
Spring free length	606 mm (23.86 in)	467 mm (18.39 in)	540.5 mm (21.28 in)

Front forks

	XJ750 RH and RJ	XJ750 J	XJ750(UK)
Type	Oil damped, air assisted, telesopic, with anti-dive	Adjustable, oil damped, air assisted telescopic	Oil damped, air assisted, telescopic, with anti-dive
Oil capacity (per leg)	309 cc (10.5 US fl oz)	257 cc (8.7 US fl oz)	312 cc (11.0 Imp fl oz)
Oil grade	SAE 20W fork oil	SAE 10W/30 motor oil or SAE 10W fork oil	
Air pressure:			
Standard	5.7 psi (0.4 kg/cm²)	5.7 psi (0.4 kg/cm²)	5.7 psi (0.4 kg/cm²)
Range	0-17 psi (0-1.2 kg/cm²)	5.7-17 psi (0.4-1.2 kg/cm²)	0-36 psi (0-2.5 kg/cm²)
Maximum	36 psi (2.5 kg/cm²)	36 psi (2.5 kg/cm²)	36 psi (2.5 kg/cm²)
Travel	150 mm (5.91 in)	150 mm (5.91 in)	150 mm (5.91 in)
Spring free length	604.9 mm (23.81 in)	458.5 mm (18.05 in)	604.9 mm (23.81 in)

Rear suspension units

Type	Oil damped, telescopic
Spring preload settings	5
Damper settings:	
750 models	4
All others	N/A
Travel	80 mm (3.15 in)
Spring free length:	
XJ650 G, L, LH, J	232 mm (9.15 in)
XJ650 RJ, 650(UK)	236 mm (9.31 in)
XJ750 models	227 mm (8.97 in)

Torque wrench settings

Component	kgf m	lbf ft
Steering stem pinch bolt	2.0	14
Crown bolt	5.4	39
Upper yoke pinch bolt	2.0	14
Lower yoke pinch bolt	2.0	14
Swinging arm LH pivot bolt	10.0	72
Swinging arm RH pivot locknut	10.0	72
Torque arm bolts	2.0	14
Brake arm pinch bolt	2.0	14
Suspension unit mountings	3.0	22

1 General description

The frame used on the Yamaha XJ650 and 750 models is of the full cradle type, in which the engine/gearbox unit is supported in duplex tubes running below the crankcase. The frame tubes run from the steering head lug above and below the engine unit, meeting at the swinging arm pivot point. An outrigger section provides support and location for the seat, rear suspension units and ancillary components. The frame is extensively braced and gusseted to minimise flexing.

The front forks are of the oil-damped telescopic type, there being several different types as discussed later in this Chapter. Some models incorporate a hydraulically actuated anti-dive system which helps to stabilise the machine during heavy braking. The fork legs are retained by a pair of fork yokes which pivot on cup and cone steering head races. Rear suspension is provided by a pivoted fork, or swinging arm, which is supported on tapered roller bearings located by adjustable stubs, and controlled by oil-damped coil spring suspension units. The left-hand longitudinal section of the swinging arm takes the form of a torque tube which contains the final drive shaft and to which the final drive housing is retained.

2 Front forks: types and applications

1 There are five different front fork arrangements used on the various XJ models, ranging from simple oil damped telescopic to sophisticated versions using various combinations of air pressure, damping adjustment and adjustable anti-dive control. To avoid further confusion the types and features are decribed below, together with the models to which each type is fitted.

XJ650(UK) and XJ650 RJ

2 These models are fitted with a simple centre-axle telescopic fork employing coil springs and non-adjustable hydraulic damping.

XJ650 G, H and LH

3 The 650 Maxim models employ a leading-axle telescopic fork with coil springs and non-adjustable hydraulic damping. The fork is generally similar to the type described above.

XJ650 J

4 The XJ650 J is equipped with a leading-axle telescopic fork, utilising coil springs supplemented by adjustable air

pressure as the suspension medium. Damping is hydraulic and non-adjustable.

XJ750 RH and RJ, UK 750 model
5   These models are fitted with leading-axle telescopic forks, featuring air-assisted coil springing and hydraulic damping. An additional refinement is an anti-dive system in which damping resistance is increased in proportion to braking effort.

XJ750 J
6   This last model differs significantly from the remaining 750 cc machines. It does not employ the anti-dive arrangement mentioned above, but does incorporate damping adjustment via a knurled knob at the top of each fork leg.

3   Front forks legs: removal and replacement – centre axle types

1   This type of fork is fitted to the XJ650(UK) and XJ650 RJ. Place the machine securely on its centre stand, leaving plenty of working area at the front and sides. Arrange wooden blocks beneath the crankcase so that the front wheel is raised clear of the ground.
2   Release the speedometer drive cable at the front wheel and lodge it clear of the forks. Slacken and remove the two bolts, spring washers and plain washers which secure each of the front brake calipers to its fork leg. As the calipers are lifted clear position some small pieces of scrap wood between the pads so that they are not expelled if the brake lever is inadvertently operated whilst the calipers are removed. Tie the calipers to the frame so that their weight is not supported by the hydraulic hoses. A similar approach should be applied to the single disc brake versions.
3   Remove the split pin from the wheel spindle nut, which can then be slackened and removed. Slacken the pinch bolt at the bottom of each fork leg. Support the wheel and displace the wheel spindle. The wheel can now be manoeuvred clear of the forks and placed to one side.
4   Remove the front mudguard mounting bolts and disengage the mudguard from the fork legs. Slacken the single pinch bolt at the upper yoke and the two pinch bolts at the lower yoke to release the fork leg, which can then be twisted and pulled

downwards to disengage it from the steering head assembly. Repeat this operation to remove the remaining fork leg.
5   Reassembly is a straightforward reversal of the removal sequence, noting that the tops of the fork stanchions should be positioned flush with the top edge of the upper yoke. Once the stanchion is in position, tighten the clamp bolts to 2.0 kgf m (14.5 lbf ft).
6   Refit the front wheel, checking that the speedometer drive gearbox engages properly on the locating lug of the fork leg. Slide the wheel spindle home and refit the castellated nut, tightening it to 10.7 kgf m (77.4 lbf ft). Once the nut is tight, fit a new split pin.
7   Refit the caliper(s), tightening the caliper bracket bolts to 2.6 kgf m (18.8 lbf ft). Operate the brake lever a few times to centralise the disc(s) between the caliper(s), and then compress the forks a few times to allow them to settle in their normal position. The spindle clamp bolts should now be tightened to 2.0 kgf m (14.5 lbf ft). Refit the front mudguard, tightening the mounting bolts to 1.0 kgf m (7.2 lbf ft) and reconnect the speedometer drive cable.

3.4a Release mounting bolts and remove the front mudguard

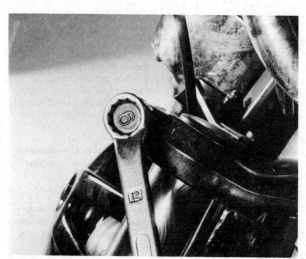

3.4b Slacken the top yoke pinch bolt ...

3.4c ... and the double pinch bolts on the lower yoke

3.4d Fork legs can be removed by pulling and twisting them downwards until clear of the yokes

4 Front fork legs: removal and replacement – leading-axle type

1   This type of fork is fitted to the XJ650 G, H, LH and J models. The procedure for removing and replacing the leading-axle type fork legs is essentially the same as that described for the standard type, and reference should be made to Section 3 for details. The following remarks should, however, be taken into consideration.

2   The single brake caliper is of the single piston type and is mounted on a cast-in lug on the left-hand lower leg. The caliper body pivots on a pin which is secured to the lug by a bolt which passes up through the pin, and a nut and spring washer. Once the nut and washer have been removed, the caliper can be lifted away and lodged clear of the forks and front wheel, having placed a wooden wedge between the brake pads.

3   A pinch bolt is fitted to the plain end of the spindle at the base of the right-hand lower leg. To free the spindle, first remove the split pin and castellated nut, then release the single pinch bolt. The spindle can now be displaced and the wheel removed.

4   All torque figures are the same as those shown in Section 3, as are the remaining removal details.

5 Front fork legs: removal and replacement – leading-axle type with adjustable air pressure and damping

1   This type of fork is fitted to the XJ750 J. Place the machine securely on its centre stand, arranging wooden blocks or similar beneath the crankcase so that the wheel is raised clear of the ground.

2   Straighten and remove the split pin which secures the wheel spindle nut, then slacken and remove the nut. Release the bolt which holds the speedometer drive cable in the drive gearbox and pull the cable clear. Slacken and remove the two bolts which secure each caliper to its fork lug. Lift the calipers away and place wooden wedges between the pads to prevent their accidental expulsion. The calipers should be tied clear of the wheel and fork, noting that they should not be left hanging by the hydraulic hoses.

3   Slacken the pinch bolt at the base of the lower right-hand fork leg to free the wheel spindle. Using a soft faced mallet, tap the threaded end of the spindle until the cross drilling on the opposite end is accessible. Insert a tommy bar or a screwdriver and twist and pull the spindle clear of the wheel and forks. It will

be found helpful if the weight of the wheel is supported at this stage. Once the spindle has been freed, lower the wheel and place it to one side. Remove the four bolts which retain the front mudguard to the fork legs, then remove the mudguard taking care not to scratch it

4   Remove the air valve cap at the top of the left-hand stanchion. Using a small screwdriver or similar, depress the valve insert to release the fork air pressure. Slacken the fork cap bolts by one turn only. **Do not** attempt to remove them at this stage. Remove the finisher trim from the bottom yoke. Slacken the top and bottom yoke pinch bolts, and slide each fork leg down by about two inches. Slide the air valve unions and hose upwards and clear of the stanchions together with their internal O-rings and the rubber spacers which fit between them and the underside of the top yoke. Using a small electrical screwdriver, peel the wire circlips, which locate the unions on the stanchions, clear of the stanchion grooves and slide them upwards and off the stanchions. Each fork leg assembly can now be removed by twisting it and pulling it downwards.

5   The fork legs are refitted by reversing the removal sequence, noting the following points. Do not omit to fit the union locating circlip to the stanchion after the latter has been fitted through the bottom yoke. Examine, and where necessary renew, the O-rings which seal the air unions on the stanchions. Coat the O-rings and stanchion with a lithium soap-based grease prior to fitting. Take care that the O-rings are not displaced as the unions are pushed down over the stanchions or air pressure leakage will occur.

6 Front fork legs: removal and replacement – leading-axle type with adjustable air pressure and anti-dive

1   This type of fork is fitted to the XJ750(UK) and XJ750 RH and RJ. Place the machine securely on its centre stand, using wooden blocks or similar beneath the crankcase to raise the front wheel well clear of the ground.

2   Straighten and remove the split pin which retains the wheel spindle nut, then slacken and remove the nut. Release the bolt which secures the speedometer drive cable in the drive gearbox and pull the cable clear, lodging it away from the wheel and forks.

3   Remove the two Allen screws which retain the actuator piston housing to the top of the anti-dive housing and lift the piston housing away together with its hydraulic hose. Slacken and remove the two caliper mounting bolts and lift away the caliper with the piston housing attached. Tie the assembly clear of the wheel, then repeat the process with the remaining caliper and anti-dive components. Place wooden wedges between the brake pads to prevent the caliper pistons from being expelled should the brake lever be squeezed inadvertently.

4   Slacken the pinch bolt at the bottom of the right-hand lower leg to free the plain end of the wheel spindle. Tap the spindle through using a soft-faced mallet, then insert a bar or screwdriver through the plain end. Twist and pull the spindle clear, whilst supporting the wheel. Manoeuvre the wheel clear of the forks and place it to one side. Remove the front mudguard retaining bolts and remove it from between the fork legs.

5   Continue removal and replacement by following the sequence described in Section 5 paragraph 4 onwards, noting that damping adjustment is not fitted to the anti-dive fork, and references to it can be ignored, and that each fork leg has its own air pressure valve on these models.

7 Steering head assembly: removal and replacement – XJ650(UK) and XJ650 RJ

1   The steering head assembly comprises the lower yoke and steering stem, the upper yoke and the two cup-and-cone

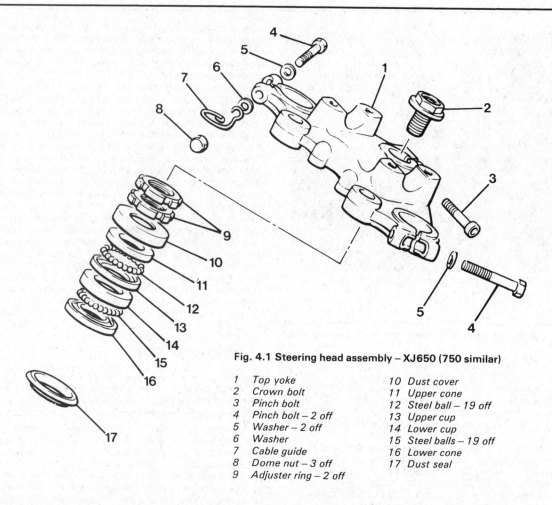

Fig. 4.1 Steering head assembly – XJ650 (750 similar)

1 Top yoke	10 Dust cover
2 Crown bolt	11 Upper cone
3 Pinch bolt	12 Steel ball – 19 off
4 Pinch bolt – 2 off	13 Upper cup
5 Washer – 2 off	14 Lower cup
6 Washer	15 Steel balls – 19 off
7 Cable guide	16 Lower cone
8 Dome nut – 3 off	17 Dust seal
9 Adjuster ring – 2 off	

bearing race sets. It is unlikely to require attention unless the bearings have become worn or damaged, or if accident damage has been sustained. If necessary, it is possible to remove the lower yoke together with the fork legs, and this approach may be required where the fork stanchions are bent and jammed in the lower yoke. In all other cases the fork legs are best removed first.

2   Start by removing the front brake calipers, the fork legs and mudguard as described in Section 3 of this Chapter. The fuel tank should be protected by covering it with a blanket or, better still, removed and placed to one side. This will prevent any risk of damage to the painted finish.

3   The procedure from this point onwards must depend on individual circumstances. For obvious reasons, the full dismantling sequence is described here, but it is quite in order to avoid as much of the dismantling as possible by careful manoeuvring of the ancillary components. As an example, the handlebar assembly can be removed as a unit, and threaded around the yokes as these are released. The same applies to the headlamp unit and instrument panel. Obviously, much depends on a commonsense approach and a measure of ingenuity on the part of the owner.

4   Remove the screws which secure the headlamp lens and reflector assembly to the headlamp shell. Disconnect the headlamp bulb connector and the parking lamp bulb (UK and European models) and place the unit to one side. Trace the various multi-pin connectors which enter the headlamp shell, making a quick sketch to show their relative positions. Separate the connectors and push them out of the headlamp shell. Remove the headlamp shell mounting nuts and release the shell, noting the disposition of the various washers and spacers.

Do not forget the headlamp adjuster bolt which passes through an elongated hole in the bracket on the rear of the shell.

5   Unscrew the knurled rings which retain the speedometer and tachometer drive cables to the underside of the instrument panel and manoeuvre the cables clear of the stee   ing head area. The instrument panel is secured by two studs which pass down through rubber bushed holes in the upper yoke. Remove the two nuts from the underside of the yoke and pull the assembly upwards to free it. Remove the plastic cover from the lower yoke by releasing its two securing screws. Once this cover has been lifted away the lower mounting of the headlamp bracket will be revealed. Remove the single bolt and lift the bracket away together with the front indicator lamps.

6   Release the ignition switch from the top yoke by removing the two securing bolts. Prise out the black plastic caps which cover the handlebar clamp Allen screws. Release the screws and lift away the clamps to free the handlebar assembly. It should not be necessary to remove the levers, master cylinder and switches from the handlebar, but make sure the master cylinder remains upright to prevent leakage from the reservoir.

7   Slacken the pinch bolt at the rear of the top yoke, then remove the steering stem top bolt. The top yoke can now be lifted away. If it proves stubborn, tap it upwards using a soft-faced mallet. Using a C-spanner, slacken and remove the steering stem lock nut.

8   Before the lower yoke is removed, it should be noted that the steering head balls will be released as it is withdrawn. It follows that great care must be taken to catch these since they will travel a remarkable distance if allowed to fall onto a hard surface. Anyone not familiar with the behaviour of small steel balls dropping from a steering head race will be astonished at

the speed with which they disappear into small nooks and crannies. To obviate a protracted and frustrating search it is advisable to place something like an old blanket beneath the steering head area to collect any errant balls. A magnet will prove invaluable when collecting them as the yoke is lowered away from the steering head.

9 Support the lower yoke with one hand, then unscrew the adjustment nut and remove the upper bearing cone. The bearing balls will usually remain in the bearing cup and can be collected with the magnet and placed in a suitable container. Things are less simple with the lower race. Slowly remove the lower yoke, keeping one hand cupped beneath it to catch any bearing balls which drop free. Remove the remainder with the magnet and place them in the container. There is a total of 38 balls in the two races.

10 The steering head components should be examined for wear as described in Section 17, prior to reassembly. Make sure that the bearing cups and cones and the bearing balls are clean before assembly commences. Wipe a thick layer of grease around the lower bearing cup, using the grease to stick the nineteen lower race balls into position. It will be noted that a small gap remains in the circle of balls. This is intentional, and allows the balls to rotate without scuffing against each other. Do not be tempted to fit an extra ball. Grease the upper bearing cup and stick the remaining nineteen balls in place in the same manner.

11 Taking great care not to dislodge the lower bearing balls, slide the steering stem into position, holding it firmly against the bearing while the upper cone, shroud and adjustment nut are fitted. It is important to ensure that the steering head is adjusted correctly. If left too loose, the small amount of free play will be much magnified at the front wheel and will cause serious handling problems. Conversely, over-tight head races will cause rapid bearing wear and make the steering stiff and unresponsive. The steering head has a fine thread which makes it quite easy to apply inadvertently a loading of several tons on the bearings.

12 The official method of adjustment is to tighten the slotted adjuster nut to 2.5 kgf m (18.0 lbf ft) and then to back it off by $\frac{1}{4}$ turn. This poses a problem since most owners will not have access to the special socket required. It is quite feasible to use an extended C-spanner, to which is attached a spring balance at a point one foot from the centre of the steering stem. If the nut is tightened by pulling on the spring balance until a reading of eighteen pounds is indicated, the specified torque will have been achieved.

13 If care is exercised, the nut can be set by feel alone. It should be borne in mind that it is necessary only to tighten the nut sufficiently to remove all trace of free play, but no more. Check that the yoke turns smoothly and easily. Once adjustment is correct, hold the nut in position, then fit and tighten the locknut. Continue assembly by fitting the top yoke. To ensure that the yokes align properly, slide the fork legs into position whilst the top bolt and pinch bolt are fitted. The correct torque figures for the latter are as follows:

Steering stem pinch bolt   2.0 kgf m (14 lbf ft)
Crown bolt   5.4 kgf m (39 lbf ft)

7.4 Note adjustable mounting on back of headlamp shell

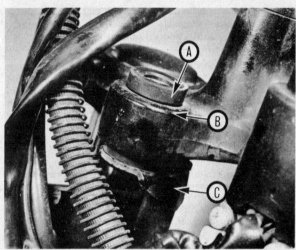
7.5 A: Rubber mounting B: Support lug C: Headlamp subframe

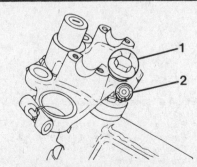
Fig. 4.2 Steering head top yoke pinch bolt and top bolt positions – XJ650(UK) and XJ650 RJ

1 Top bolt   2 Pinch bolt

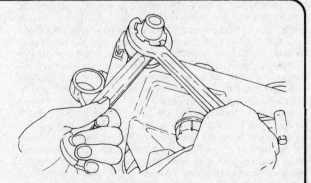
Fig. 4.3 Adjusting the steering head bearings

8   Steering head assembly: removal and replacement – XJ 650 G, H, LH and J models

1   Steering head removal and reassembly for the above models is generally similar to the procedure described in Section 7 of this Chapter, with a few minor variations. A brief summary of the sequence is given below and any procedural differences explained.

2   Start by removing the front wheel, forks and handlebar assembly, noting that it is advisable to remove the fuel tank as a precautionary measure. Release and disconnect the headlamp assembly from the shell, pushing the wiring and connectors through to allow the shell to be lifted clear. The shell can be detached together with its supporting subframe. Removal of the finisher plate on the bottom yoke will allow easy access to the subframe lower mounting. The finisher is retained by two screws fitted from the underside.

3   Free the hydraulic union from the bottom yoke by releasing its single mounting bolt. Disconnect the instrument drive cables, then release each instrument, together with its mounting bracket, from the top yoke. The rest of the dismantling sequence is the same as that described in Section 7. Reassemble by reversing the dismantling sequence, noting the following torque settings:

Steering stem pinch bolt	2.0 kgf m (14.5 lbf ft)
Steering stem top bolt	5.4 kgf m (39.1 lbf ft)

9   Steering head assembly: removal and replacemnet – XJ750(UK), XJ750 RH and RJ models

1   Preliminary dismantling, in preparation for the removal of the steering head components, differs quite markedly from that described in Sections 7 and 8 for the 650cc models. Most of the general remarks, however, can be applied, and reference should be made to the earlier Sections for this reason. The sequence described in this Section assumes that a complete strip of the steering head is required, and this may be modified to suit individual requirements as necessary.

2   Start by removing the front wheel and forks as described in Section 6. It is recommended that the fuel tank is removed to avoid any risk of damage to the paint finish. It will now be necessary to remove the various electrical and hydraulic components grouped between the top and bottom fork yokes. As owners of these machines will have noticed, there are a large number of individual items packed into this small space, and to avoid any unnecessary work it is suggested that the whole assembly is detached complete as described below.

3   Slacken fully the front brake cable adjuster and disconnect the cable at the upper end. Remove the two screws which retain the finisher to the underside of the bottom yoke and lift it clear. Examination will show that the front brake master cylinder is mounted on a subframe which also carries the front brake distributor union, the auxiliary lamp and the headlamp. The subframe is secured at the bottom by two bolts, whilst the top mounting bolts also retain the instrument panel assembly. Release the bolts, which will require some dexterous spanner work, and move the assembly as far clear of the yokes as the cables and wiring will allow.

4   Moving to the top yoke, prise out the rectangular blank which is fitted in the centre of the handlebar cover. Release the two screws beneath the blank and lift the handlebar cover clear. Remove the four handlebar clamp screws, lift away the clamp halves and move the handlebar assembly forward and clear of the yoke. Remove the two screws which retain the ignition switch/steering lock assembly and allow it to hang from its wires.

5   Access to the steering head for removal is now possible, though complicated somewhat by the components which have just been detached from it. The rest of the dismantling sequence is best carried out with the aid of an assistant, following the procedures described in Section 7, paragraph 7 onwards. Refer to Fig. 4.5 for details of cable and wiring routing during reassembly.

10 Steering head assembly: removal and replacement – XJ750 J

1   The steering head dismantling and reassembly sequence is largely the same as that described in Section 9 of this Chapter for the remaining XJ750 models. Note, though, that the procedure for removing the handlebar assembly differs in view of the adjustable bars used on the J model. Each side of the handlebar is removed independently of the other. Start by prising off the plastic caps which conceal the heads of the mounting stub cap bolts. Slacken the cap bolts and the pinch bolts to allow the handlebar sections to be pulled upwards and clear of their mounting splines. The handlebar sections can be lodged or tied clear of the steering head area during the rest of the dismantling sequence. After reassembly has been completed, refer to Section 36 of this Chapter for details on handlebar adjustment for the XJ750 J model.

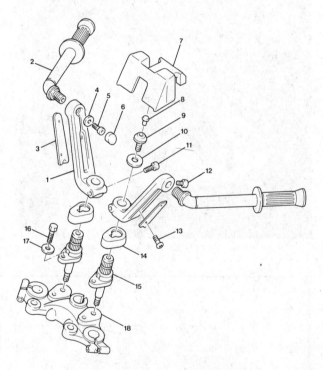

Fig. 4.4 Handlebar assembly – XJ750 J

1	Riser	10	Washer – 2 off
2	Bar end	11	Screw – 2 off
3	Finisher plate	12	Screw – 2 off
4	Washer – 2 off	13	Screw – 4 off
5	Bolt – 2 off	14	Trim – 2 off
6	Plug – 2 off	15	Stub – 2 off
7	Cover	16	Bolt – 2 off
8	Plug – 2 off	17	Washer – 2 off
9	Bolt – 2 off	18	Upper yoke

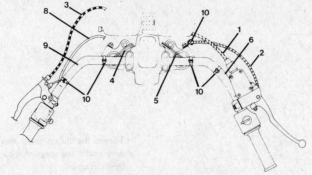

XJ650 G, H, LH and J models

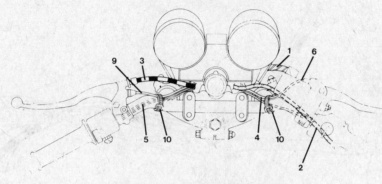

XJ650 (UK) and XJ650 RJ models

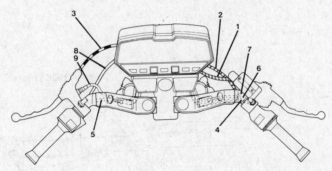

XJ750 J models

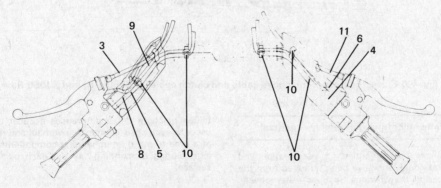

XJ750 (UK) and XJ750 RH and RJ

Fig. 4.5 Correct routing of handlebar cables

1 Front brake hose	5 Left-hand handlebar switch	8 Starter motor lead
2 Throttle cable	wiring	9 Clutch switch wire
3 Clutch cable	6 Front brake stop lamp	10 Cable tie
4 Right-hand handlebar	switch wire	11 Front brake cable
switch wiring	7 Low brake fluid sensor	
	wire	

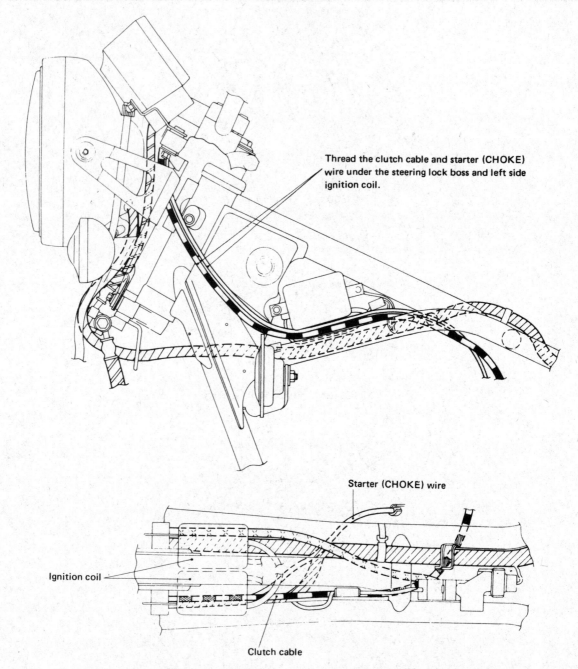

Thread the clutch cable and starter (CHOKE) wire under the steering lock boss and left side ignition coil.

Starter (CHOKE) wire

Ignition coil

Clutch cable

Fig. 4.6 Correct routing of the choke cable and clutch cable – XJ650(UK) and XJ650 RJ

11 Front forks: dismantling and reassembly – general

1   The front forks can be dismantled for examination and overhaul after the individual legs have been removed from the machine as described in the previous Sections. Alternatively, attention to all but the stanchions is possible by removing the springs and lower legs as described in the subsequent Sections, leaving the stanchions clamped in the yokes. Needless to say, it will first be necessary to remove the front wheel, brakes and mudguard.

2   The general procedure relating to fork dismantling is similar on all types. Section 11 relates specifically to the standard type fork fitted to the XJ650 Seca models, whilst the Sections which

follow highlight detail differences peculiar to the other types used. In each case, dismantle, overhaul and reassemble one leg at a time to avoid interchanging components between the two legs. During dismantling, lay out each part on a clean work surface.

12 Front forks: dismantling and reassembly – XJ650(UK) and XJ650 G, H, LH and RJ models

1   Clamp the fork stanchion firmly in a vice, but taking great care not to over-tighten the jaws and crush or distort the stanchion. It is essential to use soft jaws to prevent damage to

the stanchion surface. Hard wood, fibre or nylon types being ideal for this purpose. Prise out from the stanchion top the black plastic dust cap and place it to one side.

2   The fork top plug is under pressure from the fork spring and is located in the stanchion by an internal wire circlip (snap ring). To free the plug it will be necessary to depress it sufficiently to reveal the circlip which must then be worked out of its groove. An assistant is invaluable at this point since it may prove awkward to release the circlip whilst holding the plug down with one hand. The method is illustrated in the accompanying photograph.

3   Having removed the circlip, pressure on the plug can be released progressively, and the plug and fork spring removed. Remove the fork from the vice and invert it over a drain tray. Remove the residual damping oil by 'pumping' the fork.

4   It is now necessary to remove the damper bolt from the bottom of the lower leg. This takes the form of an Allen bolt which passes upwards through the lower leg to secure the damper rod. The head of the latter runs inside the stanchion, and effectively holds the fork assembly together. It is common to experience some difficulty when removing the bolt, because its threads are coated with Loctite during assembly and the damper rod base is not located in the lower leg. This usually results in the damper rod rotating rather than the bolt being removed.

5   Yamaha produce a holding tool, part number 90890-01300, to overcome this problem. The tool is introduced through the top of the stanchion and engages the head of the damper rod to hold it while the bolt is removed. Few owners will have access to such a tool, so an improvised version will be necessary. It helps to start by using a torch to inspect the damper rod head. It will be noted that it has an internal bi-hexagon recess, much like a socket. The method used on the machine featured in the photographs was to search through a box of odd nuts and bolts to find one which was of similar size to the damper rod bi-hexagon. The assembly shown in the photograph was then made up, using a socket as a convenient spacer. This was dropped into the stanchion and a second socket and extension bars used to hold it.

6   A more permanent version of the above can be made by brazing or welding a bolt into the end of a suitable length of steel tubing. The upper end can be drilled to take a tommy bar. Another method which has been found successful in the past is to use a length of wooden dowel of about 1 inch diameter. Form a blunt taper on one end and introduce this end down the stanchion. Clamp the lower leg in a vice, using the caliper lug and protecting it by using soft jaws. An assistant can push hard on the dowel while the bolt is slackened.

7   Once the damper bolt has been freed, disengage the dust seal from the top of the lower leg and withdraw the stanchion. The damper rod asembly can be tipped out of the stanchion and the damper rod seat displaced from the lower leg. Wipe the fork components with a clean rag to remove residual oil, and lay them out for examination on a clean surface. The fork oil seal should be renewed as a precautionary measure whenever the forks are dismantled. To remove it, first release the wire circlip which locates in a groove in the top of the lower leg, then work the seal out using a screwdriver. Note that the seal will almost invariably be damaged during removal and that care must be taken to avoid damage to the soft alloy lower leg.

8   Reassemble the forks by reversing the dismantling sequence. When fitting a new seal it should be carefully tapped into position with a flat wooden block and a hammer, making sure that it enters the bore squarely and seats fully. Grease the seal lip before the stanchion assembly is inserted. When fitting the fork springs note that the lower end of each has one small diameter coil, and that the tighter pitch coils face the top of the fork. It is convenient to put in the damping oil at this stage, having first fitted and tightened the drain screws. The correct quantity of oil for each fork leg for each model is given in the Specifications at the beginning of the Chapter. After filling, pump the fork legs a few times to expel any air, then refit the fork top plugs.

12.1 Remove plastic cap from top of stanchion

12.2 Depress plug as shown and prise out the circlip

12.3 The top plug will be displaced by spring pressure

12.5 Home-made tool was used to hold damper rod ...

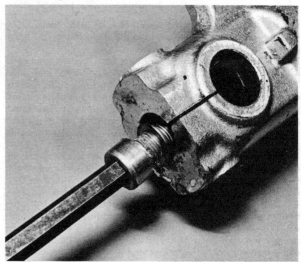

12.6 ... while the Allen-headed damper bolt was unscrewed

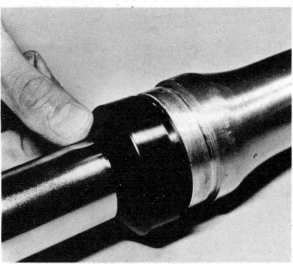

12.7a Slide the dust seal clear of the lower leg ...

12.7b ... and withdraw the stanchion and damper assembly

12.7c The damper rod and rebound spring can be tipped out

12.7d Fork oil seal is located by a wire retaining clip

12.8a Check that new seal seats squarely in lower leg

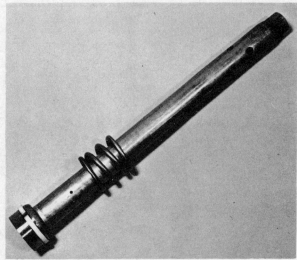
12.8b Place rebound spring over damper rod ...

12.8c ... and drop the assembly into the stanchion

12.8d Fit the oil lock piece to the damper rod end

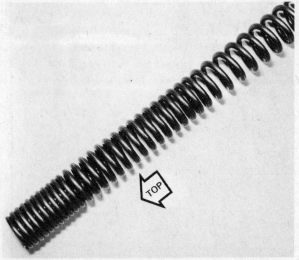
12.8e Fit fork spring with the tightly wound coils near the top

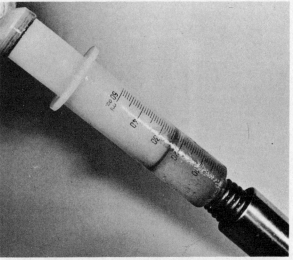
12.8f Add the prescribed quantity of damping oil ...

12.8g ... then fit top plug, securing it with its circlip

Fig. 4.7 Front forks – XJ650(UK) and XJ650 RJ

1	Dust seal – 2 off	12	Sealing washer – 2 off
2	Damper rod seat – 2 off	13	Drain screw – 2 off
3	Circlip – 2 off	14	Stanchion – 2 off
4	Oil seal – 2 off	15	Pinch bolt – 4 off
5	Rubber plug – 2 off	16	Pinch bolt – 2 off
6	Circlip – 2 off	17	Washer – 2 off
7	Plug – 2 off	18	Bolt – 2 off
8	O-ring – 2 off	19	Sealing washer – 2 off
9	Fork spring – 2 off	20	Washer – 2 off
10	Damper rod – 2 off	21	Spring washer – 2 off
11	Lower leg	22	Nut – 2 off

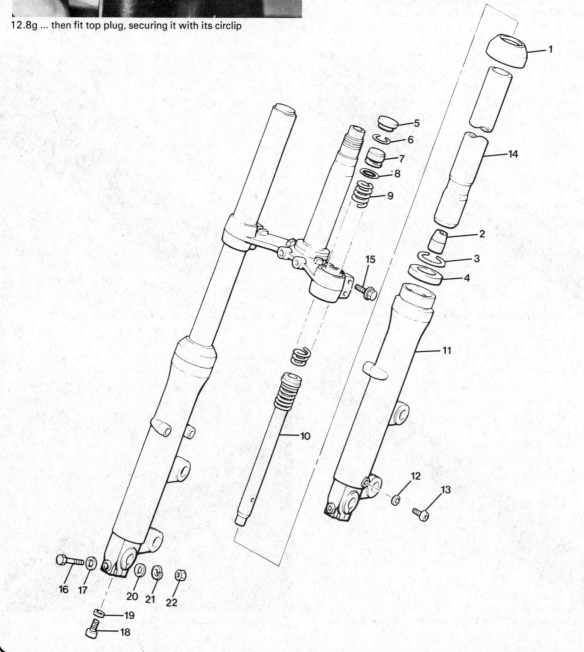

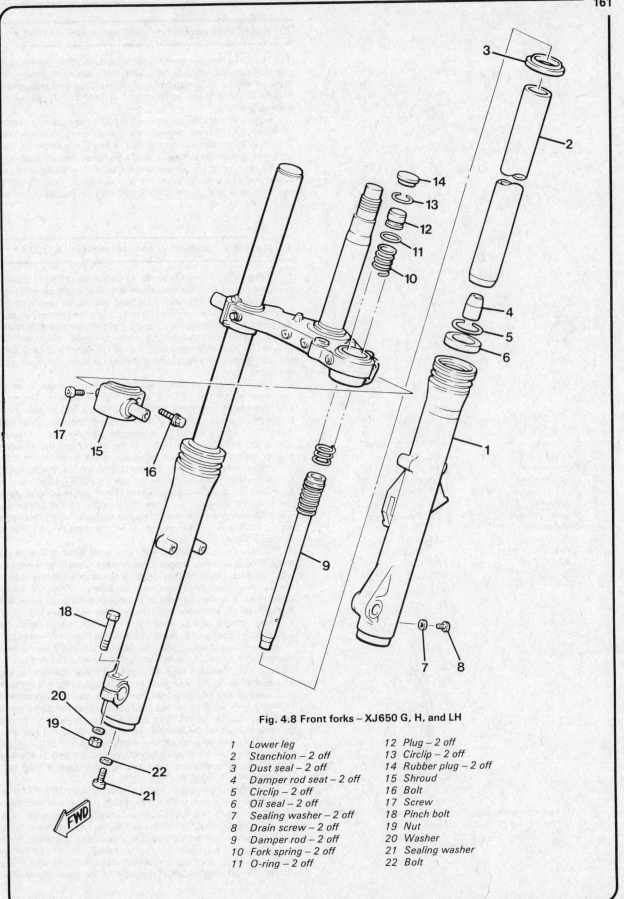

Fig. 4.8 Front forks – XJ650 G, H, and LH

1	Lower leg	12	Plug – 2 off
2	Stanchion – 2 off	13	Circlip – 2 off
3	Dust seal – 2 off	14	Rubber plug – 2 off
4	Damper rod seat – 2 off	15	Shroud
5	Circlip – 2 off	16	Bolt
6	Oil seal – 2 off	17	Screw
7	Sealing washer – 2 off	18	Pinch bolt
8	Drain screw – 2 off	19	Nut
9	Damper rod – 2 off	20	Washer
10	Fork spring – 2 off	21	Sealing washer
11	O-ring – 2 off	22	Bolt

FWD

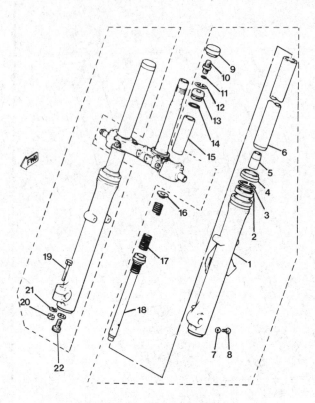

Fig. 4.9 Front forks – XJ650 J

1	*Lower leg*
2	*Oil seal – 2 off*
3	*Circlip – 2 off*
4	*Dust excluder – 2 off*
5	*Damper rod seat – 2 off*
6	*Stanchion – 2 off*
7	*Sealing washer – 2 off*
8	*Drain screw – 2 off*
9	*Rubber plug – 2 off*
10	*Air valve – 2 off*
11	*O-ring – 2 off*
12	*Circlip – 2 off*
13	*Plug – 2 off*
14	*O-ring – 2 off*
15	*Spacer – 2 off*
16	*Spring seat – 2 off*
17	*Fork spring – 2 off*
18	*Damper rod – 2 off*
19	*Pinch bolt*
20	*Nut*
21	*Washer*
22	*Bolt*

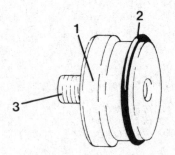

Fig. 4.10 Front fork top plug showing air valve – XJ650 J

1	*Plug*
2	*O-ring*
3	*Valve*

13 Front forks: dismantling and reassembly – XJ650 J

1   The forks fitted to these models are essentially the same as the standard leading-axle type, with the addition of Schraeder-type air valves in the top plugs. Dismantling and reassembly should be undertaken as described in Section 11, noting the following points.

2   Before attempting to remove the fork top plugs, use a piece of wire or similar to depress the valve insert to release air pressure. When depressing the plug to free the wire circlip, take care not to damage the valve. It will be noted that an additional spring seat is fitted between the top of the spring and the underside of the plug. Note that it is especially important that the fork seal is in good condition if it is to maintain fork air pressure.

14 Front fork: dismantling and reassembly – XJ750 J

1   Before any attempt is made to dismantle the front forks it must be noted that the procedure is rather more complicated than normal on these machines, and access to two Yamaha service tools will be required. It is strongly recommended that the procedure is read through and some decision made as to the practicality of undertaking this work at home. Many owners may prefer to entrust the job to a qualified Yamaha dealer. Note the dismantling and reassembly should take place on each fork leg separately.

2   Prise out and discard the fork dust seal using a small electrical screwdriver. Take care not to scratch the fork stanchion during removal. Remove the circlip which retains the oil seal in place. Unscrew and remove the fork cap bolt assembly, taking care not to damage the damper adjuster rod as it is withdrawn from the centre of the stanchion. Remove the spacer spring seat and fork spring and place them to one side.

3   Invert the fork leg over a drain tray and pump the damping oil out. Use Yamaha tool number 90890-04084 or its equivalent to prevent damper rod rotation, slacken and remove the damper securing bolt from the bottom of the lower leg. Carefully invert the fork and allow the damper rod and rebound spring to slide out. Take care not to let the damper rod fall onto a hard surface.

4   Clamp the lower leg in soft vice jaws against the wheel spindle boss. Using a propane torch, heat the top of the lower leg to expand the light alloy casting. Great care must be taken during this operation because the painted finish is easily damaged if it is overheated. Pull the stanchion sharply outwards to displace the top bush and oil seal from the lower leg.

5   After cleaning and examining the fork components as described in Section 16, reassemble in the following sequence. Place the rebound spring over the damper rod and slide the two components down inside the stanchion so that the end of the damper rod protrudes from the bottom. Place the fork spring and spacer inside the stanchion to hold the damper rod in place and arrange the stanchion so that it rests on the workbench vertically with the damper rod uppermost. Place the damper rod seat over the end of the damper rod, then carefully lower the fork lower leg over the stanchion assembly. Once the seat and damper rod are in position the damper holding bolt can be fitted. Make sure its threads are clean and dry and coat them with Loctite or a similar thread locking compound. Fit the bolt and tighten to 2.0 kgf m (14.0 lbf ft).

6   A Yamaha service tool, part number TLM-11080-10-00 or YM-08010 is required to fit the top bush, oil seal and dust seal. The tool consists of a small tubular fitting guide together with a larger sliding weight which is used to tap the component into position. Some owners may be able to approximate these items using odds and ends found in the workshop. The smaller fitting guide part of the tool has a shouldered internal bore which will accept the top bush and hold it square to the lower leg while it

is driven home. With care, an ordinary tubular drift can be made up from a section of thick-walled tube. Care must be taken to ensure that the bush enters the lower leg squarely. If this is not the case, the bush may jam in place and could prove difficult to remove without damage.

7 Lubricate the oil seal and slide it into place over the stanchion, with the bevelled face upwards. Tap it down onto the top of the bush using the special tool as described above. Once in position, fit the circlip to retain it. Next, fit the dust seal and gently tap it into place above the oil seal.

8 Fill the fork leg with 257 cc (8.7/9.0 US/Imp fl oz) of SAE 10W fork oil or SAE 10W/30 type SE motor oil. Drop the fork spring, spring seat and spacer into the top of the stanchion. Lower the cap bolt and damper adjuster rod slowly home. It is very important that the rod engages in the corresponding hole in the damper rod. If it is positioned correctly, the cap bolt will rest against the top of the spacer. If this is not the case, slowly rotate the cap bolt assembly until it is felt and seen to drop into engagement. On no account should force be used – if the rod is bent or deformed it must be renewed. Once the assembly is in position, screw the cap bolt home and tighten it to 3.0 kgf m (22 lbf ft).

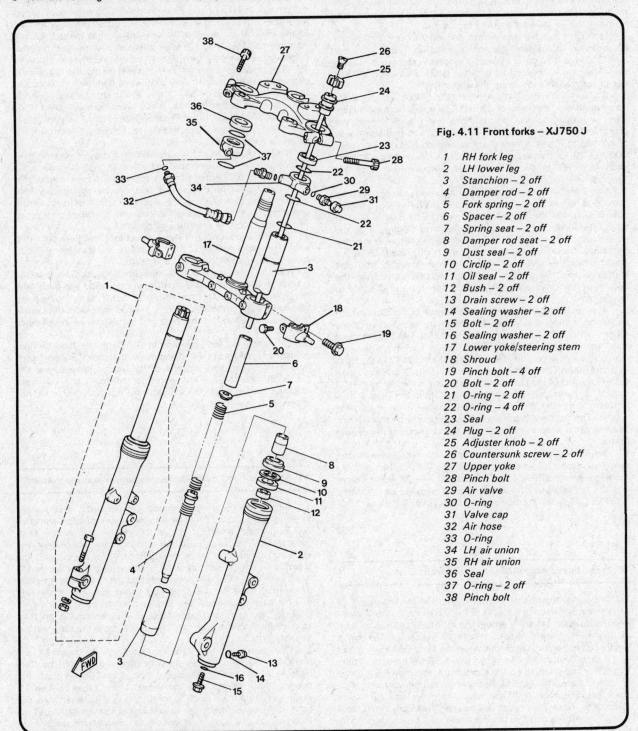

Fig. 4.11 Front forks – XJ750 J

1   RH fork leg
2   LH lower leg
3   Stanchion – 2 off
4   Damper rod – 2 off
5   Fork spring – 2 off
6   Spacer – 2 off
7   Spring seat – 2 off
8   Damper rod seat – 2 off
9   Dust seal – 2 off
10 Circlip – 2 off
11 Oil seal – 2 off
12 Bush – 2 off
13 Drain screw – 2 off
14 Sealing washer – 2 off
15 Bolt – 2 off
16 Sealing washer – 2 off
17 Lower yoke/steering stem
18 Shroud
19 Pinch bolt – 4 off
20 Bolt – 2 off
21 O-ring – 2 off
22 O-ring – 4 off
23 Seal
24 Plug – 2 off
25 Adjuster knob – 2 off
26 Countersunk screw – 2 off
27 Upper yoke
28 Pinch bolt
29 Air valve
30 O-ring
31 Valve cap
32 Air hose
33 O-ring
34 LH air union
35 RH air union
36 Seal
37 O-ring – 2 off
38 Pinch bolt

15 Front forks: dismantling and reassembly – XJ750(UK), XJ750 RH and RJ

1   Hold the stanchion in soft vice jaws, or wrap it in rag before tightening the vice. Do not overtighten the vice or the stanchion may be distorted. With the dust cap removed, depress the fork top plug with a screwdriver or similar tool, holding it down against spring pressure while the internal wire circlip is displaced. This can be done using an electrical screwdriver and will demand a certain amount of patience. Gradually release the plug, allowing it to be displaced by fork spring pressure. Remove the spring seat and the fork spring.

2   Invert the fork leg over a drain tray and allow the damping oil to drain. This process can be speeded up by pumping the fork. Slide the dust cover clear of the lower leg to expose the oil seal. The seal is retained by a circlip which should be worked out of its groove and slid up the stanchion.

3   Slacken and remove the damper holding bolt which passes up through the underside of the fork lower leg. Some difficulty may be experienced if the damper rod rotates inside the stanchion, in which case it will be necessary to find some way of holding it. Yamaha can supply a holding tool, Part number 90890-01300. Its use, and the construction of a home made equivalent, will be found in Section 12, paragraph 4 of this Chapter.

4   Once the holding bolt has been removed, the fork can be separated by jarring the top bush out of the lower leg, bringing the oil seal and the related components with it. It is best to clamp the lower leg between soft vice jaws, holding it by the caliper mounting boss. Pull the stanchion sharply outwards, repeating the procedure until the bush is displaced. Once the stanchion is free, tip out the damper rod assembly from the stanchion and the oil lock piece and related parts from the bottom of the lower leg. Lay these components out as an aid to reassembly.

5   Reassembly is a straightforward reversal of the dismantling sequence, noting that it will be necessary to contrive some means of fitting the top bush and the oil seal. The best method is to use a short length of steel tubing, having an internal diameter slightly greater than the stanchion. This can be used as a tubular drift to tap the bush home squarely. It is a good idea to heat the top of the lower leg using boiling water, or by careful use of a blowlamp flame. Remember that the fork leg is lacquered and that the finish will be damaged if too much heat is used. The heat will expand the alloy and the bush should then fit quite easily. Allow the leg to cool before pressing the oil seal home.

6   Before the top plug is fitted, fill each fork leg with the specified grade and amount of oil. The XJ750 RH and RJ models will take 309 cc (10.5 US fl oz) of SAE 20W fork oil per leg. The UK version of the XJ750 should take 312 cc (11.0 Imp fl oz) of SAE 10W/30 motor oil or SAE 10W fork oil per leg.

16 Front forks: examination and renovation

1   The parts most liable to wear over an extended period of service are the wearing surfaces of the fork stanchion and lower leg, the damper assembly within the fork tube and the oil seal at the sliding joint. Wear is normally accompanied by a tendency for the forks to judder when the front brake is applied and it should be possible to detect the increased amount of play by pulling and pushing on the handlebars when the front brake is applied fully. This type of wear should not be confused with slack steering head bearings, which can give rise to similar symptoms. Note that where air assisted forks are fitted, seal wear, and subsequent loss of air pressure, will be accelerated by excessive clearance between the stanchion and lower leg.

2   Renewal of the worn parts is quite straightforward. Particular care is necessary when renewing the oil seal, to ensure that the feather edge seal is not damaged during reassembly. Both the seal and the fork tube should be greased, to lessen the risk of damage. Note that oil seal condition is of particular significance on those forks which are air-assisted.

3   After an extended period of service, the fork springs may take a permanent set. If there is any doubt as to their condition check the free lengths against those of a new spring. If there is a noticeable difference, renew the springs as a complete set.

4   Check the outer surface of the fork tube for scratches or roughness. It is only too easy to damage the oil seal during reassembly, if these high spots are not eased down. Note that in the case of air assisted forks even a small scratch on the stanchion surface may allow air pressure leakage, and renewal of the stanchion may be the only way to rectify the problem. The fork stanchions are unlikely to bend unless the machine is damaged in an accident. Any significant bend will be detected by eye, but if there is any doubt about straightness, roll the stanchions on a flat surface. If the stanchions are bent, they must be renewed. Unless specialised repair equipment is available, it is rarely practicable to straighten them to the necessary standard.

5   The dust seals must be in good order if they are to fulfil their proper function. Replace any that are split or damaged.

6   Damping is effected by the damper units contained within each fork tube. The damping action can be controlled within certain limits by changing the viscosity of the oil used as the damping medium, although a change is unlikely to prove necessary except in extremes of climate.

7   Note that a number of the types of fork fitted are not fitted with renewable bushes. If wear develops, the stanchions and/or lower fork legs will have to be renewed.

8   Where renewable bushes are fitted, check first that the stanchion is in serviceable condition, then slide the bush over the stanchion and feel for free play. Little or no movement should be apparent if the bush is in good condition. In practice the bush will often become oval and the difference in clearance may be felt.

9   On machines with adjustable damping, the long damper adjustment rod must be checked for straightness. Support each end on V-blocks and arrange a dial gauge to rest near the centre. Rotate the rod and note the maximum and minimum readings. Half of this figure indicates the runout, which must not exceed 1.0 mm (0.04 in).

17 Steering head bearings: examination and renovation

1   Before commencing reassembly of the forks, examine the steering head races. The ball bearing tracks of the respective cup and cone bearings should be polished and free from indentations, cracks or pitting. If signs of wear are evident, the cups and cones must be renewed. In order for the straight line steering on any motorcycle to be consistently good, the steering head bearings must be absolutely perfect. Even the smallest amount of wear on the cups and cones may cause steering wobble at high speeds and judder during heavy front wheel braking. The cups and cones are an interference fit on their respective seatings and can be tapped from position with a suitable drift.

2   Ball bearings are relatively cheap. If the originals are marked or discoloured they **must** be renewed. To hold the steel balls in place during reasembly of the fork yokes, pack the bearings with grease. The upper and lower races contain 19 steel balls each. Although a small gap will remain when the balls have been fitted, on no account must an extra ball be inserted, as the gap is intended to prevent the balls from skidding against each other and wearing quickly.

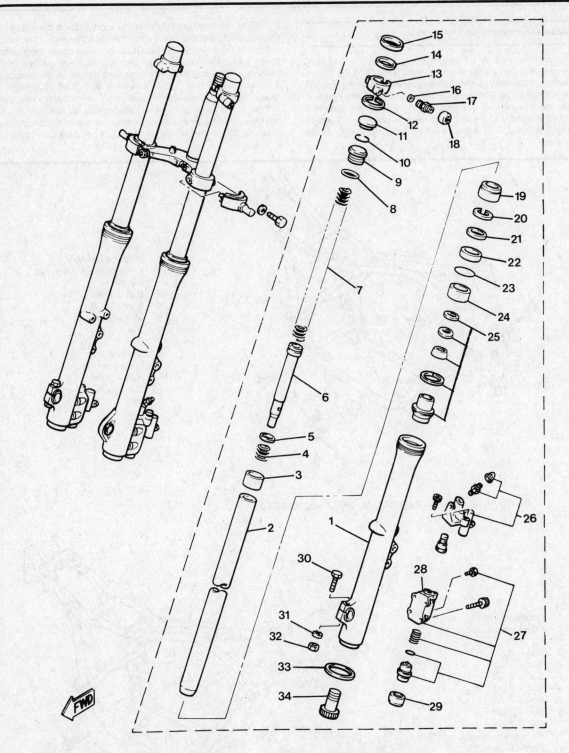

Fig. 4.12 Front forks – XJ750(UK) and XJ750 RJ and RH

1	Lower leg	10	Circlip
2	Stanchion	11	Rubber plug
3	Bush	12	Circlip
4	Rebound spring	13	Air union
5	Piston ring	14	O-ring
6	Damper rod	15	Seal
7	Fork spring	16	O-ring
8	O-ring	17	Air valve
9	Plug	18	Valve cap

19	Dust seal
20	Circlip
21	Washer
22	Oil seal
23	O-ring
24	Upper bush
25	Damper seat/oil lock assembly
26	Plunger assembly

27	Anti-dive valve assembly
28	Anti-dive valve housing
29	Dust cap
30	Pinch bolt
31	Washer
32	Nut
33	Sealing washer
34	Bolt

18 Frame: examination and renovation

1   The frame is unlikely to require attention unless accident damage has occurred. In some cases, replacement of the frame is the only satisfactory course of action if it is badly out of alignment. Only a few frame repair specialists have the jigs and mandrels necessary for resetting the frame to the required standard of accuracy and even then there is no easy means of assessing to what extent the frame may have been overstressed.

2   After the machine has covered a considerable mileage, it is advisable to examine the frame closely for signs of cracking or splitting at the welded joints. Rust can also cause weakness at these joints. Minor damage can be repaired by welding or brazing, depending on the extent and nature of the damage.

3   Remember that a frame which is out of alignment will cause handling problems and may even promote 'speed wobbles'. If misalignment is suspected, as the result of an accident, it will be necessary to strip the machine completely so that the frame can be checked and, if necessary, renewed.

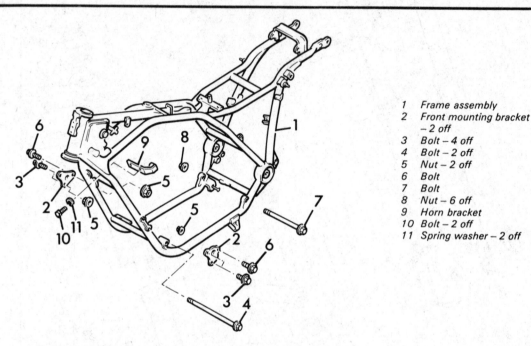

1   Frame assembly
2   Front mounting bracket
    – 2 off
3   Bolt – 4 off
4   Bolt – 2 off
5   Nut – 2 off
6   Bolt
7   Bolt
8   Nut – 6 off
9   Horn bracket
10 Bolt – 2 off
11 Spring washer – 2 off

Fig. 4.13 Frame – XJ650 G, H, LH, J and XJ750 J

1   Frame assembly
2   Front mounting bracket
    – 2 off
3   Bolt – 4 off
4   Nut – 4 off
5   Bolt – 2 off
6   Nut – 4 off
7   Bolt – 2 off
8   Bolt – 2 off
9   Nut – 2 off
10 Horn bracket – except US
11 Bolt – except US
12 Spring washer – except US

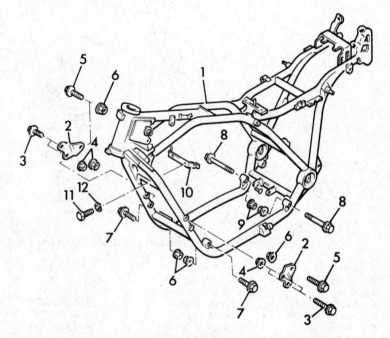

Fig. 4.14 Frame – XJ650(UK), XJ750(UK) and XJ750 RH and RJ

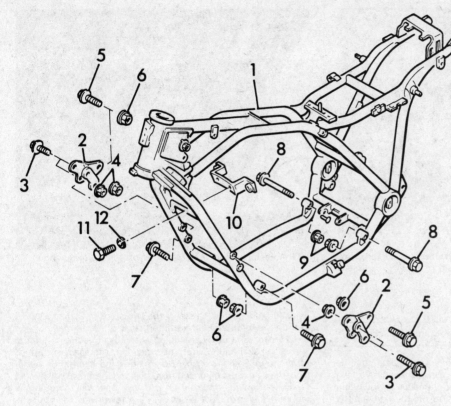

1 Frame assembly
2 Front mounting bracket
  – 2 off
3 Bolt – 4 off
4 Nut – 4 off
5 Bolt – 2 off
6 Nut – 4 off
7 Bolt – 2 off
8 Bolt – 2 off
9 Nut – 2 off
10 Horn bracket
11 Bolt
12 Spring washer

Fig. 4.15 Frame – XJ650 RJ

19 Swinging arm bearings: checking and adjustment

1   The rear swinging arm fork pivots on two tapered roller bearings, which are supported on adjustable screw stubs fitted to the lugs either side of the frame. After a period of time the tapered roller bearings will wear slightly, allowing a small amount of lateral shake at the rear wheel. This condition, even in its early stages, will have a noticeable effect on handling.

2   To check the play accurately, and if necessary to make suitable adjustments, it will be necessary to remove the rear wheel and detach the rear suspension units.

3   The machine should be placed securely on its centre stand, leaving the rear wheel clear of the ground. Unscrew and remove the rear brake adjuster nut and disengage the brake rod from the brake arm. Displace the trunnion and fit it, together with the spring, to the end of the rod, then retain them by refitting the adjuster. Straighten and remove the split pin which secures the torque arm nut. Remove the nut and displace the torque arm.

4   Remove the split pin from the wheel spindle nut, which can then be unscrewed. Slacken the wheel spindle pinch bolt, then displace and withdraw the wheel spindle to free the rear wheel. Manoeuvre the wheel clear of the frame and place it to one side. Remove the suspension unit lower mounting bolts to free the rear of the swinging arm assembly.

5   The swinging arm fork can now be checked for play. Grasp the fork at the rear end and push and pull firmly in a lateral direction. Any play will be magnified by the leverage effect. Move the swinging arm up and down as far as possible. Any roughness or a tightness at one point may indicate bearing damage. If this is suspected, the bearings should be inspected after removal of the swinging arm as described in the following Section.

6   Prise off the dust cap which covers the right-hand end of the swinging arm pivot. Slacken the adjuster stub locknut. It is

necessary to tighten the adjuster stub to a prescribed torque setting so that the bearings are under the right amount of side pressure. If a socket-type Allen key is not available it will be necessary to improvise using an extended Allen key with a spring balance attached one foot from the driving point, tightening by pulling on the spring balance until a reading of 3.6 lbs is shown. Tighten the adjuster to 0.5 – 0.6 kgf m (43 – 52 lbf in), then hold the adjuster in this position whilst the locknut is secured to 10.0 kgf m (72.3 lbf ft). Refit the dust cap and reassemble by reversing the dismantling sequence.

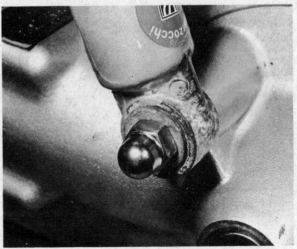

19.4 Remove domed nuts and washers to free the suspension unit lower mounts

19.5 Push the swinging arm from side to side to check for play, then up and down to check for roughness

19.6 Adjuster should be tightened to prescribed torque figure

20 Swinging arm: removal, renovation and replacement

1   If on inspection for play in the swinging arm bearings it is found that damage has occurred, or wear is excessive, the swinging arm fork should be removed. Commence by following paragraphs 3 – 5 of the preceding Section and then detach the swinging arm as follows.

2   Prise the final driveshaft rubber gaiter off the boss to the rear of the gearbox so that access to the final driveshaft flange is gained. Slacken evenly and remove the four flange bolts, turning the flange as necessary to reach the bolts.

3   The final drive box may be detached as a complete unit after removing the four flange nuts and washers. Support the weight of the casing as the nuts are removed. Drainage of the lubricating oils is not required, providing the casing is moved and stored in an upright position.

4   Prise off the dust caps on either side of the swinging arm pivot. On the left-hand side, bend back the tab washer which secures the stub, and on the right-hand side, slacken the

adjuster stub locknut. Unscrew the two stubs until the swinging arm unit is freed and can be lifted clear of the frame.

5   To each end of the swinging arm cross-member is fitted a spacer, oil seal and tapered roller bearing. The spacer is a push fit in the oil seal. To inspect or remove the bearing on either side, the seal must be prised from position. Removal will almost certainly damage the seal and a new one will therefore be required. Lever out the seals with a screwdriver. Take out the bearing inner races and clean and inspect them. Clean the bearing outer races whilst they are still in place. Check the rollers for pitting and the outer race for pitting or indentation. If the bearings need renewing, the outer races must be extracted from position using an internal puller attached to a slide hammer. Driving the bearings out is not possible because they are located in blind housings. New inner races may be driven in using a suitable tubular drift.

6   When assembling, clean and lubricate the bearings. Use a waterproof grease of the type recommended for wheel bearings.

7   Refit the swinging arm by reversing the dismantling procedure. Adjust the bearings by referring to Section 19.

20.3 Detach final drive housing by removing nuts

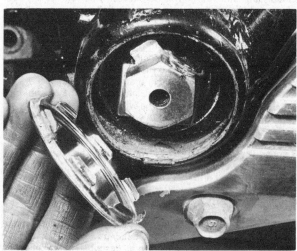

20.4a Prise off the end caps, and release locking tab

20.4b Release swinging arm by unscrewing stubs

20.5a Spacer is fitted inside seal

20.5b The lip of this seal is damaged (arrowed), requiring renewal

20.5c Inner race can be removed for cleaning and inspection. Outer race requires use of extractor

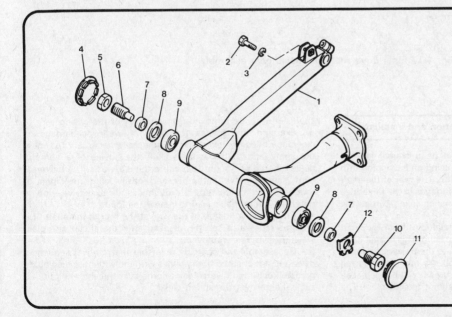

Fig. 4.16 Swinging arm

1   Swinging arm
2   Bolt
3   Spring washer
4   Cover
5   Locknut
6   Adjuster stub
7   Collar – 2 off
8   Oil seal – 2 off
9   Bearing – 2 off
10   Pivot stub
11   Cover
12   Tab washer

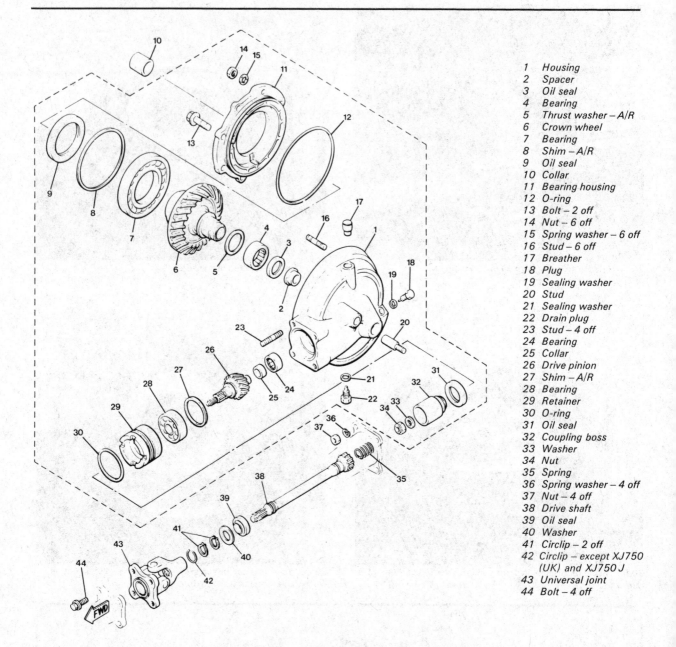

1   Housing
2   Spacer
3   Oil seal
4   Bearing
5   Thrust washer – A/R
6   Crown wheel
7   Bearing
8   Shim – A/R
9   Oil seal
10 Collar
11 Bearing housing
12 O-ring
13 Bolt – 2 off
14 Nut – 6 off
15 Spring washer – 6 off
16 Stud – 6 off
17 Breather
18 Plug
19 Sealing washer
20 Stud
21 Sealing washer
22 Drain plug
23 Stud – 4 off
24 Bearing
25 Collar
26 Drive pinion
27 Shim – A/R
28 Bearing
29 Retainer
30 O-ring
31 Oil seal
32 Coupling boss
33 Washer
34 Nut
35 Spring
36 Spring washer – 4 off
37 Nut – 4 off
38 Drive shaft
39 Oil seal
40 Washer
41 Circlip – 2 off
42 Circlip – except XJ750
     (UK) and XJ750 J
43 Universal joint
44 Bolt – 4 off

Fig. 4.17 Final drive shaft and bevel gear assembly

21 Final driveshaft: removal, renovation and installation

1   Access to the final driveshaft and its universal joint is
gained after removal of the final drive casing and swinging arm
unit as described in the preceding Section. The rear of the shaft
terminates in a machined spline which engages in the final drive
pinion, whilst at the forward end a smaller spline carries the
universal joint.
2   It is necessary to remove the universal joint from the shaft
to permit removal of the shaft from the casing. This is best done
with the aid of a slide hammer attached to the universal joint as
shown in the accompanying photograph. As the joint is jarred
off the splines the spring clip which secures it will be displaced
into its groove in the shaft, allowing the joint to slide clear. The

shaft can now be removed from the rear of the casing.
3   Examine the splines for signs of wear or damage. If
clearance becomes excessive it may prove necessary to renew
the shaft, and if this appears likely the advice of a Yamaha
dealer should be sought to confirm the diagnosis. The universal
joint should move smoothly and easily with no signs of
roughness or notchiness. If roughness or excessive slop is
noted, it will be necessary to renew the joint.
4   Check the condition of the seal at the rear of the shaft. If it
requires renewal it can be released after the circlip and plain
washer have been removed.
5   Reassemble the shaft by reversing the removal sequence,
having first lubricated both splines with molybdenum disulphide
grease. Push the universal joint home over the shaft end until it
clicks home over the spring clip.

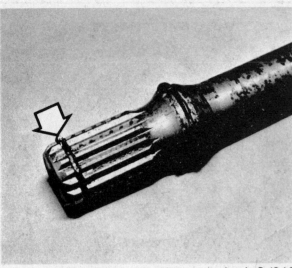

21.2b Spring ring (arrowed) locates in groove in joint

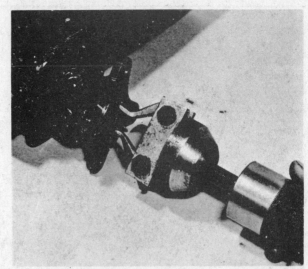

21.2a Slide hammer can be used to jar universal joint free

21.4a Circlip secures washer and seal on shaft

21.2c Shaft can be removed once housing is released

21.4c ... and check that the seal is secured

21.4b Push shaft into splined housing ...

21.5a Note spring inside splined housing

21.5b When joint is located by clip, shaft end will **just** be visible (arrowed)

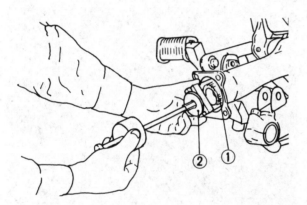

Fig. 4.18 Removal of final drive shaft universal joint

1   *Drive shaft*   2   *Slide hammer*

22 Final drive bevel gears: fault diagnosis

1   The final drive casing components are of well proven design, having been used on the XS750, 850 and 1100 models. Given regular oil changes the unit should last for the life of the motorcycle. In the event of damage or failure it should be noted that checking and overhauling the internal components requires the use of specialist equipment not normally available in the home workshop. The work is also rather complicated and for these reasons must be entrusted to a Yamaha dealer. Refer to the accompanying figure for a list of symptoms and possible causes which may be of assistance if a fault is suspected.

2   It is important to check that any suspected fault is not attributable to worn front or rear wheel bearings. These can give rise to very similar symptoms to those described above.

3   Failure of the seals either at the input shaft or output shaft will be self-evident by oil leakage.

4   Check the splines on the input shaft and output boss for wear or damage.

5   If damage to the gear teeth or bearings is suspected, drain the final drive oil into a clean container and examine it carefully

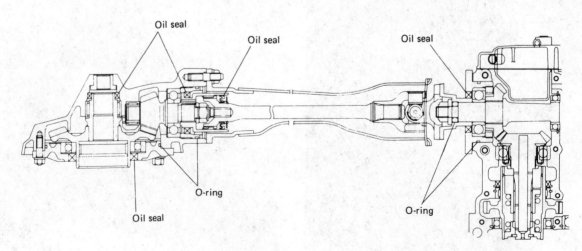

Fig. 4.19 Location of final drive assembly oil seals

Symptoms	Possible damaged areas
1. A pronounced hesitation or "jerky" movement during acceleration, deceleration, or sustained speed. (This must not be confused with engine surging or transmission characteries).	A. Damage to bearings.
	B. Improper gear lash.
	C. Gear tooth damage.
2. A "rolling rumble" noticeable at low speed; a high-pitched whine; a "clunk" from a shaft drive component or area.	D. Drive flange/universal joint bolts loose.
3. A locked-up condition of the shaft drive mechanism; no power transmitted from engine to rear wheel.	E. Broken drive-shaft.
	F. Disconnected flange/universal joint connection.
	G. Broken gear teeth.
	H. Seizure due to lack of lubrication.
	I. Small foreign object lodged between moving parts.

NOTE:
Damage areas A, B, and C above may be extremely difficult to diagnose. The symptoms are quite subtle and difficult to distinguish from normal motorcycle operating noise. If there is reason to believe component(s) are damaged, remove component(s) for specific inspection.

Fig. 4.20 Final drive assembly fault diagnosis chart

for signs of contamination by metal particles. A small amount of debris is normal, but any large particles are a good indication that all is not well.

6   If attention to the final drive bevel box is required, the complete unit should be returned to a Yamaha Service Agent, who will have the necessary tools and experience to carry out inspection and overhaul.

Important note: Investigate and rectify any suspected fault before riding the machine. Although failure is uncommon, it is possible for the final drive to lock up without warning if a piece of broken gear tooth or bearing becomes jammed between the final drive pinion and ring gear.

23 Rear suspension units: examination – XJ650 models

1   The rear suspension units fitted to the Yamaha XJ650 models are of the normal hydraulically damped type, adjustable to give 5 different spring settings. A screwdriver shaft or round metal rod should be used to turn the lower spring seat and so alter its position on the adjustment projection. When the spring seat is turned so that the effective length of the spring is shortened the suspension will become stiffer.

2   If a suspension unit leaks, or the damping efficiency is reduced in any other way the two units must be replaced as a pair. For precise roadholding it is imperative that both units react to movement in the same way. It follows that the units must always be set at the same spring loading.

3   The suspension units are of sealed construction and should be regarded as expendable components, their useful life being dependent upon mileage, terrain and individual riding style. As wear develops, there will be an increasing tendency for the units to fade during hard riding. The standard units are a necessary compromise between cost, handling quality and ride quality, and many owners will wish to fit units better suited to their own riding style as replacements. The range of choice is very wide and beyond the scope of this book, but a good guide is to speak to owners of similar machines fitted with non-standard units. When purchasing non-standard units, make sure that the spring and damping characteristics are suited to the specific machine to which they are to be fitted.

24 Rear suspension units: examination – XJ750 models

1   The XJ750 models are fitted with slightly more complicated units than the XJ650 machines, and feature 4-position damping adjustment in addition to the 5-position spring preload adjustment. The damping adjuster takes the form of a knurled ring just below the upper mounting eye. The adjuster has four click-stopped settings marked from 1 (standard) to 4 (maximum damping). Adjustment is largely a matter of personal preference, but like preload adjustment, should be identical on both units. In other respects the units are similar to those fitted to the XJ650 and the remarks in the preceding Section can be applied.

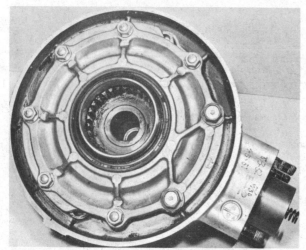

24.1a Check final drive housing for signs of leakage ...

24.1b ... or wear in splines. Do not dismantle

25 Suspension adjustment

1   The range of suspension adjustment available is dependent on the model concerned. On many of the XJ650 versions, for example, the only adjustable element is rear spring preload. On the XJ750 models things become more complex as will be seen below.

XJ650(UK), XJ650 G, H, LH and RJ models
2   Front suspension on the above models is non-adjustable. Rear spring preload can be adjusted to one of five positions to cater for additional loads imposed by a passenger or luggage, or for high-speed riding. The choice of setting is at the rider's discretion, no recommendations being given by the manufacturer.

XJ650 J model
3   Front suspension is by telescopic fork with variable air pressure to supplement the fork springs. Rear suspension is by standard oil-damped coil spring units with five-position spring preload adjustment. The accompanying table shows the manufacturer's recommendations for various conditions. Note that rear spring preload settings are quoted as positions A to E, A being the softest setting.
4   Note also the following remarks concerning front fork adjustments:

Standard air pressure   5.7 psi (0.4 kg/cm^2)
Maximum air pressure   17.0 psi (1.2 kg/cm^2)
Minimum air pressure   0 psi (0 kg/cm^2)
Maximum pressure difference
between legs   1.4 psi (0.1 kg/cm^2)

XJ750 J model
5   Front suspension is by leading axle, air assisted forks with adjustable damping. Rear suspension is by oil-damped coil-spring units with 5-position spring preload and 4-position damping adjustment. The accompanying table shows combinations of settings recommended by the manufacturer.
6   Note also the following remarks concerning front fork adjustments:

Standard air pressure   5.7 psi (0.4 kg/cm^2)
Maximum air pressure   17.0 psi (1.2 kg/cm^2)
Minimum air pressure   5.7 psi (0.4 kg/cm^2)

	Front fork	Rear shock absorber	Loading condition			
	Air pressure	Spring seat	Solo rider	With passenger	With accessory equipment	With accessory equipments and passenger
1	0.4 ~ 0.8 kg/cm^2 (5.7 ~ 11.4 psi)	A ~ C	O			
2	0.4 ~ 0.8 kg/cm^2 (5.7 ~ 11.4 psi)	A ~ C	O	O		
3	0.4 ~ 0.8 kg/cm^2 (5.7 ~ 11.4 psi)	C ~ E			O	O
4	0.8 ~ 1.2 kg/cm^2 (11.4 ~ 17.1 psi)	E			O	O

Fig. 4.21 Suspension settings table – XJ650 J

The front forks are interconnected and thus it is impossible to get a pressure difference between them.

XJ750(UK) and XJ750 RH and RJ models
7 Front suspension is by telescopic fork with adjustable air pressure and adjustable anti-dive. Rear suspension units have five-position spring preload adjustment and four-position damping adjustment. The accompanying table shows combinations of settings recommended by the manufacturer.

8 Note also the following remarks concerning front fork adjustments:

Standard air pressure	5.7 psi (0.4 kg/cm^2)
Maximum air pressure	36.0 psi (2.5 kg/cm^2)
Minimum air pressure	0 psi (0 kg/cm^2)
Maximum pressure difference between legs	1.4 psi (0.1 kg/cm^2)

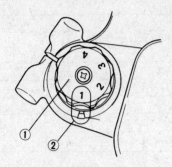

Fig. 4.22 Front fork damping adjustment – XJ750 J

Front fork		Rear shock absorber		Loading condition			
Air pressure	Damping adjuster	Spring seat	Damping adjuster	Solo rider	With passenger	With accessory equipments	With accessory equipments and passenger
39.2 ~ 78.5 kPa (0.4 ~ 0.8 kg/cm^2, 5.7 ~ 11 psi)	1	A ~ C	1	O			
	2	A ~ C	2	O	O		
	3	C ~ E	3		O	O	
78.5 ~ 118 kPa (0.8 ~ 1.2 kg/cm^2, 11 ~ 17 psi)	4	E	4			O	O

Fig. 4.23 Suspension settings table – XJ750 J

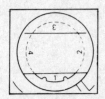

Fig. 4.24 Rear suspension unit 4-position damping adjuster – XJ750

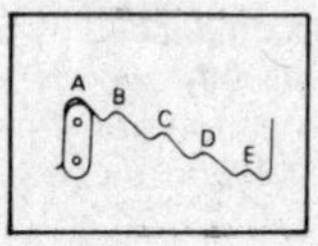

Fig. 4.25 Rear suspension unit 5-position spring preload adjuster

	Front fork	Rear shock absorber		Loading condition			
	Air pressure	Spring seat	Damping adjuster	Solo rider	With passenger	With accessory equipments	With accessory equipments and passenger
1.	0.4 ~ 0.8 kg/cm^2 (5.7 ~ 11.4 psi)	A ~ C	1	O			
2.	0.4 ~ 0.8 kg/cm^2 (5.7 ~ 11.4 psi)	A ~ C	2	O	O		
3.	0.4 ~ 0.8 kg/cm^2 (5.7 ~ 11.4 psi)	C ~ E	3		O	O	
4.	0.8 ~ 1.2 kg/cm^2 (11.4 ~ 17.1 psi)	E	4			O	O

Fig. 4.26 Suspension settings table – XJ750(UK) and XJ750 RH and RJ

Adjusting bolt position	Loading condition		
	Solo rider	With accessory equipments or passenger	With accessory equipments and passenger
1	○		
2	○	○	
3	○	○	○
4		○	○
5			○

Fig. 4.27 Front fork anti-dive settings table – XJ750(UK) and XJ750 RH and RJ

26 Front fork adjustment procedures

General remarks

1   The following sequences assume the forks to be in good condition and with sound seals. It is also important that the fork oil capacities are as specified and equal in both legs. If this is not the case it will be impossible to achieve balanced air pressure settings and handling may be affected.

Air pressure adjustment

2   Fork air pressure is increased by pumping air into the forks via the Schrader-type air valve or valves. The valve insert can be depressed to release pressure. It is safer to use a manual pump rather than compressed air since the small volume of the forks makes it difficult to judge pressure accurately enough with an air line. It is important not to exceed the maximum pressure recommendation since this will usually result in damaged fork seals. It is worth investing in a syringe-type air pump such as that marketed by S & W and specifically designed for suspension use.

3   Also needed is an accurate pressure gauge, preferably of the type which holds the reading until a rest button is pressed. Remember that a small drop in pressure will occur each time the gauge is used. With a little experience this can be allowed for when adding air.

4   Always check and adjust pressures with no weight on the front forks. This can be accomplished by placing the machine on its centre stand and placing wooden blocks beneath the crankcase. Take care to ensure that the pressure in each fork leg is equal as far as is possible, on those models where the forks are not interconnected.

Damping adjustment

5   The XJ750 J has damping adjustment controlled by a knurled knob at the top of each stanchion. The four marked settings are aligned with an index mark on the top yoke. The standard (softest) damping setting is number 1; number 4 being the hardest setting. Always ensure that the adjusters are set in the same position on each fork leg.

Anti-dive adjustment

6   The anti-dive adjuster is located at the bottom of each fork leg, and is covered by a rubber dust cap to exclude road dirt and water. With the cap removed it will be noted that the bottom of the anti-dive unit has four slots machined in it, through which the adjuster bolt is visible. A series of four lines is engraved on the adjuster bolt head, and these, together with the bottom of the bolt head, indicate the five adjustment positions. Maximum anti-dive effect is obtained with the bolt turned fully clockwise so that the bottom of the head is level with the top of the slots (position 5). As the bolt is unscrewed from this position each successive line which becomes visible denotes the next softest setting (positions 4, 3, 2 and 1). Note that once line 1 is visible a slight resistance will be felt. **Do not** attempt to unscrew the bolt further or the anti-dive unit may be damaged. Set both bolts to the same setting, as described in the preceding Section, then refit the rubber dust cap.

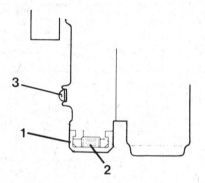

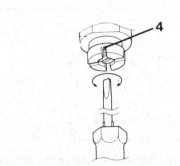

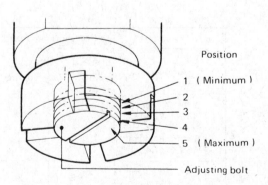

Fig. 4.28 Front fork anti-dive adjustment – XJ750(UK) and XJ750 RH and RJ

1	Rubber cap	3	Oil drain plug
2	Adjuster bolt	4	Machined slot

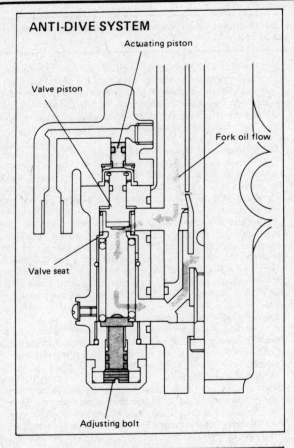

ANTI-DIVE SYSTEM

Actuating piston

Valve piston

Fork oil flow

Valve seat

Adjusting bolt

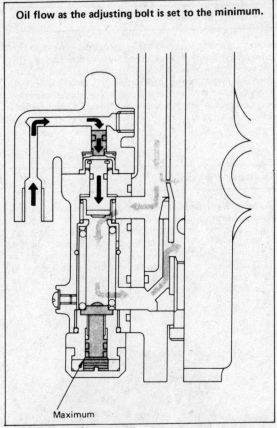

Oil flow as the adjusting bolt is set to the minimum.

Maximum

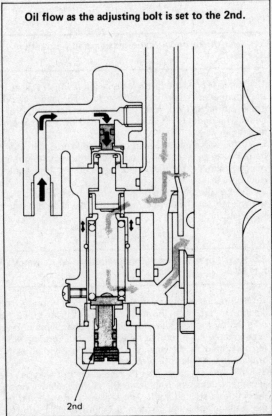

Oil flow as the adjusting bolt is set to the 2nd.

2nd

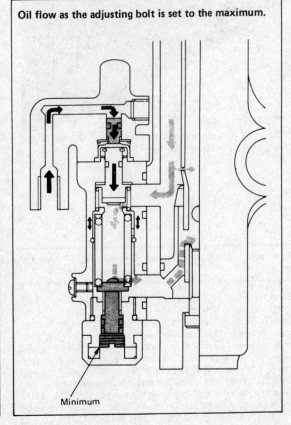

Oil flow as the adjusting bolt is set to the maximum.

Minimum

Fig. 4.29 Front fork anti-dive system method of operation

27 Centre stand: examination

1   The centre stand is attached to the machine by two bolts on the bottom of the frame. It is returned by a centre spring. The bolts and spring should be checked for tightness and tension respectively. A weak spring can cause the centre stand to 'ground' on corners and unseat the rider. For similar reasons, keep the stand pivots well lubricated and ensure that the stand is retracted fully.

28 Prop stand: examination

1   The prop stand is secured to a plate on the frame with a bolt and nut, and is retracted by a tension spring. Make sure the bolt is tight and the spring not overstretched, otherwise an accident can occur if the stand drops during cornering. In the case of machines fitted with a computer monitor system, a switch is incorporated in the stand which will operate a warning on the LCD panel if the stand is not retracted. This system is covered in detail in Chapter 6.

28.1 This stand was cracked and would have failed soon if not spotted. It was repaired by welding

29 Footrests: examination and renovation – (except XJ750 J)

1   Each footrest is an individual unit retained by a single bolt to a suitable part of the frame.
2   Both pairs of footrests are pivoted on clevis pins and spring loaded in the down position. If an accident occurs, it is probable that the footrest peg will move against the spring loading and remain undamaged. A bent peg may be detached from the mounting, after removing the clevis pin securing split pin and the clevis pin itself. The damaged peg can be straightened in a vice, using a blowlamp flame to apply heat at the area where the bend occurs. The footrest rubber will, of course, have to be removed as the heat will render it unfit for service.

30 Footrests: examination, renovation and adjustment – XJ750 J

1   The XJ750 J is equipped with similar footrests to the rest of the range, apart from the fact that the front footrests are attached via an adjuster plate. The rear brake and gearchange pedals are also adjustable to accommodate changes in position. General remarks concerning renovation are as given in Section 29. The adjustment procedure is described below.

RH footrest and brake pedal
2   Slacken the brake adjuster locknuts and the self-locking nut which retain the brake pedal and footrest. The footrest can now be moved within its adjustment range, to suit rider preference. Once the best position has been found, tighten the self-locking nut to secure the footrest.
3   Adjust the brake pedal height to suit the new footrest position. This is largely a matter of discretion, but Yamaha recommend a setting of 20 mm (0.8 in) below the top of the footrest. Adjustment is effected by turning the adjuster turnbuckle to suit. Once set correctly, tighten the adjuster locknuts. It will now be necessary to check and adjust the rear brake to give 20 – 30 mm (0.8 – 1.2 in) free play at the pedal.

LH footrest and gear change pedal
4   Slacken the gearchange pedal adjuster locknuts and the large footrest mounting bolt. Adjust the footrest to the required position, ensuring that it is in line with the right-hand item when viewed from above. Once the position has been set correctly, tighten the bolt to 5.5 kgf m (40 lbf ft) and then tighten the large securing nut (right-hand footrest) to the same figure. Set the gearchange adjuster to the desired position, sitting on the machine to make sure that it is conveniently positioned. The manufacturer recommends that the pedal is set about 20 mm (0.8 in) below the footrest. Once set, secure the adjuster locknuts.

31 Rear brake pedal: examination and renovation – except XJ750 J

1   The rear brake pedal pivots on a shaft which passes through the frame right-hand intersection lug. The shaft carrying the brake arm is splined, to engage with splines of the rear brake pedal. The pedal is retained to the shaft by a simple pinch bolt arrangement.
2   If the brake pedal is bent or twisted in an accident, it should be removed by slackening the pinch bolt and straightened in a manner similar to that recommended for the footrests in Section 29.
3   Make sure the pinch bolt is tight. If the lever is a slack fit on the splines, they will wear rapidly and it will be difficult to keep the lever in position.

32 Rear brake pedal: examination and renovation – XJ750 J

1   The rear brake pedal of the XJ750 J incorporates a remote linkage to facilitate footrest adjustment. The brake pedal is retained by the footrest mounting bolt and pivots on a bush. It is connected via a turnbuckle linkage to a short link which is in turn clamped to the splined shaft which normally carries the brake pedal on the other models.
2   If it proves necessary to dismantle the linkage, note the alignment dots which indicate the correct relative positions of the link and splined shaft. The brake adjuster should be set at the nominal position to give a length of 131 mm (5.16 in) between centres. Refer to Fig. 4.31 for details.

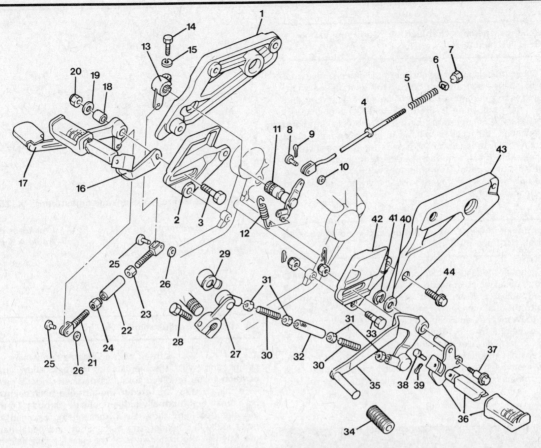

Fig. 4.30 Footrests and operating pedal assemblies – XJ750 J

1	Mounting plate	12	Return spring	23	Locknut	34	Rubber

1   Mounting plate
2   Mounting plate
3   Bolt
4   Brake rod
5   Spring
6   Trunnion
7   Adjuster nut
8   Clevis pin
9   Split pin
10 Washer
11 Brake shaft

12 Return spring
13 Link
14 Pinch bolt
15 Washer
16 RH footrest
17 Brake lever
18 Collar
19 Washer
20 Nut
21 Adjuster rod
22 Adjuster barrel

23 Locknut
24 Locknut
25 Clevis pin
26 Washer
27 Link
28 Pinch bolt
29 Boot
30 Adjuster rod
31 Lock nut – 2 off
32 Adjuster barrel
33 Bolt – 2 off

34 Rubber
35 Gearchange lever
36 LH footrest assembly
37 Bolt
38 Clevis pin
39 Split pin
40 Washer
41 E-clip
42 Mounting plate
43 Mounting plate
44 Bolt – 4 off

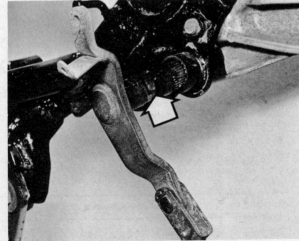

31.1 Link can be displaced inwards to permit greasing

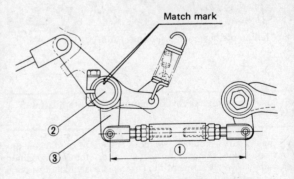

Fig. 4.31 Rear brake pedal alignment – XJ750 J

1   Standard length
    131 mm (5.16 in)

2   Splined shaft
3   Linkage

33 Gearchange pedal: examination and renovation – XJ650 G, H, LH and J

1   The above models employ a conventional gearchange pedal retained by splines and a pinch bolt to the gearchange shaft. The splined fitting permits a degree of coarse adjustment, but since the footrest position is fixed, it is unlikely that this will prove necessary. It is important that the pinch bolt is kept tight. If play develops, the splines will wear and eventually strip, necessitating the renewal of the pedal, and possibly the gearchange shaft. If accident damage occurs, it may be possible to straighten a bent pedal after heating it with a blowlamp, but since the finish will be destroyed it will usually be preferable to fit a new component.

34 Gearchange linkage: examination and renovation – all UK and all US Seca models

1   The above models are equipped with a semi-rearset gearchange pedal which connects to the gearchange shaft via a turnbuckle adjuster and a short operating link. The link is clamped to the shaft splines by a pinch bolt, whilst the pedal is retained to its pivot by an E-clip. The adjuster can be used to set the pedal height to the rider's preference. Maintenance is confined to periodic cleaning and lubrication of the moving parts to ensure smooth operation and prevent premature wear. It is worth removing the pedal and greasing the pivot pin to this end. The remaining pivots can be lubricated with a few spots of light oil.

34.1 The remote gearchange linkage (XJ650 UK)

35 Gearchange linkage: examination and adjustment – XJ750 J

1   The XJ750 J employs a remote linkage to permit realignment of the gearchange pedal after footrest position adjustment. Maintenance of the linkage is essentially the same as that described for the Seca models in Section 34. If the linkage is removed for any reason, note during reassembly that the split in the short operating link (shift arm) should align with the crankcase mark and that the angle between the pedal and adjuster should be set at 90°. Refer to Fig. 4.32 for further details.

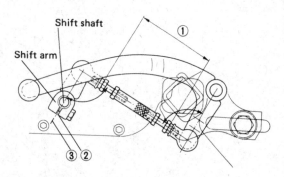

Fig. 4.32 Gearchange linkage adjustment – XJ750 J

1	Standard length 94 mm (3.70 in)	3	Crankcase mark Angle A & B 90°
2	Short operating link split		

36 Handlebar adjustment – XJ750 J

1   The XJ750 J is fitted with adjustable handlebars which, in conjunction with the adjustable footrest, allow the riding position to be modified to suit a particular rider's preference. Two splined stubs are fitted to the top yoke, each being retained by a small bolt, normally hidden beneath shrouds. To the stubs are clamped light alloy risers, these being retained by pinch bolts and Allen-headed top bolts. Short individual handlebars are fitted to the upper ends of the risers, again held by pinch bolts and stop bolts. The various component parts are shown in Fig. 4.4.

Vertical adjustment

2   Vertical position and angle of the handlebar ends can be set in one of three positions, giving one position higher and lower from the standard centre position. To effect adjustment, remove the two wiring finisher plates from the outer faces of the risers, each being secured by two cross-head screws. Prise out the blanking plugs which hide the stop bolts, then slacken them and the two pinch bolts. Set both bars to the chosen position, noting that movement is restricted to the three positions, the standard position being marked by punched dots. Once set, tighten the bolts to the following torque settings and refit the trim to complete.

Stop bolts	1.6 kgf m (11.0 lbf ft)
Pinch bolts	1.6 kgf m (11.0 lbf ft)

Horizontal adjustment

3   Remove the handlebar cover and top bolt blanking plugs, then slacken the top bolts and pinch bolts. Lift the risers slightly and re-position them either side of the standard centre position. This is denoted by alignment dots. Tighten the top bolts and pinch bolts to the torque figures quoted below, then refit the blanking plugs and the handlebar cover.

Top bolts	2.3 kgf m (17.0 lbf ft)
Pinch bolts	3.0 kgf m (22.0 lbf ft)

37 Dualseat: removal and replacement

All Seca models

1   Insert the key into the lock, turn it through 90° clockwise to unlock it, then slide the catch rearwards. The seat can now be

hinged upward. The seat will lock automatically when it is closed. If it is wished to remove the seat completely, slacken and remove the nuts and spring washers which secure the seat base to its hinges. When refitting the seat, check that the latch engages correctly before final tightening of the hinge nuts.

All Maxim variants (except XJ750 J)

2   Insert the key and turn it through 90° anticlockwise, then pull the catch lever rearwards. Lift the rear of the seat clear of the frame, then pull it back to free the front mounting lug. To refit the seat, engage the front mounting and lower the seat into position. Pull the catch rearwards while the seat is fitted then forwards to hold it in place. Turn the key clockwise to lock it and remove the key.

XJ750 J only

3   This is similar to the other Maxim models except for the lock mechanism. Note that the key moves through 45° and that two catch levers are fitted, one on each side. These should be pushed down to release the seat.

Lock renewal

4   In the event of failure of the seat lock/helmet lock unit it will be necessary to renew it. It is not practicable to effect repairs to the assembly. The lock is retained by screws to a frame lug.

38 Helmet lock: general description

1   All models are equipped with a helmet lock which is built into the seat lock mechanism. The lock is opened by turning the key in the opposite direction to that used to release the seat. To lock it, the shackle bar is pushed home into the lock body. The lock mechanism is unlikely to give problems, but must be renewed in the event of failure (see Section 37).

39 Steering lock: location and renewal

1   A steering lock mechanism is incorporated in the ignition switch assembly to facilitate the locking of the steering on full left or right lock. The lock is operated by depressing the key in the 'Off' position and turning it anticlockwise to the 'Lock' position.

2   If the steering lock or the ignition switch malfunction, it will be necessary to renew the unit. Repair is not practicable. The switch/lock assembly is bolted to the underside of the top yoke and can be removed after the instrument panel has been detached as described in Chapter 6.

40 Security chain: location and renewal

1   A security chain, operated by the machine's ignition key, is fitted in a small holder below the left-hand side panel. It is normally locked to a frame projection when being carried and is intended to immobilise the machine by locking the rear wheel to a convenient immovable object when parked. If the lock mechanism fails it will be necessary to purchase a new chain complete with key. Should this occur when the chain is locked, either to secure the machine or locked to the frame for storage, some difficulty may be experienced. It is recommended that the assistance of a Yamaha dealer or locksmith is sought.

41 Instrument panel: general

1   A wide variety of instrumentation is fitted to the Yamaha XJ range, ranging from the UK XJ650 (US XJ650 Seca) with

mechanical speedometer and tachometer plus conventional warning lamps, to the 1982 XJ750 Maxim's fully electronic CYCOM display and computer monitor system. Since the bias is towards electronic, rather than mechanical operation, details on the instrument panel will be found in Chapter 6. Information on mechanical drives, where applicable, will be found in the subsequent Sections of this Chapter.

40.1 Security chain in storage holder

41.1 Instrument panel layout (XJ650 UK)

42 Speedometer drive gearbox and cable: examination

1   The speedometer drive gearbox is fitted on the front wheel spindle where it is driven internally by the left-hand side of the wheel hub.

2   The gearbox rarely gives trouble if it is lubricated with grease at regular intervals. This can only be done after the wheel has been removed and the gearbox has been detached since no external grease nipple is fitted. The gearbox can be pulled from position after wheel removal.

3   The drive cable can often give rise to speedometer faults,

ranging from complete failure to jerky or erratic operation due to a kinked inner cable. In the event of malfunction a new cable must be fitted, repair or dismantling being impracticable.

43 Tachometer drive gearbox and cable: examination — mechanical tachometer models

1   Where a mechanical tachometer is fitted, drive is taken from a gear machined on the exhaust camshaft. This arrangement is kept well lubricated by engine oil and thus does not need to be dismantled for regular greasing. In the event of failure, the driven gear and housing may be detached after removing the single bolt and retainer plate which secures it to the cylinder head.

2   The housing is sealed by an O-ring which is fitted to a groove machined in its outer diameter. An internal seal is fitted to prevent oil leakage between the driven gear and the housing. This can be removed after the gland nut which retains it has been unscrewed. A slot is provided to facilitate removal.

3   The flexible drive cable is the most likely source of problems and can be dealt with in the same way as is described for speedometer cable in the preceding Section.

42.2 Speedometer drive can be removed for greasing

Chapter 5 Wheels, brakes and tyres

Refer to Chapter 7 for information on the 1983 US models

Contents

Specifications

Tyres

	XJ650 G, H, L LH, J and XJ750 J	XJ650(UK) models	All other models
Front	3.25H19-4PR Tubeless	3.25H19-4PR Tubed	3.25H19-4PR Tubeless
Rear	130/90-16 67H Tubeless	120/90-18 65H Tubed	120/90-18 65H Tubeless

Tyre pressures (cold)

	Front	Rear
XJ650 (UK), XJ650 RJ:		
Up to 198 lb (90 kg)	26 psi (1.8 kg/cm^2)	28 psi (2.0 kg/cm^2)
198-331 lb (90-150 kg)	28 psi (2.0 kg/cm^2)	33 psi (2.3 kg/cm^2)
331-478 lb (150-217 kg)	28 psi (2.0 kg/cm^2)	40 psi (2.8 kg/cm^2)
High speed riding	33 psi (2.3 kg/cm^2)	36 psi (2.5 kg/cm^2)
XJ650 G, H, LH, J:		
Up to 198 lb (90 kg)	26 psi (1.8 kg/cm^2)	28 psi (2.0 kg/cm^2)
198-353 lb (90-160 kg)	28 psi (2.0 kg/cm^2)	33 psi (2.3 kg/cm^2)
353-507 lb (160-230 kg)	28 psi (2.0 kg/cm^2)	40 psi (2.8 kg/cm^2)
High speed riding	33 psi (2.3 kg/cm^2)	36 psi (2.5 kg/cm^2)
XJ750 (UK), XJ750 RH, RJ:		
Up to 198 lb (90 kg)	26 psi (1.8 kg/cm^2)	28 psi (2.0 kg/cm^2)
198-474 lb (90-215 kg)	28 psi (2.0 kg/cm^2)	33 psi (2.3 kg/cm^2)
High speed riding	33 psi (2.3 kg/cm^2)	36 psi (2.5 kg/cm^2)
XJ750 J:		
Up to 198 lb (90 kg)	26 psi (1.8 kg/cm^2)	28 psi (2.0 kg/cm^2)
198-507 lb (90-230 kg)	28 psi (2.0 kg/cm^2)	33 psi (2.3 kg/cm^2)
High speed riding	28 psi (2.0 kg/cm^2)	33 psi (2.3 kg/cm^2)

Note: *Pressures are for original equipment tyres only; check with tyre supplier whether different pressures are to be used if non-standard tyres are fitted. Loads are weight of rider, passenger and accessories or luggage.*

Wheels

	XJ650 G, H, LH, J and XJ750 J	All other models
Type	Cast aluminium alloy	Cast aluminium alloy
Rim size:		
Front	MT 1.85 x 19	MT 1.85 x 19
Rear	MT 3.00 x 16	MT 2.15 x 18
Vertical runout (max)	2.0 mm (0.08 in)	2.0 mm (0.08 in)
Lateral runout (max)	2.0 mm (0.08 in)	2.0 mm (0.08 in)

Brakes

Front:

	XJ650 G, H, L, LH, J	XJ650 RJ XJ650(UK)	XJ750(UK), RJ, RH	XJ750 J
Type	Single hydraulic disc	Double hydraulic disc	Double hydraulic disc	Double hydraulic disc
Pad minimum overall thickness	6.5 mm (0.26 in)	6.0 mm (0.24 in)	4.0 mm (0.16 in)	–
Pad material minimum thickness	–	1.5 mm (0.061 in)	–	0.8 mm (0.03 in)
Maximum disc warpage	0.15 mm (0.006 in)	0.15 mm (0.006 in)	0.15 mm (0.006 in)	0.15 mm (0.006 in)
Minimum disc thickness	6.5 mm (0.26 in)	4.5 mm (0.18 in)	4.5 mm (0.18 in)	–
Hydraulic fluid	DOT 3 (US) or SAE J1703 (UK)			

Rear:

All models

Type	Single leading shoe, drum
Lining thickness (min)	2.0 mm (0.08 in)
Pedal free play	20-30 mm (0.8-1.2 in)
Drum diameter	200 mm (7.87 in)
Spring free length	68 mm (2.68 in)

Torque wrench settings

Component	kgf m	lbf ft
Front wheel spindle	10.7	77
Spindle pinch bolt	2.0	14
Brake disc bolt	2.0	14
Caliper mounting:		
650 models	2.6	19
750 models	4.5	32
Hose union bolts	2.6	19
Rear wheel spindle nut	10.7	77
Torque arm bolts	2.0	14
Brake arm pinch bolt	2.0	14

1 General description

All models in the XJ650/750 range are fitted with cast alloy 'Italic' wheels. The UK market XJ650 models use tubed tyres, whereas all other models use tubeless tyres. All models use a 3.25H 19-4PR front tyre. UK models and US Seca models have a 120/90-18-65H rear tyre, whilst the US Maxim variants employ a 130/90-16-67H rear tyre.

Braking on all models is by hydraulic disc at the front wheel and by a rear drum brake. The US XJ650 Maxim models have a single front disc brake, all other models having twin front discs.

2 Front wheel: examination and renovation

1   Carefully check the complete wheel for cracks and chipping, particularly at the spoke roots and the edge of the rim. As a general rule a damaged wheel must be renewed as cracks will cause stress points which may lead to sudden failure under heavy load. Small nicks may be radiused carefully with a fine file and emery paper (No 600 – No 1000) to relieve the stress. If there is any doubt as to the condition of a wheel, advice should be sought from a Yamaha repair specialist.
2   Each wheel is covered with a coating of lacquer, to prevent corrosion. If damage occurs to the wheel and the lacquer finish is penetrated, the bared aluminium alloy will soon start to corrode. A whitish grey oxide will form over the damaged area, which in itself is a protective coating. This deposit however, should be removed carefully as soon as possible and a new protective coating of laquer applied.
3   Check the lateral run out at the rim by spinning the wheel and placing a fixed pointer close to the rim edge. If the maximum run out is greater than 2.0 mm (0.080 in), Yamaha recommend that the wheel be renewed. This is, however, a counsel of perfection; a run out somewhat greater than this can probably be accommodated without noticeable effect on steering. No means is available for straightening a warped wheel without resorting to the expense of having the wheel skimmed on all faces. If warpage was caused by impact during an accident, the safest measure is to renew the wheel complete. Worn wheel bearings may cause rim run out. These should be renewed as described in Section 21 of this Chapter.
4   Note that impact damage or serious corrosion on models fitted with tubeless tyres has wider implications in that it could lead to a loss of pressure from the tubeless tyres. If in any doubt as to the wheel's condition, seek professional advice.

3 Front wheel: removal and replacement

1   The approach to front wheel removal is dependent on the model concerned and the reason for wheel removal. Where a

single disc brake is fitted, the procedure is quite straightforward, but is rather less so where twin disc brakes are used. The main difficulty lies in the fact that it is not possible to manoeuvre the wheel rim and tyre past the calipers. If attention to the calipers or pads is necessary, one of the calipers can be removed to gain the necessary clearance. Alternatively, if the wheel spindle and front mudguard are released it is possible to turn the fork legs so that the calipers are swung clear of the wheel.The above should be taken into consideration if wheel removal becomes necessary.

2   Start by placing the machine on the centre stand, ensuring that it is stable and leaving working space to the sides and at the front. Raise the front wheel clear of the ground by placing wooden blocks beneath the front of the crankcase. Make sure that the arrangement is secure and that the load on the underside of the crankcase is evenly distributed.

XJ650 G, H, LH and J models
3   Remove the split pin which locks the wheel spindle nut, and remove the nut. Release the knurled ring and free the speedometer drive cable, lodging it clear of the wheel. Slacken the pinch bolt and withdraw the wheel spindle. This is facilitated by using a tommy bar or screwdriver passed through the hole in the spindle, allowing it to be twisted free.

XJ650(UK) and XJ650 RJ models
4   Remove the split pin which secures the wheel spindle nut, then slacken and remove the nut. Unscrew the knurled ring which retains the speedometer drive cable, which can then be pulled clear of the wheel and forks. Slacken the pinch bolt at the bottom of each fork leg.
5   If the calipers are to be left in position, remove the four mudguard retaining bolts and lift the mudguard clear of the forks. Withdraw the wheel spindle and lower the wheel until the calipers are clear of the disc. Turn each lower leg so that the calipers are swung outwards allowing the wheel to be lifted clear.
6   Alternatively, remove the two bolts that secure one of the calipers to the fork leg. Withdraw the wheel spindle and lower the wheel, which can now be manoeuvred clear of the forks. Note that either caliper can be removed, and should be tied to the frame to avoid damaging the hydraulic hose.

XJ750(UK), XJ750 RH and RJ models
7   The above models are equipped with anti-dive suspension, and it is usually preferable to avoid disturbing this unless the calipers or anti-dive units require overhaul. Remove the split pin which secures the wheel spindle nut and remove the nut. Release the speedometer cable holding bolt and free the cable. Slacken the pinch bolt at the bottom of the right-hand lower leg.
8   If the calipers are to be left in position, remove the four mudguard retaining bolts and lift the mudguard clear of the forks. Withdraw the wheel spindle and lower the wheel until the calipers are clear of the disc. Turn each lower leg so that the calipers are swung outwards allowing the wheel to be lifted clear.
9   Alternatively, remove the two bolts that secure one of the calipers to the fork leg. Withdraw the wheel spindle and lower the wheel, which can now be manoeuvred clear of the forks. Note that either caliper can be removed, and should be tied clear of the wheel. It should not be necessary to disconnect the hydraulic hoses to the caliper or between the caliper and anti-dive unit, but care should be taken to avoid placing undue strain on the hoses or unions.

XJ750 J model
10 The front wheel of the XJ750 J can be removed by the same method described for the other XJ750 models, noting that it does not have the added complication of anti-dive braking, and thus the caliper may be removed quite easily if this approach is chosen.

All models
11   Note that once the wheel has been removed there is a distinct risk that the brake pads will be expelled should the brake lever be squeezed accidentally. This can be guarded against by slipping a strip of plywood or similar between the pads until the wheel is refitted.
12 Reassembly is a straightforward reversal of the removal sequence, noting the various torque settings given in the specifications. Make sure that the wheel aligns properly and that the discs are correctly positioned in the calipers. Check that the speedometer gearbox is located over the stop on the fork leg.

3.4a Remove split pin and unscrew wheel spindle nut

3.4b Free speedometer cable by unscrewing knurled ring

3.4c Slacken lower leg pinch bolts

3.6a Caliper can be released to allow wheel removal

3.6b Withdraw wheel spindle and lift wheel clear of forks

3.6c During installation, check that speedometer drive gearbox locates as shown

3.6d Tighten wheel spindle nut and secure with **new** split pin

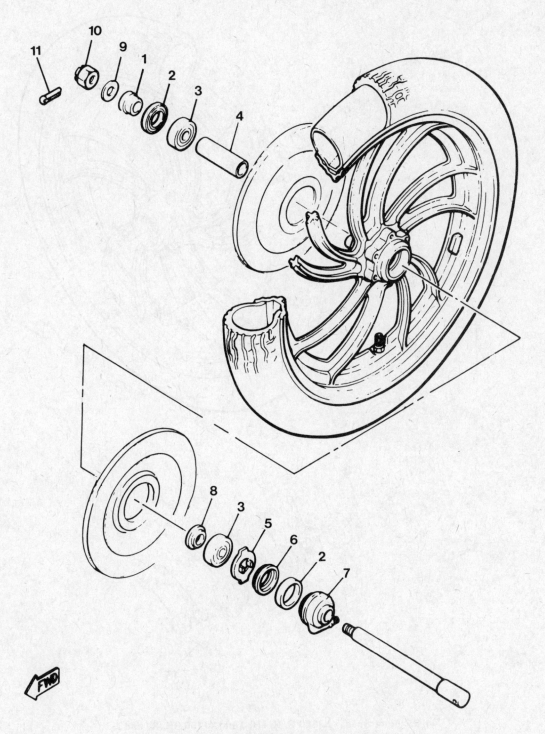

Fig. 5.1 Front wheel – XJ650 and 750(UK) and XJ650 RJ

1	Spacer	5	Drive dog	9	Washer
2	Oil seal - 2 off	6	Retainer	10	Nut
3	Bearing - 2 off	7	Speedometer gearbox	11	Split pin
4	Spacer	8	Spacer		

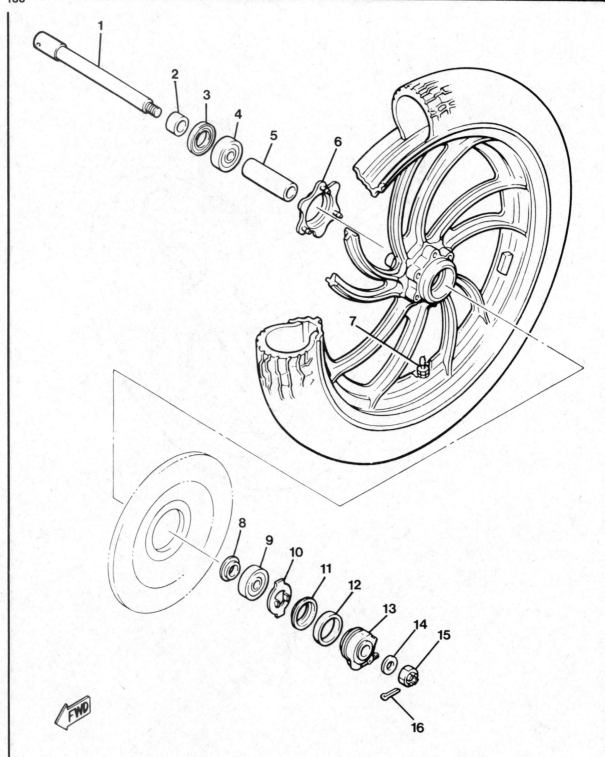

Fig. 5.2 Front wheel – XJ650 G, H, LH, J and XJ750 RH, RJ and J

1	Spindle	5	Spacer	9	Bearing	13	Speedometer gearbox
2	Spacer	6	Cover	10	Drive dog	14	Washer
3	Oil seal	7	Valve	11	Retainer	15	Nut
4	Bearing	8	Spacer	12	Oil seal	16	Split pin

4 Front disc brake: checking and renewing the pads – XJ650(UK) and XJ650 RJ

1 To facilitate the checking of brake pad wear, each caliper is provided with an inspection window closed by a small cover. Prise the cover from position and inspect both pads. Each pad has a red wear limit line on its periphery. If either pad has worn down to or past the line, both pads in that set should be renewed. In practice, it is probable that both sets of pads will wear at a similar rate, and therefore the two sets will require renewal at the same time.

2 Removal of the pads is straightforward. The procedure is identical for both calipers, and does not require the hydraulic hose to be disconnected. Slacken and remove the single bolt which retains the caliper unit to the support bracket. Next, working from the inner face of the caliper, remove the single pad locating screw (see photograph). The caliper body can now be pulled clear of the support bracket, leaving the pads in position in the support bracket. Remove the pads, taking note of which pad is fitted to each side of the caliper, and the relative positions of the backing shims.

3 Fit new pads by reversing the dismantling sequence. If difficulty is encountered when fitting the caliper over the brake disc, due to the reduced distance between the new pads, use a wooden lever to push the pad on the piston side inwards.

4 In the interests of safety, always check the function of the brakes before taking the machine on the road.

4.1a A: Pad backing metal B: Friction material C:Disc

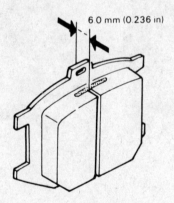

Fig. 5.3 Brake pad wear limit inspection – XJ650(UK)

6.0 mm (0.236 in)

4.1b Pad removed to show wear range. A: Area visible through caliper inspection window B: Usable range

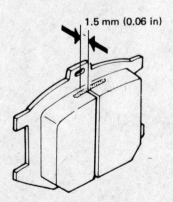

1.5 mm (0.06 in)

Fig. 5.4 Brake pad wear limit inspection – XJ650 RJ

4.2a Remove locating screw (arrowed) from inner face of caliper

4.2b Remove support bolt and pull off caliper unit

4.2c Note anti-rattle shim (arrowed) inside caliper unit

4.2d Lift away the pads ...

4.2e ... noting position of shims

4.2f Note pad guide shims on support bracket lugs

4.3 Check correct shim position during assembly

5 Front disc brake: checking and renewing the pads – XJ650 G, H, LH and J

1   The US XJ650 Maxim models make use of a different type of caliper unit than that described in Section 4 of this Chapter. The caliper body is attached to the fork leg via a pivot bolt and sleeve. This allows the caliper to move through a small arc in order to maintain even pressure from each pad against the disc. This caliper arrangement requires a different approach for the purposes of pad inspection and renewal.

2   Pad condition may be checked by viewing from the front of the caliper unit. Each pad incorporates a wear indicator in its friction material, and renewal will be necessary when either pad has worn down to this mark. Note that the design of the caliper means that the pads will probably appear to wear unevenly. This is quite normal, but it is important to renew the pads as a pair, regardless of appearance.

3   It is not necessary to remove the wheel or the caliper in order that the pads may be renewed. The pads are retained by a pin which is secured by a small coil spring in a locating groove. To free the pin use a pair of pointed-nose pliers to squeeze the spring ends together. The pin can now be displaced and withdrawn to free the pads. Grasp the projecting portion of the pads and pull them out of the caliper. The pad retainer and anti-rattle spring will also be displaced.

4   Brush out the accumulated dust from the caliper opening and check visually for signs of leakage. **Do not** inhale the dust from the caliper unit; it contains asbestos and can cause respiratory disorders. Clean the anti-rattle spring and pad retainer. Reassemble in the reverse of the dismantling order, using new pads. It may prove necessary to push the piston inwards slightly to accommodate the new, thicker pads. Refit the retaining pin, ensuring that the small coil spring engages in its groove.

6 Front disc brake: checking and renewing the pads – XJ750(UK), XJ750 RH and RJ

1   The maximum wear limit of the pads is indicated by a small raised tang on the backing metal. The pads should be renewed, as a pair, when worn to the point where the tang is about to contact the disc. The friction material also has a central groove, the bottom of which denotes that the pad has worn to its limit

thickness of 4 mm (0.16 in) overall, including the backing metal.

2   Access to the pads may be gained after the front wheel has been removed, together with the front mudguard. See Section 3 for details. The pads are retained by a pin which is pushed through the caliper body and through projections of the pad backing metal. The pin is locked in place by a safety clip. Unhook the end of the clip and pull it out of the pin. The pin can now be removed from the outer face of the caliper by pulling it clear with pliers. Flats are provided on the head of the pin to facilitate removal.

3   Withdraw the pads, anti-rattle spring and shim. Clean and inspect the latter, and also the inside of the caliper opening as described in Section 5.4. Reassemble, using new pads, by reversing the removal sequence. Note that the pad which backs onto the piston is fitted with a shim. It is important to fit this to the new pad in the same position. This will be evident from its outline on the old pad. Yamaha recommend that, along with the pads, the anti-rattle spring, shim, pad retaining pin and the safety clip are all renewed.

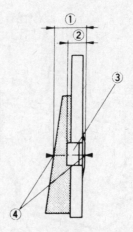

Fig. 5.5 Brake pad wear limit inspection – XJ650 G, H, LH and J

1	Pad thickness	3	Wear indicator
2	Wear limit	4	Measurement points

4.0 mm (0.16 in)

Wear indicator

Fig. 5.6 Brake pad wear limit inspection – XJ750(UK) and XJ750 RH and RJ

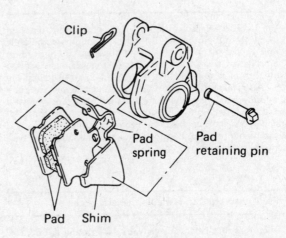

Clip

Pad spring

Pad retaining pin

Pad   Shim

Fig. 5.7 Pad assembly – XJ750(UK) and XJ750 RH and RJ

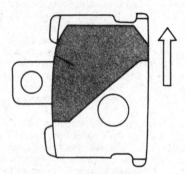

Fig. 5.8 Correct fitting of brake pad shim – XJ750(UK) and
XJ750 RH and RJ

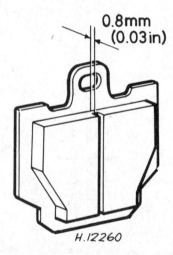

0.8mm
(0.03in)

H.12260

Fig. 5.9 Brake pad wear limit inspection – XJ750 J

7 Front disc brake: checking and renewing the pads – XJ750 J

1　The XJ750 J use a fourth type of caliper, differing
somewhat from the other XJ750 types. Although a completely
different caliper mounting arrangement is used, the procedure
for pad renewal is virtually identical to that described in Section
6. Note, however, that a long Allen-headed bolt is used in place
of the pad retaining pin and safety clip and that the pads are
shaped differently, as is the shim. The pads should be renewed
when there is 0.8 mm (0.03 in) or less of **friction material**
remaining; do not confuse this with the overall thickness
measurement, including backing metal, that applies to the type
described in the previous Section.

8 Front disc brake: overhaul – general notes

1　There are four basic caliper types used on the XJ650/750
range, and thus a slightly different approach is necessary when
overhauling the various units. Apart from details in the dismantling sequence, however, one caliper is much the same as

another in its construction. To avoid repetition, Section 9 gives
a full description of the procedure for a specific caliper type, the
remaining Sections describing detail differences on the
remaining models.

2　Irrespective of the type of caliper, collect together the
following before the overhaul commences. A can of DOT 3 or
SAE J1703 specification hydraulic fluid, a set of new caliper
seals, a length of 4.5 mm ($\frac{3}{16}$ in) bore clear plastic tubing for
bleeding and a supply of clean lint-free rag. Prepare a clean
work area, if necessary laying out an old sheet or similar on the
workbench.

9 Front disc brake: removing, renovating and replacing the caliper units – XJ650(UK) and XJ650 RJ

1　Before either caliper assembly can be removed from the
fork leg upon which it is mounted, it is first necessary to drain
off the hydraulic fluid. Disconnect the brake pipe at the union
connection it makes with the caliper unit and allow the fluid to
drain into a clean container. It is preferable to keep the front
brake lever applied throughout this operation, to prevent the
fluid from leaking out of the reservoir. A thick rubber band cut
from a section of inner tube will suffice, if it is wrapped tightly
around the lever and the handlebars. Alternatively, empty the
entire hydraulic system by placing the end of the hose in a
suitable container and squeezing the brake lever to pump the
fluid out. Unless the fluid is known to be almost new it is
recommended that this last approach is adopted.

2　Note that brake fluid is an extremely efficient paint stripper.
Take care to keep it away from any paintwork on the machine
or from any clear plastic, such as that sometimes used for
instrument glasses.

3　When the fluid has drained off, remove the caliper
mounting bolts, separate the two main caliper components and
remove the pads as described in Section 4. Note that from this
point onwards it is best to work on each caliper separately, to
avoid interchanging components between the two.

4　To displace the piston, apply a blast of compressed air to
the brake fluid inlet. Take care to catch the piston as it emerges
from the bore – if dropped or prised out with a screwdriver a
piston may suffer irreparable damage. Before removing the
piston, displace the dust seal which is retained by a circlip.

5　Remove the sleeve and protective boot upon which the
caliper unit slides. If play has developed between the sleeve and
the caliper, the former must be renewed. Check the condition of
the boot, renewing it if necessary.

6　The parts removed should be cleaned thoroughly, using
only brake fluid as the liquid. Petrol, oil or paraffin will cause the
various seals to swell and degrade, and should not be used
under any circumstances. When the various parts have been
cleaned, they should be stored in polythene bags until reassembly, so that they are kept dust free.

7　Examine the pistons for score marks or other imperfections.
If they have any imperfections they must be renewed, otherwise
air or hydraulic fluid leakage will occur, which will impair
braking efficiency. With regard to the various seals, it is
advisable to renew them all, irrespective of their appearance. It
is a small price to pay against the risk of a sudden and complete
front brake failure. It is standard Yamaha practice to renew the
seals every two years, even if no braking problems have
occurred.

8　Reassemble under clinically-clean conditions, by reversing
the dismantling procedure. Apply a small quantity of graphite
grease to the slider sleeve before fitting the boot. Reconnect the
hydraulic fluid pipe and make sure the union has been tightened
fully. Before the brake can be used, the whole system must be
bled of air, by following the procedure described in Section 18
of this Chapter.

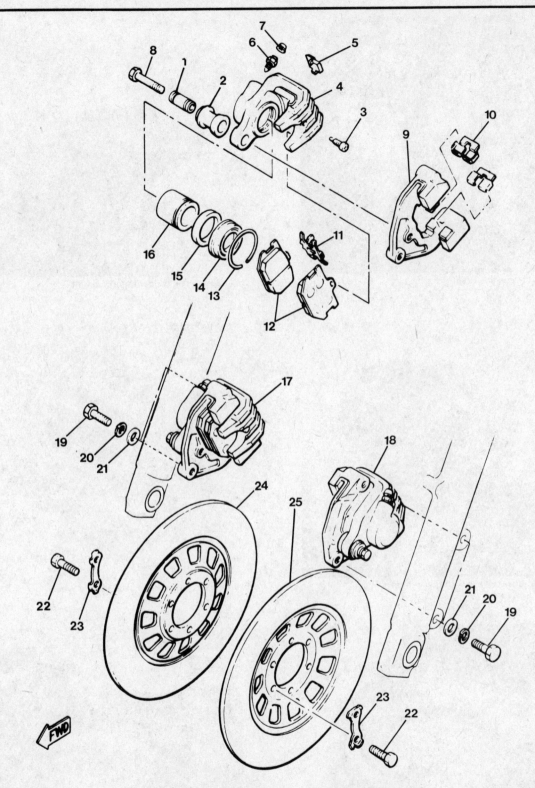

Fig. 5.10 Front brake caliper – XJ650(UK) and XJ650 RJ

1	Sleeve	8	Bolt	15	Seal	21	Washer - 2 off
2	Boot	9	Support bracket	16	Piston	22	Bolt - 12 off
3	Screw	10	Anti-rattle spring	17	RH caliper assembly	23	Tab washer
4	Caliper body	11	Anti-rattle spring	18	LH caliper assembly		- 6 off
5	Inspection window	12	Pad set	19	Bolt - 2 off	24	RH disc
6	Bleed nipple	13	Circlip	20	Spring washer	25	LH disc
7	Cap	14	Boot		- 2 off		

9.1 Disconnect hose and drain system

9.5a Use new seal on caliper mounting bolt ...

9.5b ... and lubricate sleeve with brake grease

9.7a Examine piston for signs of scoring or wear

9.7b Seals should be renewed irrespective of condition

9.8a Clean components, then slide piston into place

9.8b Fit dust seal over retaining lip ...

9.8c .. and secure with wire circlip

10 Front disc brake: removing, renovating and replacing the caliper unit – XJ650G, H, LH and J

1   Detach the hydraulic hose and drain the braking system as described in paragraphs 1 and 2 of Section 9. Remove the caliper by releasing the pivot bolt which retains it to the fork leg. Displace the washer, O-ring and pivot sleeve. Remove the pads, shim and anti-rattle spring as described in Section 5. Continue

dismantling and overhaul as described in Section 9, paragraph 4 onwards.
2   When fitting the caliper pivot bolt, lubricate it with graphite or molybdenum grease before sliding it into position. De-grease the threads which protrude beyond the washer and coat them with a thread locking compound such as Loctite. Offer up the caliper and fit the pivot bolt nut and spring washer, then tighten to 2.6 kgf-m (18.8 lbf ft).

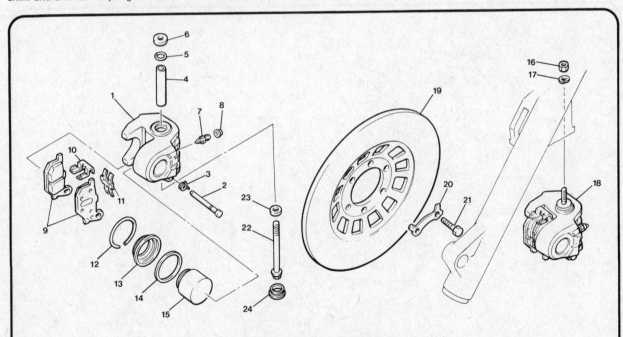

Fig. 5.11 Front brake caliper – XJ650 G, H, LH and J

1 Caliper body	7 Bleed nipple	13 Boot	19 Disc
2 Pad retaining pin	8 Cap	14 Seal	20 Tab washer - 3 off
3 Spring	9 Pad set	15 Piston	21 Bolt - 6 off
4 Pivot sleeve	10 Anti-rattle spring	16 Nut	22 Pivot bolt
5 O-ring	11 Anti-rattle spring	17 Spring washer	23 Washer
6 Washer	12 Circlip	18 LH caliper assembly	24 Dust cap

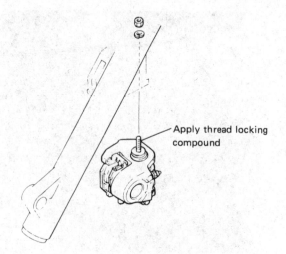

Fig. 5.12 Front brake caliper pivot bolt thread locking compound application

Apply thread locking compound

11 Front disc brake: removing, renovating and replacing the caliper units – XJ750(UK), XJ750 RH and RJ

1   Working with one caliper at a time, slacken the cross head screw which retains the plastic dust cover (where fitted) to the caliper body and place it to one side. Remove the brake hose union bolt to free the master cylinder and anti-dive hoses. Place the former in a suitable container, and pump the brake lever to expel the hydraulic fluid. Place the end of the anti-dive hose in a plastic bag and secure it with an elastic band.

2   Remove the two bolts which secure the caliper mounting bracket to the fork leg and lift the caliper assembly clear of the disc and fork leg. Remove the retaining pin, pads, shim and anti-rattle spring as described in Section 6, paragraph 2.

3   The piston may be driven out of the caliper body by an air jet – a foot pump if necessary. Remove the piston seal and dust seal from the caliper body. Under no circumstances should any attempt be made to lever or prise the piston out of the caliper. If the compressed air method fails, temporarily reconnect the caliper to the flexible hose, and use the handlebar lever to displace the piston hydraulically. Wrap some rag around the caliper to catch the inevitable shower of brake fluid.

4   The caliper mounting bracket can be freed from the caliper body by withdrawing the support pin. This is retained by a small split pin which should be withdrawn and discarded; a new split should be fitted during reassembly. Remove the bracket and lift off the retainer and shim.

5   Clean each part carefully, using only clean hydraulic fluid. On no account use petrol, oil or paraffin as these will cause the seals to degrade and swell. Keep all components dust free.

6   Examine the piston surface for scoring or pitting, any imperfection will necessitate renewal. The seals should be renewed as a matter of course, re-using an old seal is a false economy. Remember that the safety of the machine is very much dependent on seal and piston condition.

7   Reassemble, again ensuring absolute cleanliness, by reversing the dismantling procedure. Use clean hydraulic fluid as lubricant. Replace the caliper unit on the machine and reconnect the hydraulic hose. Remember that the system will need bleeding before use, by following the instructions given in Section 18 of this Chapter.

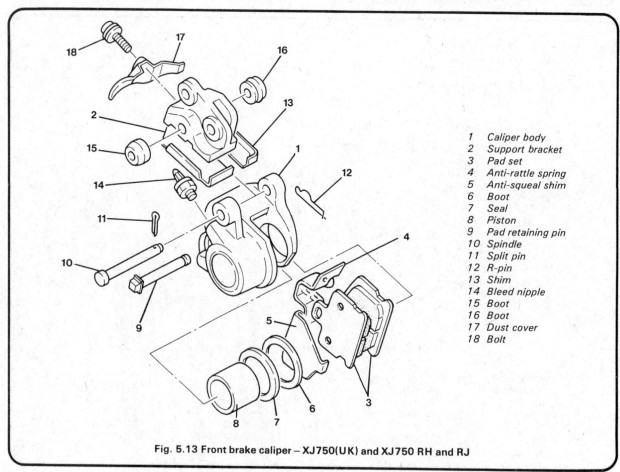

1	Caliper body
2	Support bracket
3	Pad set
4	Anti-rattle spring
5	Anti-squeal shim
6	Boot
7	Seal
8	Piston
9	Pad retaining pin
10	Spindle
11	Split pin
12	R-pin
13	Shim
14	Bleed nipple
15	Boot
16	Boot
17	Dust cover
18	Bolt

Fig. 5.13 Front brake caliper – XJ750(UK) and XJ750 RH and RJ

Tyre changing sequence - tubeless tyres

Deflate tyre. After releasing beads, push tyre bead into well of rim at point opposite valve. Insert lever adjacent to valve and work bead over edge of rim.

Use two levers to work bead over edge of rim. Note use of rim protectors.

When first bead is clear, remove tyre as shown.

Before fitting, ensure that tyre is suitable for wheel. Take note of any sidewall markings such as direction of rotation arrows.

Work first bead over the rim flange.

Use a tyre lever to work the second bead over rim flange.

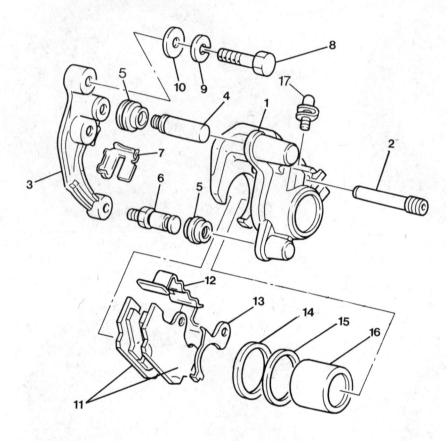

Fig. 5.14 Front brake caliper – XJ750 J

1 Caliper body
2 Pad retaining pin
3 Support bracket
4 Pin
5 Boot
6 Pin
7 Anti-rattle shim
8 Bolt - 2 off
9 Spring washer - 2 off
10 Washer - 2 off
11 Pad set
12 Anti-rattle spring
13 Anti-squeal shim
14 Boot
15 Seal
16 Piston

12 Front disc brake: removing, renovating and replacing the caliper units – XJ750 J

1   Working on one caliper unit at a time, slacken and remove the brake hose union bolt to free the hose at the caliper. Place the hose in a suitable container and pump the front brake lever to expel the hydraulic fluid. Slacken and remove the caliper mounting bolts and pull the assembly clear of the brake disc and wheel. Remove the pad retaining bolt and lift away the pads, shim and anti-rattle spring. The caliper body and mounting bracket can now be separated by pulling the two apart.

2   The piston may be driven out of the caliper body by an air jet – a foot pump if necessary. Remove the piston seal and dust seal from the caliper body. Under no circumstances should any attempt be made to lever or prise the piston out of the caliper. If the compressed air method fails, temporarily reconnect the caliper to the flexible hose, and use the handlebar lever to displace the piston hydraulically. Wrap some rag around the caliper to catch the inevitable shower of brake fluid.

3   Clean each part carefully, using only clean hydraulic fluid. On no account use petrol, oil or paraffin as these will cause the seals to degrade and swell. Keep all components dust free.

4   Examine the piston surface for scoring or pitting, any imperfection will necessitate renewal. The seals should be renewed as a matter of course, re-using an old seal is a false economy. Remember that the safety of the machine is very much dependent on seal and piston condition.

5   Check the condition of the pins upon which the caliper body slides. If wear has occurred the excessive free play may cause the caliper to 'chatter' when the brake is operated. The only remedy is to renew the pins, and if necessary, the caliper body. The pins are fitted with dust seals to prevent the ingress of water or road dirt, and it is advisable to renew these as a

precautionary measure before the unit is reassembled. The pins and bores should be cleaned and greased during reassembly.

6   Reassemble, again ensuring absolute cleanliness, by reversing the dismantling procedure. Use clean hydraulic fluid as lubricant. Replace the caliper unit on the machine and reconnect the hydraulic hose. Remember that the system will need bleeding before use, by following the instructions given in Section 17 of this Chapter.

13 Handlebar – mounted master cylinder: overhaul – All models except XJ750(UK), XJ750 RH and RJ

1   The master cylinder and hydraulic fluid reservoir takes the form of a combined unit mounted on the right-hand side of the handlebars, to which the front brake lever is attached.

2   Before the master cylinder unit can be removed and dismantled, the system must be drained. Place a clean container below each brake caliper unit and attach a plastic tube from the bleed screw of each caliper unit to the container. Lift off the master cylinder cover (cap), gasket and diaphragm, after removing the four countersunk retaining screws. Open the bleed screws one complete turn and drain the system by operating the brake lever until the master cylinder reservoir is empty. Close the bleed screws and remove the tube.

3   Before dismantling the master cylinder, it is essential that a clean working area is available on which the various component parts can be laid out. Use a sheet of white paper, so that none of the smaller parts can be overlooked.

4   Disconnect the stop lamp switch and front brake lever, taking care not to misplace the brake lever return spring. The stop lamp switch is a push fit in the lever stock. The lever pivots

on a bolt retained by a single nut. Remove the brake hose by unscrewing the banjo union bolt. Take the master cylinder away from the handlebars by removing the two bolts that clamp it to the handlebars. Take care not to spill any hydraulic fluid on the paintwork or on plastic or rubber components.

5   Withdraw the rubber boot that protects the end of the master cylinder and remove the snap ring that holds the piston assembly in position, using a pair of circlip pliers. The piston assembly can now be drawn out, followed by the return valve, spring cup and return spring.

6   The spring cup can now be separated from the end of the return valve spring and the main cup prised off the piston.

7   Examine the piston and the cylinder cup very carefully. If either is scratched or has the working surface impaired in any

other way, it must be renewed without question. Reject the various seals, irrespective of their condition, and fit new ones in their place. It often helps to soften them a little before they are fitted by immersing them in a container of clean brake fluid.

8   When reassembling, follow the dismantling procedure in reverse, but take great care that none of the component parts is scratched or damaged in any way. Use brake fluid as the lubricant whilst reassembling. When assembly is complete, reconnect the brake fluid pipe and tighten the banjo union bolt.

9   Use two new sealing washers at the union so that the banjo bolt does not require overtightening to effect a good seal. Refill the master cylinder with DOT 3 or SAE J1703 brake fluid and bleed the system of air by following the procedure described in Section 17 of this Chapter.

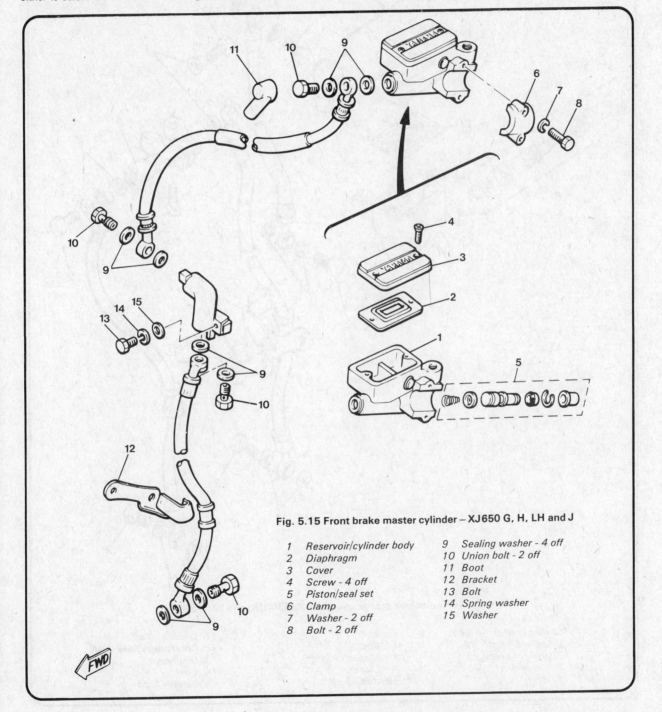

Fig. 5.15 Front brake master cylinder – XJ650 G, H, LH and J

1	Reservoir/cylinder body	9	Sealing washer - 4 off
2	Diaphragm	10	Union bolt - 2 off
3	Cover	11	Boot
4	Screw - 4 off	12	Bracket
5	Piston/seal set	13	Bolt
6	Clamp	14	Spring washer
7	Washer - 2 off	15	Washer
8	Bolt - 2 off		

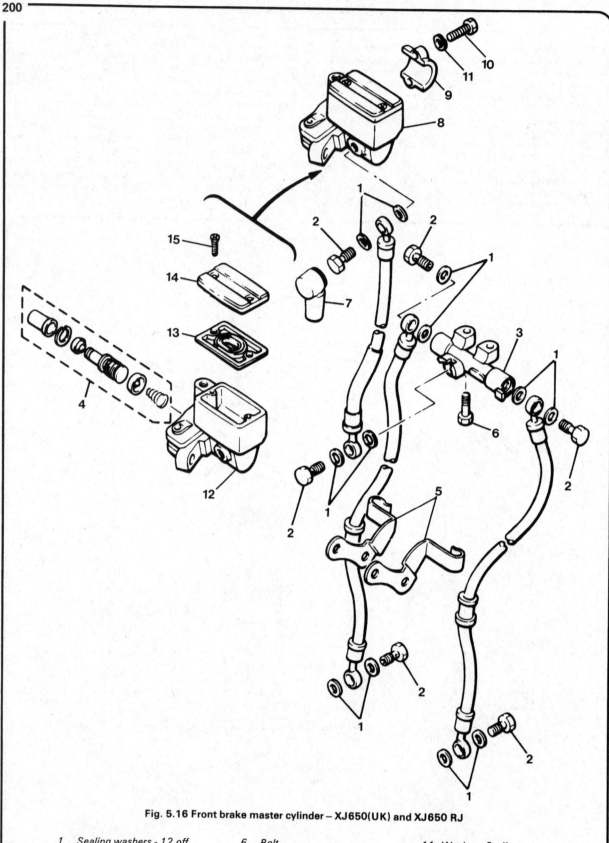

Fig. 5.16 Front brake master cylinder – XJ650(UK) and XJ650 RJ

1	Sealing washers - 12 off	6	Bolt	11	Washer - 2 off
2	Union bolt - 6 off	7	Boot	12	Reservoir/cylinder-body
3	Distributor union	8	Master cylinder	13	Diaphragm
4	Piston/seal set	9	Clamp	14	Cover
5	Brackets	10	Bolt - 2 off	15	Screw - 2 off

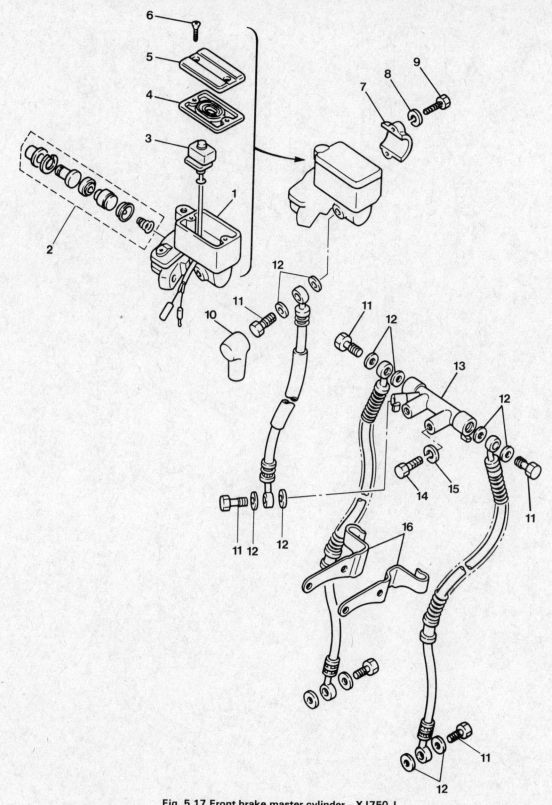

Fig. 5.17 Front brake master cylinder – XJ750 J

1	Reservoir/cylinder body	5	Cover	9	Bolt - 2 off	13	Distributor union
2	Piston/seal set	6	Screw - 2 off	10	Boot	14	Bolt - 2 off
3	Level switch	7	Clamp	11	Union bolt - 6 off	15	Washer - 2 off
4	Diaphragm	8	Washer - 2 off	12	Sealing washer - 12 off	16	Brackets

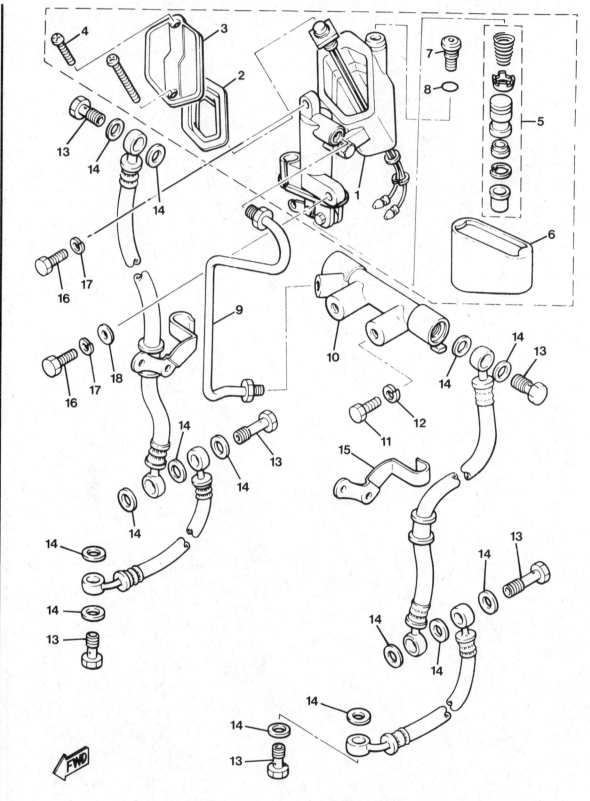

Fig. 5.18 Front brake master cylinder – XJ750(UK) and XJ750 RJ and RH

1	Reservoir/cylinder body	6	Cover	11	Bolt - 2 off
2	Diaphragm	7	Filler plug	12	Washer – 2 off
3	Cover	8	Sealing washer	13	Union bolt
4	Screw - 2 off	9	Pipe	14	Sealing washer
5	Piston/seal set	10	Distributor union		

15	Brackets
16	Bolt - 2 off
17	Spring washer - 2 off
18	Washer

14 Frame-mounted master cylinder: overhaul – XJ750(UK), XJ750 RH and RJ

1   The remotely-mounted master cylinders of the above models are well hidden amongst the various ancillary parts attached to the steering head, making access rather difficult. To reach the master cylinder it will first be necessary to remove the headlamp and auxiliary lamp units. To minimise the amount of preliminary work it is suggested that each unit is detached but left connected to the wiring harness. In this way the two lamps can be tied or propped clear of the working area.

2   Start by removing the two screws which secure the headlamp unit to the moulded plastic nacelle. Lift the unit away and disconnect the wiring from the back of the bulb. Slacken and remove the two bolts which hold the nacelle on the subframe arms and catch the captive nuts as they drop free inside the nacelle. Ease the assembly forward as far as the wiring will allow and check whether access to the master cylinder is now feasible. If not, it will be necessary to disconnect the wiring connectors, and also the horn leads on US models. Push the connectors and wiring through the cutouts at the back of the nacelle and lift it away.

3   The auxiliary lamp is mounted by a single stud on its underside and may be removed after the retaining nut and spring washer have been removed. If necessary, trace back and disconnect the leads to allow complete removal of the unit.

4   Before the master cylinder unit can be removed and dismantled, the system must be drained. Place a clean container below each brake caliper unit and attach a plastic tube from the bleed screw of each caliper unit to the container. Open the bleed screws one complete turn and drain the system by operating the brake lever until the master cylinder reservoir is empty. Close the bleed screws and remove the tube.

5   Slacken the gland nuts which retain the rigid metal pipe between the master cylinder and the distributor block. The latter is located beneath the bottom yoke and can be reached after the chromium plated finisher has been removed. This is retained by two screws which pass upwards into the underside of the bottom yoke. Remove the pipe taking care not to allow hydraulic fluid to drip onto any painted or plastic parts.

6   Slacken fully the brake cable adjuster and disconnect the cable from the master cylinder. Trace back and disconnect the hydraulic fluid level indicator leads. Slacken and remove the two bolts which retain the master cylinder and lift it clear of the subframe.

7   Before dismantling the master cylinder, it is essential that a clean working area is available on which the various component parts can be laid out. Use a sheet of white paper, so that none of the smaller parts can be overlooked. Clean the external surfaces of the master cylinder using a rag moistened with hydraulic fluid only.

8   The fluid reservoir can be removed from the master cylinder, where necessary. It is retained by a single union bolt. The reservoir cover can be removed, after releasing its two retaining screws, and the diaphragm lifted out. If the fluid level sensor requires attention, clean the area around its wiring leads very carefully, then remove the circlip which retains it in the reservoir. Displace the switch into the reservoir and remove it.

9   To gain access to the internal hydraulic components free the actuating arm from the underside of the master cylinder. Withdraw the rubber boot that protects the end of the master cylinder and remove the snap ring that holds the piston assembly in position, using a pair of circlip pliers. The piston assembly can now be drawn out, followed by the return valve, spring cup and return spring.

10   The spring cup can now be separated from the end of the return valve spring and the main cup prised off the piston.

11   Examine the piston and the cylinder cup very carefully. If either is scratched or has the working surface impaired in any other way, it must be renewed without question. Reject the various seals, irrespective of their condition, and fit new ones in their place. It often helps to soften them a little before they are fitted by immersing them in a container of clean brake fluid.

12   When reassembling, follow the dismantling procedure in reverse, but take great care that none of the component parts is scratched or damaged in any way. Use brake fluid as the lubricant whilst reassembling. When assembly is complete, refit the master cylinder and related parts by reversing the dismantling sequence. Refilling the reservoir is not easy because access is severely limited. It is suggested that a local doctor or veterinary surgeon is approached and an unwanted syringe obtained. Fit a length of small bore clear plastic tubing and use this to introduce fluid through the tiny Allen-headed filler plug at the back of the reservoir. A small funnel could be used in place of the syringe, but would not allow excess fluid to be sucked out. Always use new DOT 3 or SAE J1703 hydraulic fluid, and remember that the system must now be bled as described in Section 18 of this Chapter.

15 Anti-dive actuating valve housing: overhaul – XJ750(UK), XJ750 RH and RJ

1   The braking system of the above models is complicated by the anti-dive arrangement. The system is best dealt with in two parts; the anti-dive valve assembly, which is part of the suspension system and thus is covered in Chapter 4, and the actuating valve assembly. The latter is an appendage of the standard braking system and can be considered similar to a small secondary caliper unit for the purposes of overhauling. Like the master cylinder and the caliper, wear will develop over the years, and air or fluid leaks are a possibility. This should be remembered when diagnosing braking faults.

2   The actuating housing is mounted on top of the anti-dive housing and is retained by two Allen bolts. Commence removal by removing the fork air valve cap and releasing fork air pressure by depressing the valve insert. Obtain a jar or similar container into which hydraulic fluid can be drained. Slacken the brake hose union bolt and free the hose from the actuating housing, allowing any fluid from the hose to drain into the jar. It is sound practice to place the open end of the hose into a plastic bag, securing the end with an elastic band. This will keep dirt out and will prevent drips of hydraulic fluid falling on the painted parts.

3   Remove the two Allen screws and lift away the actuating valve housing and the white piston separator. Remove the circlip which retains the actuating piston which can then be displaced from the housing for examination.

4   Examine the valve and O-ring for signs of wear or damage. Note that the O-ring is not supplied separately and thus should not be removed from the valve. If wear is evident it will be necessary to renew the valve and O-ring as an assembly. The housing bore should also be checked for wear and renewed if necessary.

5   Clean the valve and housing prior to reassembly, and lubricate both components with clean brake fluid only. Reassemble by reversing the dismantling sequence. Note that the hydraulic system must be bled before the machine is used on the road. See Section 18 for details.

16 Hydraulic brake hoses and pipes: examination

1   External brake hoses and pipes are used to transmit the hydraulic pressure to the caliper units when the front brake or rear brake is applied. The brake hose is of the flexible type, fitted with an armoured surround. It is capable of withstanding pressures up to 350 kg/cm^2. The brake pipe attached to it is made from double steel tubing, zinc plated to give better corrosion resistance.

2   When the brake assembly is being overhauled, check the condition of both the hose and the pipe for signs of leakage or

scuffing, if either has made rubbing contact with the machine whilst it is in motion. The union connections at either end must also be in good condition, with no stripped threads or damaged sealing washers.

17 Bleeding the hydraulic system – All models except XJ750(UK), XJ750 RH and RJ

1   As mentioned earlier, brake action is impaired or even rendered inoperative if air is introduced into the hydraulic system. This can occur if the seals leak, the reservoir is allowed to run dry or if the system is drained prior to the dismantling of any component part of the system. Even when the system is refilled with hydraulic fluid, air pockets will remain and because air will compress, the hydraulic action is lost.

2   Check the fluid content of the reservoir and fill almost to the top. Remember that hydraulic brake fluid is an excellent paint stripper, so beware of spillage, especially near the petrol tank.

3   Place a clean glass jar below the brake caliper unit and attach a clear plastic tube from the caliper bleed screw to the container. Place some clean hydraulic fluid in the container so that the pipe is always immersed below the surface of the fluid.

4   Unscrew the bleed screw one complete turn and simultaneously operate the handlebar lever slowly. When the lever reaches the full extent of its travel, hold it in that position and close the bleed screw. Release the lever and repeat the process. Watch the plastic tube during this operation and note whether air bubbles are present in the expelled fluid. As the fluid is ejected from the bleed screw the level in the reservoir will fall. Take care that the level does not drop too low whilst the operation continues, otherwise air will re-enter the system, necessitating a fresh start.

5   Continue the pumping action with the lever until no further air bubbles emerge from the end of the plastic pipe. Hold the brake lever against the handlebars and tighten the caliper bleed screw. Remove the plastic tube **after** the bleed screw is closed. Where the front brakes are being bled, attach the pipe to the second caliper and repeat the sequence.

6   Check the brake action for sponginesss, which usually denotes there is still air in the system. If the action is spongy, continue the bleeding operation in the same manner, until all traces of air are removed.

7   Bring the reservoir up to the correct level of fluid and replace the diaphragm, sealing gasket and cap. Check the entire system for leaks. Recheck the brake action.

8   Note that fluid from the container placed below the brake caliper unit whilst the system is bled, should not be reused, as it will have become aerated and may have absorbed moisture.

18 Bleeding the hydraulic system – XJ750(UK), XJ750 RH and RJ

1   The procedure for bleeding the hydraulic system on machines equipped with anti-dive arrangements is essentially the same as that described in Section 17 of this Chapter, except that the anti-dive units present two additional bleed screws to be dealt with. Commence bleeding, starting with the anti-dive units. This will often clear the entire system of air, but it may prove necessary to repeat the operation at the calipers if air remains trapped in the system. Occasionally, bleeding may prove troublesome, and in these instances it will usually help if the operation is left for a few hours to allow air bubbles to collect in one place.

2   Perhaps the main problem in dealing with the above machines is how to get hydraulic fluid into the minute and awkwardly sited filler orifice. The method discussed in Section 14 of this Chapter may help with this point.

17.3 A specially made brake bleeding tool, as shown, is useful

19 Removing and replacing the brake disc

1   It is unlikely that either disc will require attention until a considerable mileage has been covered, unless premature scoring of the disc has taken place thereby reducing braking efficiency. To remove each disc, first detach the front wheel as described in Section 3. Each disc is bolted to the front wheel by six bolts, which are secured in pairs by a common tab washer. Bend back the tab washers and remove the bolts to free the disc.

2   The brake disc can be checked for wear and for warpage whilst the front wheel is still in the machine. Using a micrometer, measure the thickness of the disc at the point of greatest wear. If the measurement is much less than the recommended service limit, the disc should be renewed.

3   Check the warpage of the disc by setting up a suitable pointer close to the outer periphery of the disc and spinning the front wheel slowly. If the total warpage is more than 0.15 mm (0.006 in) the disc should be renewed. A warped disc, apart from reducing the braking efficiency, is likely to cause juddering during braking and will also cause the brake to bind when it is not in use.

19.1 Brake disc is secured by six bolts and tab washers

20 Front wheel bearings: examination and renovation

1   The front wheel bearings can be removed for examination, repacking or renewal after the front wheel has been removed. Clean any road dirt from the hub area, then remove the speedometer drive gearbox from the left-hand end of the hub. On the right-hand side of the hub, remove the short spacer which runs inside the oil seal. Pass a long steel drift through the centre of the right-hand bearing, then move it at an angle to the hub so that the internal spacer is displaced. Start tapping around the inner edge of the left hand bearing, gradually moving the drift around so that the bearing is driven out squarely. Check that as the bearing, speedometer drive dog and oil seal are displaced, there is adequate room for them to emerge. If necessary, support the wheel clear of the bench surface.

2   When the left-hand bearing drops clear it will bring with it the speedometer drive dog, a retainer and the oil seal, and will be followed by the internal bearing spacer. The wheel can now be turned over and the other bearing and oil seal removed.

3   Remove all the old grease from the hub and bearings, giving the latter a final wash in petrol. Check the bearings for signs of play or roughness when they are turned. If there is any doubt about the condition of a bearing, it should be renewed.

4   Before replacing the bearings, first pack the hub with new grease. Then drive the bearings back into position, not forgetting the distance piece that separates them. Take great care to ensure that the bearings enter the housings perfectly squarely otherwise the housing surface may be broached. Fit replacement oil seals and any dust covers or spacers that were also displaced during the original dismantling operation.

21 Rear wheel: examination, removal and replacement

1   Check the condition and alignment of the rear wheel in the same manner as described for the front wheel in Section 2 of this Chapter. Note that what may appear to be wheel bearing wear may in fact be wear in the final drive casing bearings, though this is not a common occurrence. See Chapter 4 for details.

2   To remove the rear wheel, place the machine securely on its centre stand so that the wheel is raised clear of the ground. Unscrew and remove the rear brake adjuster nut and disengage the brake rod from the brake arm. Displace the trunnion and fit it, together with the spring, to the end of the rod, then retain them by refitting the adjuster. Straighten and remove the split pin which secures the torque arm nut. Remove the nut and displace the torque arm.

3   Remove the split pin from the wheel spindle nut, which can then be unscrewed. Slacken the wheel spindle pinch bolt, then displace and withdraw the wheel spindle to free the rear wheel. Pull the wheel to the right until it drops clear of the driving splines, then manoeuvre it clear of the frame.

4   When installing the wheel, grease the oil seal lips before offering up the wheel to the driving splines. Ensure that these engage properly and that the wheel is pushed fully home. Continue reassembly, noting the following torque settings.

Rear wheel spindle   10.7 kgf m (77.4 lbf ft)
Rear wheel spindle pinch bolt   0.6 kgf m (4.5 lbf ft)

20.1a Remove speedometer gearbox from hub

20.1b Use a long drift to remove wheel bearings

20.4a Large socket ensures that bearings are fitted squarely

20.4b Drop speedometer drive dog into hub recess ...

20.4c ... followed by the steel retaining ring ...

20.4d ... and grease seal

20.4e Do not omit spacer from between bearings

20.4f Bearings must be greased before fitting

20.4g Fit a new oil seal ...

20.4h ... tapping it home until flush with the hub

21.2 Release brake rod and torque arm

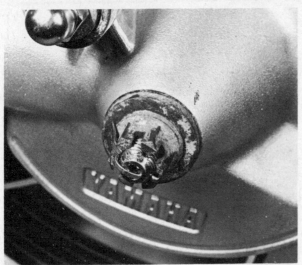
21.3a Remove split pin and spindle nut ...

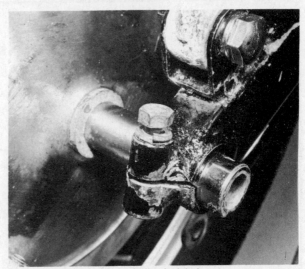
21.3b ... and slacken the spindle pinch bolt

21.3c Withdraw the spindle ...

21.3d ... and manoeuvre the wheel clear of frame

21.4a Check O-ring and renew if damaged

21.4b Lift wheel into place and align driving splines

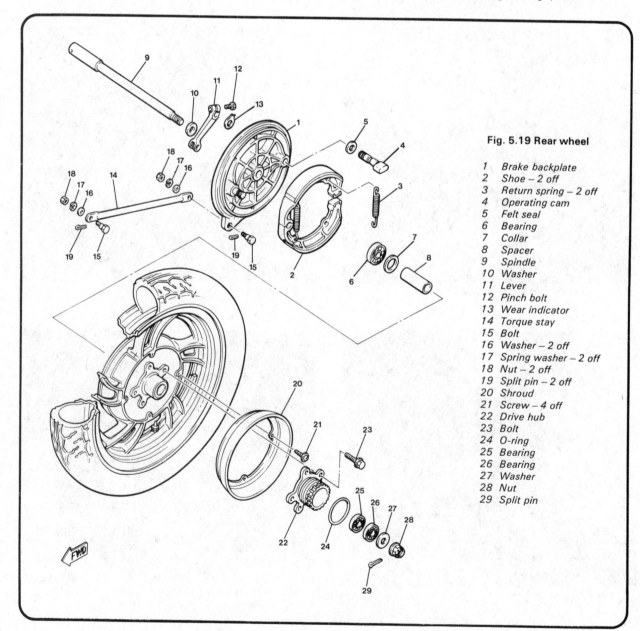

Fig. 5.19 Rear wheel

1 Brake backplate
2 Shoe – 2 off
3 Return spring – 2 off
4 Operating cam
5 Felt seal
6 Bearing
7 Collar
8 Spacer
9 Spindle
10 Washer
11 Lever
12 Pinch bolt
13 Wear indicator
14 Torque stay
15 Bolt
16 Washer – 2 off
17 Spring washer – 2 off
18 Nut – 2 off
19 Split pin – 2 off
20 Shroud
21 Screw – 4 off
22 Drive hub
23 Bolt
24 O-ring
25 Bearing
26 Bearing
27 Washer
28 Nut
29 Split pin

FWD

22 Rear wheel bearings: examination and replacement

1   The procedure for the removal and examination of the rear wheel bearings is similar to that given for the front wheel bearings. A heavy dust cover and oil seals are fitted on both sides of the hub. Commence by drifting out the right-hand wheel bearing and bearing spacer. Two bearings placed side by side are fitted on the left of the hub. These should be drifted out together. The double sealed bearing should be fitted on the outside.

23 Rear drum brake: examination and renovation

1   The rear brake backplate is fitted with a wear indicator which shows the usable range of the lining material. When this has worn down to the minimum mark it will be necessary to remove the brake backplate assembly for further investigation. Remove the rear wheel as described in Section 21, then lift the brake backplate assembly clear of the drum. Note that from this point onward care must be taken to avoid inhaling brake dust. This contains asbestos which has been shown to be toxic. Accumulated dust should be wiped away with a rag moistened with methylated spirit or petrol. On no account use compressed air to remove the dust.

2   Check the assembly for signs of oil or grease contamination, noting that the brake shoes will have to be removed if badly affected, having first located and rectified the source of the contamination. Measure lining wear by checking the overall diameter of the assembled shoes. Take measurements at several points across the shoes using a vernier caliper, then compare the smallest reading with the dimensions shown in the Specifications. If below the service limit, renew the shoes as a pair.

3   If the shoes require renewal they should be removed from the brake backplate by lifting the ends of the shoes clear of the pivot pin and 'folding' the shoes inwards as shown in the accompanying photograph. Once spring tension has been relieved the assembly can be lifted clear. Before the new shoes are fitted it is a good idea to detach the brake operating arm so that the brake cam can be displaced for cleaning and re-greasing. Mark the cam and arm as a guide during reassembly. A thin film of grease can be applied to the pivot pin as well. Fit the new shoes by reversing the removal sequence.

4   Where old shoes are to be reused, remove the glazed surface of the friction material with coarse sand paper, taking care not to inhale the dust. Examine the surface of the brake drum after cleaning it with a rag soaked in solvent. The drum should be free from excessive scoring, but any fine scoring can be removed by judicious use of fine abrasive paper.

22.1a Splined flange can be renewed if worn, but does not normally obstruct access to bearings

22.1b Drive side is fitted with inner bearing ...

22.1c ... and outer bearing as shown

22.1d Drop spacer into centre of hub

22.1e Note direction of sealed faces on bearings

22.1f Again, socket makes a good drift

25.3 Remove shoes by 'folding' them inwards

24 Adjusting the rear drum brake

1   Adjustment of the rear brake is correct when there is 20 - 30 mm ($\frac{3}{4}$″ - 1″ approx) up and down movement measured at the rear brake pedal foot piece, between the fully off and on position. Adjustment is carried out by turning the nut on the brake rod.

2   The height of the brake pedal when at rest may be adjusted by means of the stop bolt which passes through a plate welded to the pedal shank. Loosen the locknut before making the adjustment and tighten it when adjustment is complete.

3   Either adjustment may require the brake pedal operated stop lamp switch to be re-adjusted.

4   Note that the XJ750 J model is equipped with an adjustable brake linkage to compensate for changes in footrest position. Reference should be made to Chapter 4 for further details.

25 Tyres: removal and refitting – tubeless tyres

1   It is strongly recommended that should a repair to a tubeless tyre be necessary, the wheel is removed from the machine and taken to a tyre fitting specialist who is willing to do the job or taken to an official dealer. This is because the force required to break the seal between the wheel rim and tyre bead is considerable and considered to be beyond the capabilities of an individual working with normal tyre removing tools. Any abortive attempt to break the rim to bead seal may also cause damage to the wheel rim, resulting in an expensive wheel replacement. If, however, a suitable bead releasing tool is available, and experience has already been gained in its use, tyre removal and refitting can be accomplished as follows.

2   Remove the wheel from the machine by following the instructions for wheel removal as described in the relevant Section of this Chapter. Deflate the tyre by removing the valve insert and when it is fully deflated, push the bead of the tyre away from the wheel rim on both sides so that the bead enters the centre well of the rim. As noted, this operation will almost certainly require the use of a bead releasing tool.

3   Insert a tyre lever close to the valve and lever the edge of the tyre over the outside of the wheel rim. Very little force should be necessary; if resistance is encountered it is probably due to the fact that the tyre beads have not entered the well of the wheel rim all the way round the tyre. Should the initial problem persist, lubrication of the tyre bead and the inside edge and lip of the rim will facilitate removal. Use a recommended lubricant, a diluted solution of washing-up liquid or french chalk. Lubrication is usually recommended as an aid to tyre fitting but its use is equally desirable during removal. The risk of lever damage to wheel rims can be minimised by the use of proprietary plastic rim protectors placed over the rim flange at the point where the tyre levers are inserted. Suitable rim projectors may be fabricated very easily from short lengths (4-6 inches) of thick-walled nylon petrol pipe which have been split down one side using a sharp knife. The use of rim protectors should be adopted whenever levers are used and, therefore, when the risk of damage is likely.

4   Once the tyre has been edged over the wheel rim, it is easy to work around the wheel rim so that the tyre is completely free on one side.

5   Working from the other side of the wheel, ease the other edge of the tyre over the outside of the wheel rim, which is furthest away. Continue to work around the rim until the tyre is freed completely from the rim.

6   Refer to the following Section for details relating to puncture repair and the renewal of tyres. See also the remarks relating to the tyre valves in Section 27.

7   Refitting of the tyre is virtually a reversal of removal

procedure. If the tyre has a balance mark (usually a spot of coloured paint), as on the tyres fitted as original equipment, this must be positioned alongside the valve. Similarly, any arrow indicating direction of rotation must face the right way.

8   Starting at the point furthest from the valve, push the tyre bead over the edge of the wheel rim until it is located in the central well. Continue to work around the tyre in this fashion until the whole of one side of the tyre is on the rim. It may be necessary to use a tyre lever during the final stages. Here again, the use of a lubricant will aid fitting. It is recommended strongly that when refitting the tyre only a recommended lubricant is used because such lubricants also have sealing properties. Do not be over generous in the application of lubricant or tyre creep may occur.

9   Fitting the upper bead is similar to fitting the lower bead. Start by pushing the bead over the rim and into the well at a point diametrically opposite the tyre valve. Continue working round the tyre, each side of the starting point, ensuring that the bead opposite the working area is always in the well. Apply lubricant as necessary. Avoid using tyre levers unless absolutely essential, to help reduce damage to the soft wheel rim. The use of the levers should be required only when the final portion of bead is to be pushed over the rim.

10   Lubricate the tyre beads again prior to inflating the tyre, and check that the wheel rim is evenly positioned in relation to the tyre beads. Inflation of the tyre may well prove impossible without the use of a high pressure air hose. The tyre will retain air completely only when the beads are firmly against the rim edges at all points and it may be found when using a foot pump that air escapes at the same rate as it is pumped in. This problem may also be encountered when using an air hose on new tyres which have been compressed in storage and by virtue of their profile hold the beads away from the rim edges. To overcome this difficulty, a tourniquet may be placed around the circumference of the tyre, over the central area of the tread. The compression of the tread in this area will cause the beads to be pushed outwards in the desired direction. The type of tourniquet most widely used consists of a length of hose closed at both ends with a suitable clamp fitted to enable both ends to be connected. An ordinary tyre valve is fitted at one end of the tube so that after the hose has been secured around the tyre it may be inflated, giving a constricting effect. Another possible method of seating beads to obtain initial inflation is to press the tyre into the angle between a wall and the floor. With the airline attached to the valve additional pressure is then applied to the tyre by the hand and shin, as shown in the accompanying illustration. The application of pressure at four points around the tyre's circumference whilst simultaneously applying the airhose will often effect an initial seal between the tyre beads and wheel rim, thus allowing inflation to occur.

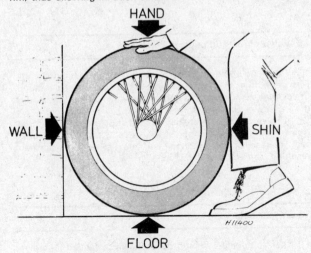

Fig. 5.20 Method of seating the tyre beads

11   Having successfully accomplished inflation, increase the pressure to 40 psi and check that the tyre is evenly disposed on the wheel rim. This may be judged by checking that the thin positioning line found on each tyre wall is equidistant from the rim around the total circumference of the tyre. If this is not the case, deflate the tyre, apply additional lubrication and reinflate. Minor adjustments to the tyre position may be made by bouncing the wheel on the ground.

12   Always run the tyre at the recommended pressures and never under or over-inflate. The correct pressures for various weights and configurations are given in the Specifications Section of this Chapter.

26 Puncture repair and tyre renewal – tubeless tyres

1   The primary advantage of the tubeless tyre is its ability to accept penetration by sharp objects such as nails etc without loss of air. Even if loss of air is experienced, because there is no inner tube to rupture, in normal conditions a sudden blow-out is avoided.

2   If a puncture of the tyre occurs, the tyre should be removed for inspection for damage before any attempt is made at remedial action. The temporary repair of a punctured tyre by inserting a plug from the outside should not be attempted. Although this type of temporary repair is used widely on cars, the manufacturers strongly recommend that no such repair is carried out on a motorcycle tyre. Not only does the tyre have a thinner carcass. which does not give sufficient support to the plug, the consequences of a sudden deflation is often sufficiently serious that the risk of such an occurrence should be avoided at all costs.

3   The tyre should be inspected both inside and out for damage to the carcass. Unfortunately the inner lining of the tyre – which takes the place of the inner tube – may easily obscure any damage and some experience is required in making a correct assessment of the tyre condition.

4   There are two main types of tyre repair which are considered safe for adoption in repairing tubeless motorcycle tyres. The first type of repair consists of inserting a mushroom-headed plug into the hole from the inside of the tyre. The hole is prepared for insertion of the plug by reaming and the application of an adhesive. The second repair is carried out by buffing the inner lining in the damaged area and applying a cold or vulcanised patch. Because both inspection and repair, if they are to be carried out safely, require experience in this type of work, it is recommended that the tyre be placed in the hands of a repairer with the necessary skills, rather than repaired in the home workshop.

5   In the event of an emergency, the only recommended 'get-you-home' repair is to fit a standard inner tube of the correct size. If this course of action is adopted, care should be taken to ensure that the cause of the puncture has been removed before the inner tube is fitted. It will be found that the valve in the rim is considerably larger than the diameter of the inner tube valve stem. To prevent the ingress of road dirt, and to help support the valve, a spacer should be fitted over the valve.

6   In the event of the unavailability of tubeless tyres, ordinary tubed tyres fitted with inner tubes of the correct size may be fitted. Refer to the manufacturer or a tyre fitting specialist to ensure that only a tyre and tube of equivalent type and suitability is fitted, and also to advise on the fitting of a valve nut to the rim hole.

27 Tyre valves: description and renewal – tubeless tyres

1   It will be appreciated from the preceding Sections, that the adoption of tubeless tyres has made it necessary to modify the valve arrangement, as there is no longer an inner tube which can carry the valve core. The problem has been overcome by

fitting a separate tyre valve which passes through a close-fitting hole in the rim, and which is secured by a nut and locknut. The valve is fitted from the rim well, and it follows that the valve can be removed and replaced only when the tyre has been removed from the rim. Leakage of air from around the valve body is likely to occur only if the sealing seat fails or if the nut and locknut become loose.

2   The valve core is of the same type as that used with tubed tyres, and screws into the valve body. The core can be removed with a small slotted tool which is normally incorporated in plunger type pressure gauges. Some valve dust caps incorporate a projection for removing valve cores. Although tubeless tyre valves seldom give trouble, it is possible for a leak to develop if a small particle of grit lodges on the sealing face. Occasionally, an elusive slow puncture can be traced to a leaking valve core, and this should be checked before a genuine puncture is suspected.

3   The valve dust caps are a significant part of the tyre valve assembly. Not only do they prevent the ingress of road dirt into the valve, but also act as a secondary seal which will reduce the risk of sudden deflation if a valve core should fail.

28 Tyres: removal, repair and refitting – tubed tyres

1   At some time or other the need will arise to remove and replace the tyres, either as a result of a puncture or because replacements are necessary to offset wear. To the inexperienced, tyre changing represents a formidable task, yet if a few simple rules are observed and the technique learned, the whole operation is surprisingly simple.

2   To remove the tyre from either wheel, first detach the wheel from the machine. Deflate the tyre by removing the valve core, and when the tyre is fully deflated, push the bead away from the wheel rim on both sides so that the bead enters the centre well of the rim. Remove the locking ring and push the tyre valve into the tyre itself.

3   Insert a tyre lever close to the valve and lever the edge of the tyre over the outside of the rim. Very little force should be necessary; if resistance is encountered it is probably due to the fact that the tyre beads have not entered the well of the rim, all the way round. If aluminium rims are fitted, damage to the soft alloy by tyre levers can be prevented by the use of plastic rim protectors.

4   Once the tyre has been edged over the wheel rim, it is easy to work round the wheel rim, so that the tyre is completely free from one side. At this stage the inner tube can be removed.

5   Now working from the other side of the wheel, ease the other edge of the tyre over the outside of the wheel rim that is furthest away. Continue to work around the rim until the tyre is completely free from the rim.

6   If a puncture has necessitated the removal of the tyre, reinflate the inner tube and immerse it in a bowl of water to trace the source of the leak. Mark the position of the leak, and deflate the tube. Dry the tube, and clean the area around the puncture with a petrol soaked rag. When the surface has dried, apply rubber solution and allow this to dry before removing the backing from the patch, and applying the patch to the surface.

7   It is best to use a patch of self vulcanizing type, which will form a permanent repair. Note that it may be necessary to remove a protective covering from the top surface of the patch after it has sealed into position. Inner tubes made from a special synthetic rubber may require a special type of patch and adhesive, if a satisfactory bond is to be achieved.

8   Before replacing the tyre, check the inside to make sure that the article that caused the puncture is not still trapped inside the tyre. Check the outside of the tyre, particularly the tread area to make sure nothing is trapped that may cause a further puncture.

9   If the inner tube has been patched on a number of past occasions, or if there is a tear or large hole, it is preferable to

discard it and fit a replacement. Sudden deflation may cause an accident, particularly if it occurs with the rear wheel.

10   To replace the tyre, inflate the inner tube for it just to assume a circular shape but only to that amount, and then push the tube into the tyre so that it is enclosed completely. Lay the tyre on the wheel at an angle, and insert the valve through the rim tape and the hole in the wheel rim. Attach the locking ring on the first few threads, sufficient to hold the valve captive in its correct location.

11   Starting at the point furthest from the valve, push the tyre bead over the edge of the wheel rim until it is located in the central well. Continue to work around the tyre in this fashion until the whole of one side of the tyre is on the rim. It may be necessary to use a tyre lever during the final stages.

12   Make sure there is no pull on the tyre valve and again commencing with the area furthest from the valve, ease the other bead of the tyre over the edge of the rim. Finish with the area close to the valve, pushing the valve up into the tyre until the locking ring touches the rim. This will ensure that the inner tube is not trapped when the last section of bead is edged over the rim with a tyre lever.

13   Check that the inner tube is not trapped at any point. Reinflate the inner tube, and check that the tyre is seating correctly around the wheel rim. There should be a thin rib moulded around the wall of the tyre on both sides, which should be an equal distance from the wheel rim at all points. If the tyre is unevenly located on the rim, try bouncing the wheel when the tyre is at the recommended pressure. It is probable that one of the beads has not pulled clear of the centre well.

14   Always run the tyres at the recommended pressures and never under or over inflate. The correct pressures are given in the Specifications Section of this Chapter.

15   Tyre replacement is aided by dusting the side walls, particularly in the vicinity of the beads, with a liberal coating of french chalk. Washing up liquid can also be used to good effect, but this has the disadvantage, where steel rims are used, of causing the inner surface of the wheel rim to rust.

16   Never replace the inner tube and tyre without the rim tape in position. If this precaution is overlooked there is a good chance of the ends of the spoke nipples chafing the inner tube and causing a crop of punctures.

17   Never fit a tyre that has a damaged tread or sidewalls. Apart from legal aspects, there is a very great risk of a blowout, which can have very serious consequences on a two wheeled vehicle.

18   Tyre valves rarely give trouble, but it always advisable to check whether the valve itself is leaking before removing the tyre. Do not forget to fit the dust cap, which forms an effective extra seal.

29 Valve cores and caps: tubed tyres

1   Valve cores seldom give trouble, but do not last indefinitely. Dirt under the seating will cause a puzzling 'slow-puncture'. Check that they are not leaking by applying spittle to the end of the valve and watching for air bubbles.

2   A valve cap is a safety device, and should always be fitted. Apart from keeping dirt out of the valve, it provides a second seal in case of valve failure, and may prevent an accident resulting from sudden deflation.

30 Wheel balancing

1   The front wheel should be statically balanced, complete with tyre. An out of balance wheel can produce dangerous wobbling at high speed.

2   Some tyres have a balance mark on the sidewall. This must be positioned adjacent to the valve. Even so, the wheel still requires balancing.

3   With the front wheel clear of the ground, spin the wheel several times. Each time, it will probably come to rest in the same position. Balance weights should be attached diametrically opposite the heavy spot, until the wheel will not come to rest in any set position, when spun.

4   Machines fitted with cast aluminium wheels require special balancing weights which are designed to clip onto the centre rim flange, much in the way that weights are affixed to car wheels. When fitting these weights, take care not to affix any weight nearer to the radial centre line of any spoke than is recommended. Refer to the accompanying diagram.

5   It is possible to have a wheel dynamically balanced at some dealers. This requires its removal.

6   Although the rear wheel is more tolerant to out-of-balance forces than is the front wheel, ideally this too should be balanced if a new tyre is fitted. Because of the drag of the final drive components the wheel must be removed from the machine and placed on a suitable free-running spindle before balancing takes place. Balancing can then be carried out as for the front wheel.

30.4 Correct fitting of balance weight to front wheel

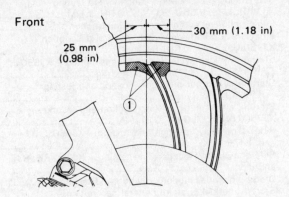

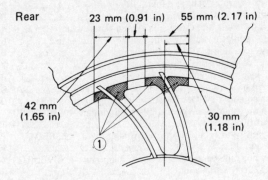

Fig. 5.21 Correct positioning of wheel balance weight

1   Do not install weight on these shaded areas

Chapter 6 Electrical system

Refer to Chapter 7 for information on the 1983 US models

Contents

Specifications

Battery

	XJ750 RH, RJ	All others
Make	Yuasa	GS
Type	YB14L-A2	12NI2A-4A
Voltage	12 volts	12 volts
Capacity	14Ah	12Ah
Earth (ground)	Negative	Negative
Charging rate	10 hours @ 1.4A	10 hours @ 1.2A

Alternator

Make	Hitachi
Type	LD119-08
Output	14V, 19A @ 5000 rpm
Field coil resistance	4.0 ohm ± 10% @ 20°C (68°F)
Stator coil resistance	0.46 ohm ± 10% @ 20°C (68°F)
Brush length	17 mm (0.67 in)
Service limit	10 mm (0.39 in)
Spring pressure	360g (12.7 oz)

Regulator/rectifier

Type	Integrated circuit
Make	Shibaura or Toshiba
Model	S8534
No-load regulated voltage	14.5 ± 0.3V
Max amperage	3A
Rectifier capacity	15A

Starter motor

Make	ND
Type	Constant mesh
Output	0.6 KW
Armature coil resistance	0.014 ohm ± 10% @ 20% (68°F)
Brush length	12 mm (0.47 in)
Service limit	8.5 mm (0.33 in)
Brush spring pressure	800 ± 150g (28.22 ± 5.29 oz)
Commutator diameter	28 mm (1.1 in)
Service limit	27 mm (1.06 in)
Mica undercut	0.6 mm (0.024 in)

Flasher relay

	XJ650(UK), XJ750(UK)	All other models
Type	Condenser	Condenser
Make	ND	ND
Flash frequency	85 ± 10 cycle/min	85 ± 10 cycle/min
Wattage	21Wx2 + 3.4W	27Wx2 + 3.4W

Bulbs

	XJ650 G, H, LH, J	XJ650 RJ	XJ650(UK)	XJ750 RH, RJ	XJ750 J	XJ750(UK)
Headlamp	50/40W	60/55W	60/55W	60/55W	60/55W	60/55W
Tail/stop lamp	8/27W (3/32cp)	8/27W (3/32cp)	5/21W	8/27W (3/32cp)	8/27W (3/32cp)	5/21Wx2
Turn indicator	27W (32cp)x4	27W (32cp)x4	21Wx4	27W (32cp)x4	27W (32cp)x4	21Wx4
License plate	8W (3cp)x2	N/A	N/A	8W (3cp)	N/A	10W
Panel lamps:						
Turn	3.4Wx2	3.4Wx2	3.4Wx2	3.4Wx2	3.4Wx2	3.4Wx2
High beam	3.4W	3.4W	3.4W	3.4W	3.4W	3.4W
Neutral	3.4W	3.4W	3.4W	3.4W	3.4W	3.4W
Oil level warning	3.4W	3.4W	3.4W	N/A	N/A	N/A
Meter illumination	3.4Wx2	3.4Wx2	3.4Wx2	3.4Wx2	3.4Wx2	3.4Wx2
Parking (auxiliary lamp)	N/A	N/A	3.4W	35W	35W	4W
CMS warning lamp	N/A	N/A	N/A	3.4W	3.4W	3.4W

Fuses

	XJ750(US)	XJ750(UK)	All 650 models
Main	30A	20A	20A
Headlamp	10A	10A	10A
Signal	10A	10A	10A
Ignition	10A	10A	10A
Tail	5A	5A	N/A

1 General description

The Yamaha XJ650 and XJ750 models are equipped with a comprehensive electrical system powered by an engine driven three-phase alternator carried on a shaft to the rear of the crankshaft. Power from the alternator is converted to dc and controlled by an electronic regulator/rectifier unit before being fed to the battery.

An electric starter is fitted to all models, there being no provision for mechanical starting. A wide range of accessories is fitted to the various models, details of which will be found under the appropriate Section heading.

2 Electrical system testing: general information

1 As already mentioned, the Yamaha XJ650/750 models feature an unusually sophisticated electrical system incorporating a number of electronic sub-assembies. These two

factors make any testing a rather exacting process which will invariably require the use of some form of test equipment. Simple continuity checks may be made using a dry battery and bulb arrangement, but for most of the tests in this Chapter a pocket multimeter can be considered essential. Many owners will already possess one of these devices, but if necessary they can be obtained from electrical specialists, mail order companies or can be purchased from a Yamaha Service Agent as a 'pocket tester', Part number 90890-03104.

2   Care must be taken when performing any electrical test, because some of the electronic assemblies can be destroyed if they are connected incorrectly or inadvertently shorted to earth. Instructions regarding meter probe connections are given for each test, and these should be read carefully to preclude any accidental damage during the test. Note that separate amp, volt and ohm meters may be used in place of the multimeter if necessary, noting that the appropriate test ranges will be required.

3   Where test equipment is not available, or the owner feels unsure of the procedure described, it is recommended that professional assistance is sought. Do not forget that a simple error can destroy a component such as the regulator/rectifier, resulting in expensive replacements being necessary.

4   A certain amount of preliminary dismantling will be necessary to gain access to the components to be tested. Normally, removal of the seat and side panels will be required, with the possible addition of the fuel tank and headlamp unit to expose the remaining components.

3   Charging system: output check

1   Set the multimeter to the 0-20 volts dc scale and connect the red positive (+) probe to the positive battery terminal and the black negative (−) probe to the negative battery terminal.
2   Start the engine, and raise the engine speed to about 2000 rpm or slightly more, noting the meter reading. The nominal charging voltage is 14.5 ± 0.3 volts, so a reading of 14.2 - 14.8 volts will indicate that the system is functioning correctly. If the voltage reading is significantly lower than that shown above, carry out the alternator resistance tests described in Section 4 of this Chapter. **Important note:** on no account should the engine be run when the battery leads are disconnected, because the resultant open voltage can destroy the rectifier diodes.

4   Alternator: rotor and stator resistance tests

1   If the charging system output has been found to be inadequate (see Section 3) it will be necessary to make a resistance check on the alternator rotor and stator windings. Trace the alternator leads back to the two-pin and three-pin connectors, noting that the two-pin connector serves the field coil windings inside the rotor, whilst the three-pin connector controls the stator connections. The test can be made with the alternator cover in position.
2   Set the multimeter to the ohms x 1 scale. Separate the two connectors and measure the resistance between the green and brown field coil leads at the two-pin connector. A reading of 4.0 ohms ± 10% at 68°F (20°C) should be obtained (3.6 - 4.4 ohms). Note that for the purpose of resistance checks, the polarity of the probe connections can be ignored.
3   Measure the resistance between the three white wires at the three-pin connector, a total of three tests. In each case, a reading of 0.46 ohms ± 10% at 68°F (20°C) should be shown (0.41 - 0.51 ohms). If any of the windings have failed, it will normally show up as a short circuit (zero resistance) or an open circuit (infinite resistance). In either case, the rotor or stator must be renewed unless the fault is caused by broken wiring connections. If the alternator appears to be functioning normally, attention can be turned to the regulator/rectifier unit as described in Section 5.

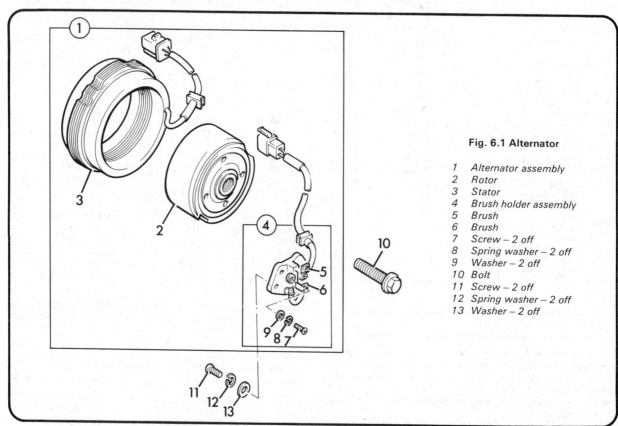

Fig. 6.1 Alternator

1   Alternator assembly
2   Rotor
3   Stator
4   Brush holder assembly
5   Brush
6   Brush
7   Screw – 2 off
8   Spring washer – 2 off
9   Washer – 2 off
10 Bolt
11 Screw – 2 off
12 Spring washer – 2 off
13 Washer – 2 off

5   Voltage regulator/rectifier unit: testing

1   The voltage regulator/rectifier unit is a small integrated circuit (IC) housed in a finned alloy casing mounted behind the left-hand side panel. Its function is to convert the alternator output to direct current (dc) from alternating current (ac), this part of the function being executed by the rectifier stage. The regulator stage monitors the drain on the electrical system and controls the alternator voltage output to suit by adjusting the effective power of the electromagnetic rotor. The unit is normally very reliable, there being no possibility of mechanical failure, but it can become damaged in the event of a short circuit in the electrical system or by poor or intermittent battery or earth connections.

Voltage regulator test

2   Remove the side panel to expose the unit, then trace the wiring back to the eight-pin connector. Note that the test is made with the connectors assembled and the electrical system intact. It is essential that the battery is fully charged during the tests because it will otherwise affect the results obtained. To this end, check that the specific gravity is at 1.260 or more, and if necessary remove the battery for recharging. For full details on the above, refer to Sections 6 and 7. Note that two multimeters are required for the regulator test, each set on the 0-20 volts dc scale.

3   Connect the meter probes to the appropriate terminals of the assembled connector as shown in Fig. 6.2. Take great care not to allow the probes to short circuit to each other or to earth during the test, ensuring that they are pushed firmly home through the back of the connector block. Note that the test should be performed with all lights and accessories off. To this end, it will be necessary to disconnect the headlamp circuit on US machines where the headlamp comes on when the ignition is switched on. This can be accomplished by removing the headlamp circuit fuse.

4   Switch on the ignition and note the reading on the V_2 meter. This should be less than 1.8 volts. Start the engine and check that the V_2 reading gradually increases to 9-11 volts as the engine speed rises. The V_1 reading should rise to 14.2-14.8 volts when the engine is started and should stabilise at this level despite variations in engine speed. The accompanying graph shows the relationship of the two voltage readings at various engine speeds. If the readings obtained are significantly outside these limits, the regulator stage must be considered defective and the unit renewed. There is no provision for adjustment or repair.

Rectifier test

5   The rectifier stage consists of an arrangement of diodes whose function can be likened to that of a one way valve. The object of the test is to ensure that each diode will pass current in one direction only. Refer to Fig. 6.3 which shows a schematic view of the rectifier connections and coded wires which will correspond with those shown in the accompanying table. Work through the test connections shown in the table, noting that if any one diode has failed, the unit must be renewed. Again, no form of repair is possible.

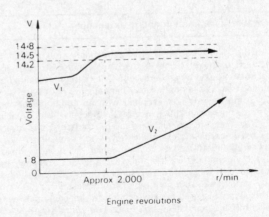

5.1 Regulator/rectifier unit location

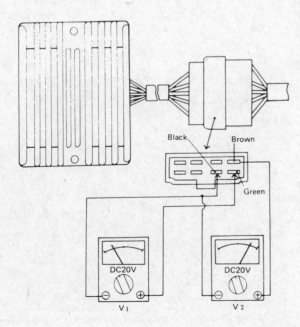

Fig. 6.2 Voltage regulator test

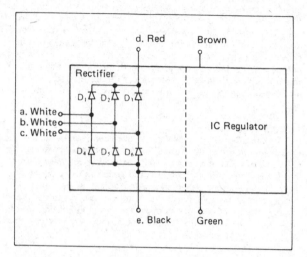

| Checking element | Pocket tester connecting point | | Good | Replace (element shorted) | Replace (element opened) |
	(+) (red)	(−) (black)			
D1	d	a	○	○	x
	a	d	x	○	x
D2	d	b	○	○	x
	b	d	x	○	x
D3	d	c	○	○	x
	c	d	x	○	x
D4	a	e	○	○	x
	e	a	x	○	x
D5	b	e	○	○	x
	e	b	x	○	x
D6	c	e	○	○	x
	e	c	x	○	x

○ : Continuity
x : Discontinuity (∞)

Fig. 6.3 Rectifier test

6 Battery: examination and maintenance

1 A GS or Yuasa battery is fitted as standard equipment. It is connected to the electrical system by a metal strap to the starter solenoid and by an earth lead. The battery is retained by a rubber strap to a mounting tray beneath the right-hand side panel. Some models are equipped with an additional battery lead from a battery condition sensor (see Section 44 for details).

2 The translucent plastic case of the battery permits the upper and lower levels of the electrolyte to be observed when the battery is lifted from its housing behind the right-hand side panel. Maintenance is normally limited to keeping the electrolyte level between the prescribed upper and lower limits and by making sure the vent pipe is not blocked.

3 Unless acid is spilt, as may occur if the machine falls over, the electrolyte should always be topped up with distilled water, to restore the correct level. If acid is spilt on any of the machine, it should be neutralised with an alkali such as washing soda and washed away with plenty of water, otherwise serious corrosion

will occur. Top up with sulphuric acid of the correct specific gravity (1.260-1.280) only when spillage has occurred. Check that the vent pipe is well clear of the frame tubes or any of the other cycle parts, for obvious reasons. Note the instructions on topping up which are fixed to the top of the battery.

4 If battery problems are experienced, the following checks will determine whether renewal is required. A battery can normally be expected to last for about 3 years, but this life can be shortened dramatically by neglect. In normal use, the capacity for storage will gradually diminish, and a point will be reached where the battery is adequate for all but the strenous task of starting the engine. It follows that renewal will be necessary at this stage, a situation best avoided where no kickstarter is available.

5 Remove the flat battery and examine the cell and plate condition near the bottom of the casing. This may prove rather difficult with the translucent-cased original battery. An accumulation of white sludge around the bottom of the cells indicates sulphation, a condition which indicates the imminent demise of the battery. Little can be done to reverse this process, but it may help to have the electrolyte drained, the battery flushed and then refilled with new electrolyte. Most electrical wholesalers have facilities for this work.

6 Warping of the plates or separators is also indicative of an expiring battery, and will often be evident in only one or two of the cells. It can often be caused by old age, but a new battery which is overcharged will show the same failure. There is no cure for the problem and the need to avoid overcharging cannot be overstressed.

7 Try charging the suspect battery as described in Section 7. If the battery fails to accept a full charge, and in particular, if one or more cells show a low hydrometer reading, the battery is in need of renewal.

8 A hydrometer will be required to check the specific gravity of the electrolyte, and thus the state of charge. Any small hydrometer will do, but avoid the very large commercial types because there will be insufficient electrolyte to provide a reading. When fully charged, each cell should read 1.280, with little discrepancy between cells.

9 *Important note* On machines equipped with computer monitor systems, a battery condition sensor is fitted in place of one of the usual cell caps. It is essential that it is fitted to the fourth cell from the negative terminal. Failure to observe this precaution may lead to damage to the microprocessor.

6.2 Battery electrolyte level is visible through translucent case

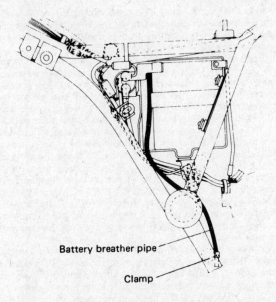

Fig. 6.4 Correct routing of battery breather pipe

7 Battery: charging procedure

1   The safe charge rate for any given battery is dependent on its capacity and will be found in the Specifications at the beginning of this Chapter. It is inadvisable to exceed this rate since it will shorten the effective life of the battery. In particular, avoid the 'quick charge' services offered by many garages. These high charge rates can damage even large car batteries and may halve the life of a motorcycle battery. In practice, most home battery chargers are not sufficiently sophisticated for the charge rate to be preset. When using a charger of this type, the initial charge rate will be rather higher than that specified, so watch the battery carefully until the rate subsides to a safe level. If the battery becomes hot, disconnect it and allow it to cool before resuming charging, otherwise the plates may become distorted.
2   Before the battery is charged, disconnect the battery leads and also the battery condition sensor, where fitted. This will preclude any chance of damage to the electrical system. Slacken the cell caps and sensor to allow gas to escape during charging. If this is not done there is a risk of gas pressure bursting the battery case. Whilst it is possible to charge the battery on the machine it is always preferable to remove it completely. Check that the electrolyte level is correct, and top up as required. When the battery is reconnected to the machine, the black lead must be connected to the negative terminal and the red lead to positive. This is most important, as the machine has a negative earth system. If the terminals are inadvertently reversed, the electrical system will be damaged permanently. The rectifier will be destroyed by a reversal of the current flow.
3   A word of caution concerning batteries. Sulphuric acid is extremely corrosive and must be handled with great respect. Do not forget that the outside of the battery is likely to retain traces of acid from previous spills, and the hands should always be washed promptly after checking the battery. Remember too that battery acid will quickly destroy clothing. In the author's experience, acid seems partial to nearly new jeans in particular, and therefore has shown that it is best to keep well clear of batteries unless old clothing is being worn. Note the following rules concerning battery maintenance.

Do not allow smoking or naked flames near batteries.
Do avoid acid contact with skin, eyes and clothing.
Do keep battery electrolyte level maintained.

Do avoid over-high charge rates.
Do avoid leaving the battery discharged.
Do avoid freezing.
Do use only distilled or demineralised water for topping up.

8 Fuses: location, function and renewal

1   A bank of four fuses is mounted in a plastic fuse box beneath the dualseat. Each fuse's function and rating is marked and a spare fuse is included. The main fuse is rated at 30 amps (XJ750, US) or 20 amps (all other models) and is designed to protect the entire system from damage in the event of a short circuit. The remaining fuses are rated at 10 amps for the protection of the headlamp, signal and ignition circuits. On XJ750 models an additional 5 amp fuse is incorporated in the tail lamp circuit, contained in an in-line holder.
2   Fuses are fitted to protect the electrical system in the event of a short circuit or sudden surge; they are, in effect, an intentional 'weak link' which will blow, in preference to the circuit burning out.
3   Before replacing a fuse that has blown, check that no obvious short circuit has occcured, otherwise the replacement fuse will blow immediately it is inserted. It is always wise to check the electrical circuit thoroughly, to trace the fault and eliminate it.
4   When a fuse blows while the machine is running and no spare is available, a 'get you home' remedy is to remove the blown fuse and wrap it in silver paper before replacing it in the fuseholder. The silver paper will restore the electrical continuity by bridging the broken fuse wire. This expedient should **never** be used if there is evidence of a short circuit or other major electrical fault, otherwise more serious damage will be caused. Replace the 'doctored' fuse at the earliest possible opportunity, to restore full circuit protection. It follows that spare fuses that are used should be replaced as soon as possible to prevent the above situation from arising.

8.1 Fuse box is located beneath the seat

9 Starter motor: removal, examination and testing

1   Starting on the XJ650 and 750 models is achieved by means of an electric starter motor mounted on the upper crankcase to the rear of the cylinder block. The system used is well proven and reliable, a desirable state of affairs since no provision is made for kick starting. In view of this it is important that the battery and the starter motor are in good condition. When the starter switch is pressed, a heavy solenoid switch is

brought into operation, switching the heavy current to the motor, cranking the engine via a primary shaft mounted roller clutch. As soon as the engine starts the clutch freewheels, disconnecting the starter drive. The clutch will rarely require any attention, a fortunate situation since a considerable amount of dismantling work would be required in the event of its failure. The starter clutch and drive are covered in Chapter 1.

2   In the event of partial or complete starter failure, check the condition of the battery, which should be fully charged, and ensure that the starter solenoid is working (See Section 10). Make sure that all switch and wiring connections are sound. If this fails to effect a cure, proceed as follows.

3   Remove the two bolts which pass through the left hand end of the starter motor, which can now be withdrawn by pulling it to the left and lifting clear of the casing recess. As the motor comes clear, release the heavy starter motor cable from its terminal.

4   Remove the two retaining screws which pass along the length of the motor, then remove the end casing which holds the reduction gearing. Prise out the motor end cover and slide it clear of the armature. The motor body and field coil windings can now be slid off, leaving the armature, end cover and brush gear. Unscrew and remove the terminal nut and washer, and ease the cover clear of the brush gear. Push the terminal into the cover so that the cover can be removed completely. The brush gear, comprising the backplate, holders, springs and brushes can now be slid off the commutator.

5   Lift up the spring clips which bear on the end of each brush and remove the brushes from their holders. Each brush should have a length of 12.0 mm (0.47 in). The minimum allowable brush length is 8.5 mm (0.33 in). If the brush is shorter it must be renewed.

6   Before the brushes are replaced, make sure that the commutator is clean. The commutator is the copper segments on which the brushes bear. Clean the commutator with a strip of glass paper. Never use emery cloth or 'wet-and-dry' as the small abrasive fragments may embed themselves in the soft copper of the commutator and cause excessive wear of the brushes. Finish off the commutator with metal polish to give a smooth surface and finally wipe the segments over with a methylated spirits soaked rag to ensure a grease free surface. Check that the mica insulators, which lie between the segments of the commutator, are undercut. The standard groove depth is 0.4-0.8 mm (0.02-0.03 in), but if the average groove depth is less than this the armature should be renewed or returned to a Yamaha dealer for re-cutting. Recutting can be undertaken at home, provided that great care is taken; new armatures are expensive. Find a broken hacksaw blade and grind the sides of the teeth so that it is the same width as the commutator

grooves, then wrap some PVC tape around the other end to form a handle. Carefully re-cut the grooves to the specified depth, taking care to keep the grooves parallel and even. Do not remove an excessive amount of material.

7   Replace the brushes in their holders and check that they slide quite freely. Make sure the brushes are replaced in their original positions because they will have worn to the profile of the commutator. Replace and tighten the end cover, then replace the starter motor and cable in the housing, tighten down and remake the electrical connection to the solenoid switch.

8   If the motor has given indications of a more serious fault, the armature and field coil winding should be checked using a multimeter set on the resistance scale. To check the armature, set the meter on the ohms x 1 scale, and measure the resistance between each pair of commutator segments. The correct figure is 0.014 ohms at 68°F (20°C). In practice, the tests will identify any dead segment. Check for armature insulation faults between each segment and the metal of the armature body. An insulation failure will require the renewal of the armature.

9   Yamaha do not provide field coil resistance figures, so a precise test of these windings is not possible. It is possible, however, to check for open or short circuits between the starter motor terminal and the brush leads. Infinite or zero resistance is indicative of a fault in the field coil windings.

9.4a Remove screws and lift away reduction gearbox

9.4b Remove end cover and motor casing

9.4c Remove terminal nut and washer (arrowed)

9.4d Pull end cover clear and displace terminal

9.4e Brush gear and support plate can now be removed

9.5a Unhook end of brush spring and displace brush ...

9.5b ... to allow it to be checked for wear

9.7a Note locating slot in brush support plate

9.7b Check and grease reduction gears before assembly

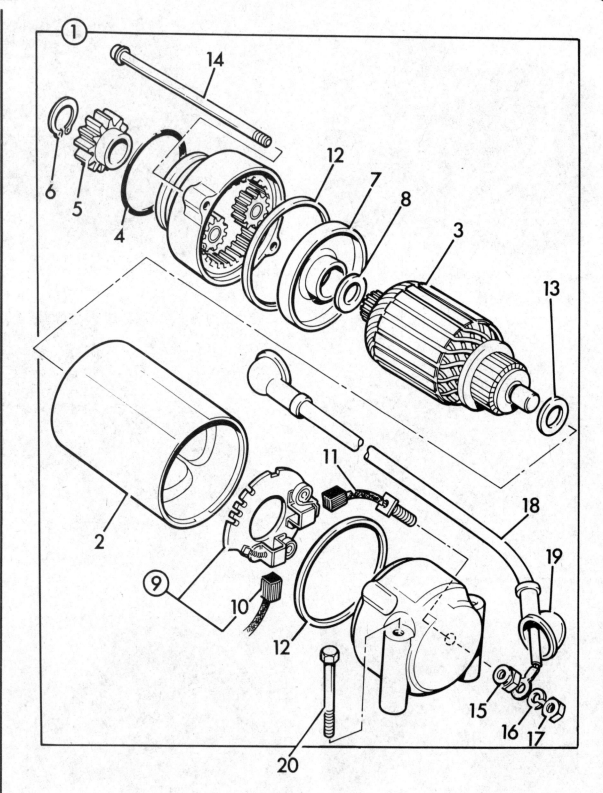

Fig. 6.5 Starter motor

1	Starter motor	6	Circlip	11	Brush	16	Spring washer
2	Stator	7	End plate	12	O-ring	17	Nut
3	Armature	8	Washer	13	Washer	18	Cable
4	O-ring	9	Brush holder	14	Bolt – 2 off	19	Boot
5	Pinion	10	Brush	15	Nut	20	Bolt – 2 off

10 Starter solenoid switch: function and location

1 The starter motor switch is designed to work on the electromagnetic principle. When the starter motor button is depressed, current from the battery passes through windings in the switch solenoid and generates an electro-magnetic force which causes a set of contact points to close. Immediately the points close the starter motor is energised and a very heavy current is drawn from the battery.

2 This arrangement is used for two reasons. Firstly, the current drawn by the starter motor is very high which requires the use of proportionally heavy cables to supply current from the battery to the motor. Running such heavy cables directly to the conveniently placed handlebar start switch would be cumbersome and impractical. Second because the demands of the starter motor are so high, as short a cable run as possible is used to minimise volt drop in the circuit. If the starter will not operate first suspect a discharged battery. This can be checked by trying the horn or switching on the lights. If this check shows the battery to be in good shape suspect the starter switch which should come into action with a pronounced click. It is located close to the battery, to which it is connected by a heavy duty metal link. Before condemning the starter solenoid, carry out the following tests.

3 To test the operation of the solenoid, disconnect the starter motor cable at the solenoid terminal and attach multimeter probe leads to each of the solenoid terminals. Set the meter on the ohms x 1 scale. Check that the machine is in neutral gear and that the kill switch is set to 'run'. Switch on the ignition and press the starter button. The solenoid should operate with an audible click and the meter needle should swing across to read zero ohms (continuity). If this is not the case the solenoid can be considered defective.

4 If the solenoid cannot be heard operating, check the battery feed wire (blue/white) and the lead from the starter button on the right-hand switch cluster. If these are undamaged and the handlebar switch works satisfactorily, check the resistance between the two wires. This will show the solenoid winding resistance which should be no more than 3.5 ohms.

11 Headlamp: bulb renewal and beam alignment – XJ650(UK) and XJ650 RJ Seca

1 Access to the headlamp bulb is gained by removing the headlamp reflector assembly from the shell. The unit is secured by two screws which pass into lugs attached to the chrome rim. Remove the screws and lift the unit clear of the shell. Pull off the headlamp bulb connector and, on UK models only, twist and pull the parking lamp bulbholder out of the reflector.

2 Place the unit face down on a workbench and peel away the rubber dust seal which protects the back of the bulb and bulbholder. Displace the end of the spring wire retainer, and swing it clear of the bulb flange. The bulb can now be lifted out of the bulb holder, noting the precautions below.

3 **Note:** On no account touch the quartz envelope of the bulb, or allow it to come into contact with oil. Any such contamination will stain or etch the surface, reducing light output and causing local overheating which may cause the envelope to fracture. Any accidental contamination must be removed promptly using alchohol or a similar de-greasing solvent. Note also that the bulb runs at a very high temperature. Always allow it to cool properly before handling.

4 Fit the bulb in the reverse of the removal sequence, being careful to handle the metal cap or shroud only. When purchasing a replacement bulb, it is worth noting that it is of the standard H4 type, in common use in the UK and Europe, and becoming more common in the US. As such, possible sources of new bulbs include auto-accessory shops and many car dealers who may be able to offer the bulb at a reduced price. Do take care to check that the new bulb is of the same construction as the original though, since this varies with different manufacturers. Some automotive types may have a reduced life when subjected to the higher vibration levels experienced on motorcycles. Recommended types are Phillips H4 12342/99 or Osram Bilux H4 64193. In the case of the latter, check that it is of the correct type since two different versions may be supplied under the same part number. The correct type has a support wire running the length of the envelope and embedded in it as shown in Fig. 6.7.

5 The headlamp can be adjusted for both vertical and horizontal alignment, the former by a lock screw which passes through a bracket on the rear of the headlamp shell, and the latter by means of the small cross-head screw in the headlamp rim.

6 In the UK, regulations stipulate that the headlamp must be arranged so that the light will not dazzle a person standing at a distance greater than 25 feet from the lamp, whose eye level is not less than 3 feet 6 inches above that plane. It is easy to approximate this setting by placing the machine 25 feet away from a wall, on a level road, and setting the dip beam height so that it is concentrated at the same height as the distance of the centre of the headlamp from the ground. The rider must be seated normally during this operation and also the pillion passenger, if one is carried regularly.

7 Most other areas have similar regulations controlling headlamp beam alignment, and these should be checked before any adjustment is made.

10.2 **A** Battery terminal **B** Starter motor lead **C** Switch lead

11.1a Headlamp assembly is retained by two screws

11.1b Release parking lamp (UK models)

11.1c Pull connector clear of bulb terminals

11.2a Disengage bulb retaining clip ...

11.2b ... and remove bulb. **Do not touch quartz envelope**

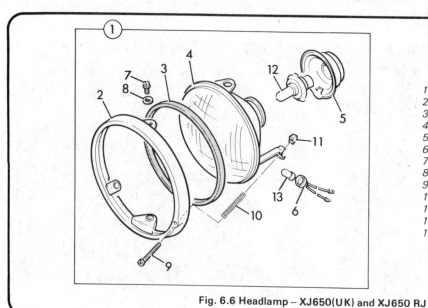

1 Headlamp assembly
2 Outer rim
3 Inner rim
4 Reflector unit
5 Rubber dust seal
6 Parking lamp bulb holder – UK only
7 Screw – 2 off
8 Spring washer – 2 off
9 Adjusting screw
10 Spring
11 Nut
12 Headlamp bulb
13 Parking lamp – UK only

Fig. 6.6 Headlamp – XJ650(UK) and XJ650 RJ

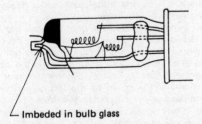

Antivibratory bulb

— Imbeded in bulb glass

Fig. 6.7 Headlamp bulb construction – XJ650(UK) and XJ650 RJ

12 Headlamp: unit renewal and beam alignment – XJ650 G, H, LH and J

1   The above models are equipped with sealed beam headlamp units. Should renewal prove necessary, start by releasing the sealed beam unit and rim assembly from the headlamp shell by removing the two screws which secure it. These are located just to the rear of the rim. Lift the assembly clear of the shell and unplug the wiring connector from the rear of the sealed beam unit. Further dismantling can now take place on a bench.

2   The sealed beam is clamped between a two-piece pressed steel carrier which pivots on two lugs on the rim, positioned at the top and bottom. A spring-loaded adjuster screw is located at 90° to the pivots. Note its position so that it can be approximated during reassembly, then remove the screw and spring noting that the captive nut into which it screws may drop free. Remove the two pivot screws and separate the rim from the carrier assembly.

3   Make a careful note of the position of the sealed beam unit within the carrier and then remove the three screws which retain the two carrier halves. Lift out the defective unit and place the new one in the same relative position. Join the carrier halves, then refit the rim and pivot screws. Fit the adjuster screw in the same position as was noted during removal, reconnect the wiring connector and install the assembly in the headlamp shell.

4   The headlamp should now be adjusted to rider preference and to comply with local laws on alignment. Horizontal alignment is set by turning the spring-loaded adjuster screw in the rim, turning it clockwise to move the beam to the right and anticlockwise to move it to the left. Vertical alignment is adjusted by slackening the screw which passes through the elongated slot in the bracket on the underside of the shell.

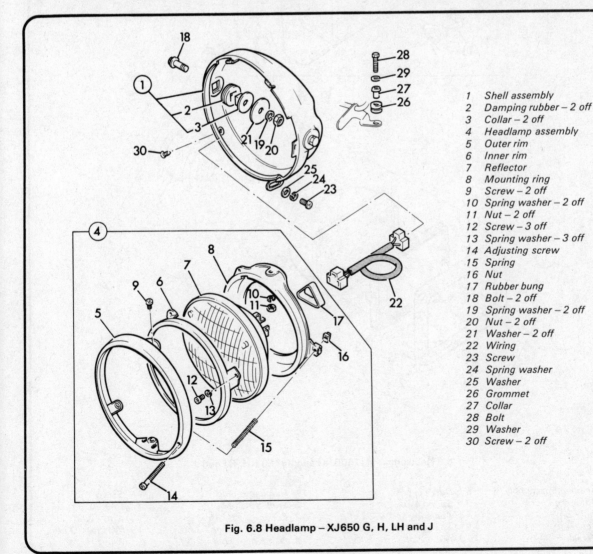

1   Shell assembly
2   Damping rubber – 2 off
3   Collar – 2 off
4   Headlamp assembly
5   Outer rim
6   Inner rim
7   Reflector
8   Mounting ring
9   Screw – 2 off
10 Spring washer – 2 off
11 Nut – 2 off
12 Screw – 3 off
13 Spring washer – 3 off
14 Adjusting screw
15 Spring
16 Nut
17 Rubber bung
18 Bolt – 2 off
19 Spring washer – 2 off
20 Nut – 2 off
21 Washer – 2 off
22 Wiring
23 Screw
24 Spring washer
25 Washer
26 Grommet
27 Collar
28 Bolt
29 Washer
30 Screw – 2 off

Fig. 6.8 Headlamp – XJ650 G, H, LH and J

13 Headlamp: bulb renewal and beam alignment – XJ750(UK), XJ750 RH, RJ and J

1   The procedure for bulb renewal and headlamp adjustment is broadly similar to that given for the models with round quartz halogen units in Section 11 of this Chapter. Where differences arise because of the rectangular unit, these are detailed below. The cautionary remarks regarding the handling and replacement of quartz halogen bulbs should be applied to the 750 cc models.
2   To remove the unit for bulb renewal, release the two holding screws which are located at the lower corners of the rim. Lift the unit away from the shell and pull off the wiring connector at the bulb terminals. Pull off the dust cover which shrouds the rear of the bulb to expose the retainer. To release the bulb, twist the retainer anticlockwise and lift it away. The bulb can now be removed, remembering not to touch the envelope; handle it by the metal parts only. Reassemble by reversing the removal sequence.
3   The unit can be adjusted for both vertical and horizontal alignment. When viewed from the front it will be noted that there are two small screws projecting from the rim edge. The upper screw controls horizontal alignment and the lower screw vertical alignment. The horizontal alignment screw is turned clockwise to move the beam to the left, whilst turning the vertical alignment screw clockwise will lower the beam. Remember to check local laws governing headlamp settings before making any adjustment. Refer to Section 11.6 for details of UK alignment regulations.

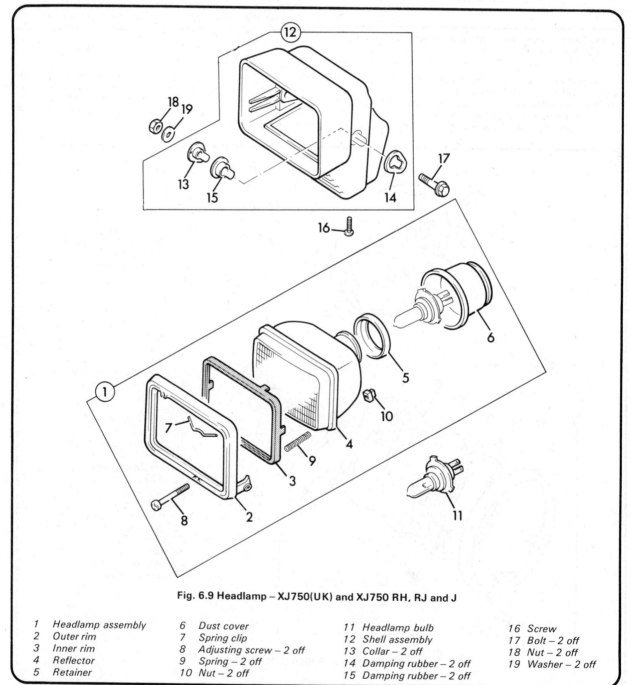

Fig. 6.9 Headlamp – XJ750(UK) and XJ750 RH, RJ and J

1 Headlamp assembly	6 Dust cover	11 Headlamp bulb	16 Screw
2 Outer rim	7 Spring clip	12 Shell assembly	17 Bolt – 2 off
3 Inner rim	8 Adjusting screw – 2 off	13 Collar – 2 off	18 Nut – 2 off
4 Reflector	9 Spring – 2 off	14 Damping rubber – 2 off	19 Washer – 2 off
5 Retainer	10 Nut – 2 off	15 Damping rubber – 2 off	

14 Auxiliary lamp: bulb renewal and beam alignment – XJ750(UK), XJ750 RH and RJ

1 The above models are equipped with an auxiliary riding lamp mounted below the main headlamp unit. It is fitted with a single-filament quartz halogen bulb rated at 35W (US models) or a 4W parking bulb (UK) and can be used as a daylight riding lamp or as a fog lamp, or simply to supplement the coverage provided by the headlamp.

2 To gain access to the bulb, first remove the two screws which retain the finisher strip to the bottom yoke. Remove the strip, then release the single screw which passes up through the lower edge of the rim to retain the unit to the shell. Lift the reflector unit away and disconnect the connecting leads at their terminals. The bulb can be removed by springing apart the ends of the wire retainer. As is the case with the headlamps of these models, note that on US models the bulb is of the quartz halogen type in which a quartz glass envelope is used to permit higher operating temperatures and, therefore, brighter light emission for a given wattage. **Never touch the glass envelope with the fingers,** because the oils and acids on the skin will contaminate the glass surface leading to the formation of hot-spots and a reduced bulb life. If contamination does occur de-natured alcohol should be used as a cleaning agent.

3 Adjustment can be made, where necessary, by slackening the lamp mounting nut and moving the entire unit to the desired position. Tighten the mounting nut to secure the adjustment.

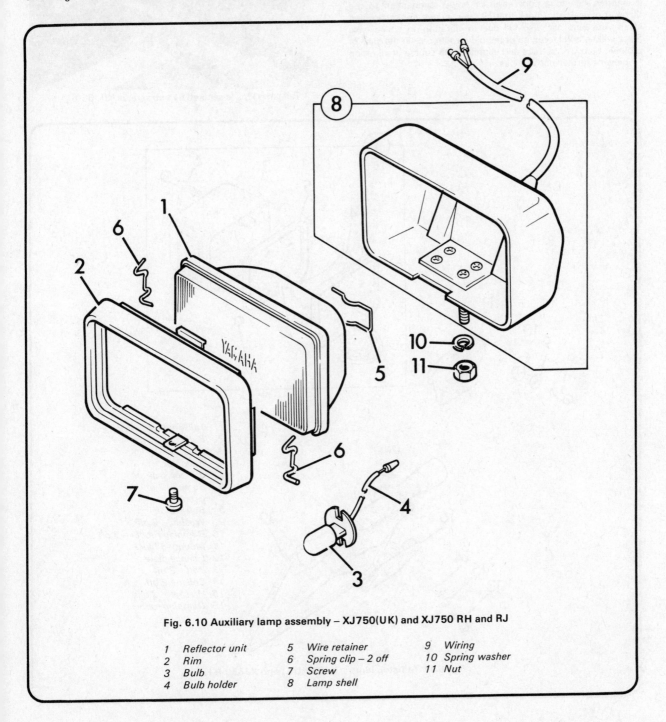

Fig. 6.10 Auxiliary lamp assembly – XJ750(UK) and XJ750 RH and RJ

1	Reflector unit	5	Wire retainer	9	Wiring
2	Rim	6	Spring clip – 2 off	10	Spring washer
3	Bulb	7	Screw	11	Nut
4	Bulb holder	8	Lamp shell		

15 Stop/tail lamp: bulb renewal – XJ650(UK) and XJ650 RJ

1   In the event of bulb failure, renewal is a straightforward matter involving the removal of the plastic lens which is retained by two screws. The bulb is a bayonet fitting, having offset pins to ensure that it is positioned correctly. It is good practice to check that both the tail and brake filaments operate properly before setting off on a journey. This should be an habitual check made whilst the engine is warmed up, particularly on single bulb models.
2   If problems with constantly blowing bulbs are experienced it is often due to a poor earth or power connection to the filament concerned. Check that all connections are secure and clean. The bulb can also fail due to vibration, in which case there will be no blackening of the old envelope as in the case of a blown bulb. Try to trace and eliminate the source of vibration to prevent further occurrences.

15.1 Tail lamp lens is secured by two screws (UK 650)

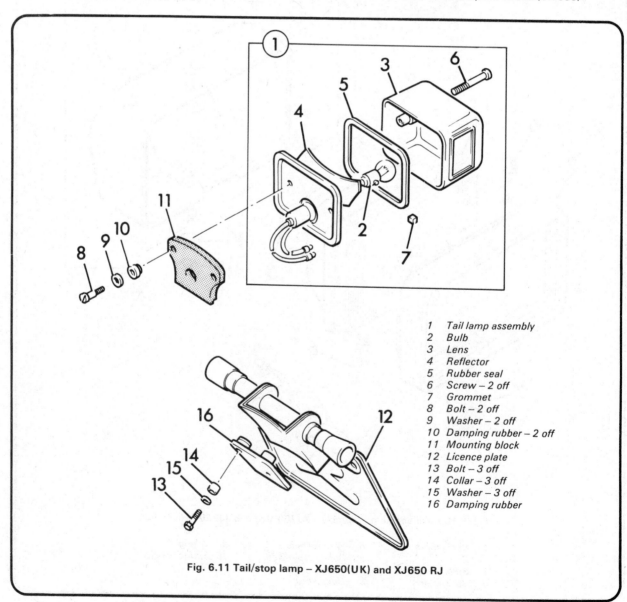

1   Tail lamp assembly
2   Bulb
3   Lens
4   Reflector
5   Rubber seal
6   Screw – 2 off
7   Grommet
8   Bolt – 2 off
9   Washer – 2 off
10 Damping rubber – 2 off
11 Mounting block
12 Licence plate
13 Bolt – 3 off
14 Collar – 3 off
15 Washer – 3 off
16 Damping rubber

Fig. 6.11 Tail/stop lamp – XJ650(UK) and XJ650 RJ

16 Stop/tail lamp: bulb renewal – XJ650 G, H, LH and J

1   The procedure for bulb renewal is generally similar to that described for the other XJ650 models in Section 15. The lens is retained by two cross-head screws and the bulb is of the offset pin bayonet fitting type. Note that the number (license) plate is illuminated by a separate lamp unit located below the main tail lamp assembly. Bulb renewal is described in Section 18 of this Chapter.

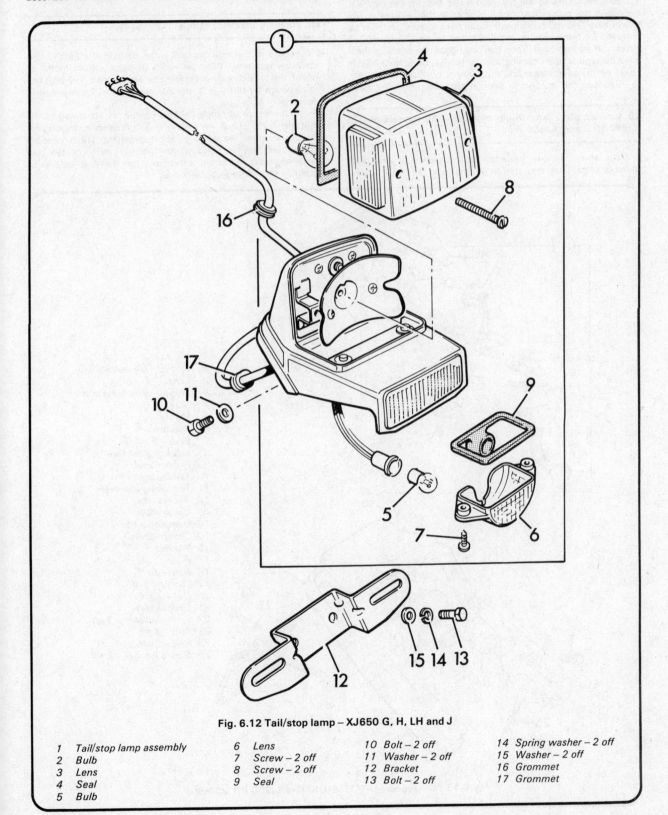

Fig. 6.12 Tail/stop lamp – XJ650 G, H, LH and J

1	Tail/stop lamp assembly	6	Lens	10	Bolt – 2 off	14	Spring washer – 2 off
2	Bulb	7	Screw – 2 off	11	Washer – 2 off	15	Washer – 2 off
3	Lens	8	Screw – 2 off	12	Bracket	16	Grommet
4	Seal	9	Seal	13	Bolt – 2 off	17	Grommet
5	Bulb						

17 Stop/tail lamp: bulb renewal – XJ750(UK), XJ750 RH, RJ and J

1   To gain access to the tail lamp bulbs, first lift the dualseat then open the access flap which is fitted in the tail fairing. The two bulbholders will be visible through the aperture, and may be removed by turning them anti-clockwise by about $\frac{1}{8}$ turn. The bulbs can be removed from the bulbholder by pushing them inwards slightly, then turning anti-clockwise. Each bulb has an offset pin bayonet fitting and thus cannot be fitted incorrectly.
2   See also the remarks in Section 15, paragraphs 2 and 3.

18 License plate lamp: bulb renewal – All models except XJ650(UK) and XJ650 RJ

1   The above models make use of a separate license plate (number plate) lamp mounted below the main tail lamp. To gain access to the bulb, remove the two self-tapping screws which retain it. The bulbholder is fitted with a single-filament bayonet cap bulb.

19 Flashing indicator lamps: bulb renewal

1   Access to the indicator bulbs can be gained after the lens has been removed. The lens varies in pattern from model to model, but in each case is retained by two screws. The bulb can be removed by pushing it inwards slightly and turning it anti-clockwise.
2   In the event of failures which cannot be attributed to the indicator bulb, check the wiring and connections, looking for corrosion, water, fractures or short circuiting. Other possible causes are malfunctions of the flasher relay or the self cancelling unit or circuit. Information on these areas will be found in the subsequent sections.

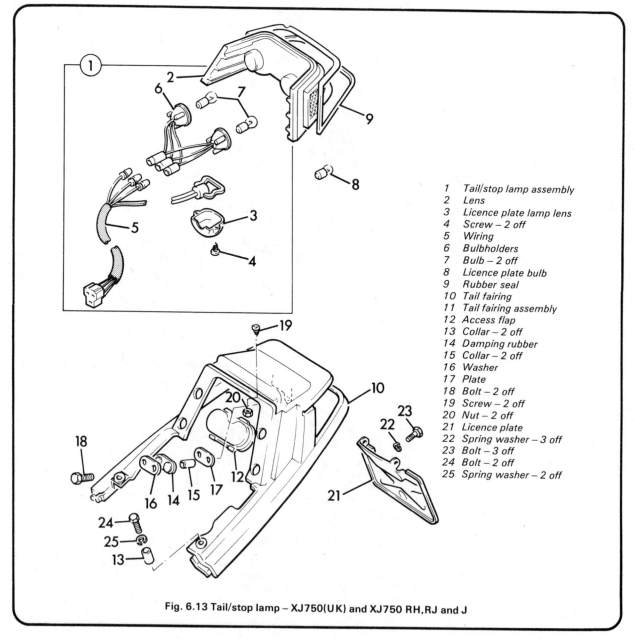

1	Tail/stop lamp assembly
2	Lens
3	Licence plate lamp lens
4	Screw – 2 off
5	Wiring
6	Bulbholders
7	Bulb – 2 off
8	Licence plate bulb
9	Rubber seal
10	Tail fairing
11	Tail fairing assembly
12	Access flap
13	Collar – 2 off
14	Damping rubber
15	Collar – 2 off
16	Washer
17	Plate
18	Bolt – 2 off
19	Screw – 2 off
20	Nut – 2 off
21	Licence plate
22	Spring washer – 3 off
23	Bolt – 3 off
24	Bolt – 2 off
25	Spring washer – 2 off

Fig. 6.13 Tail/stop lamp – XJ750(UK) and XJ750 RH,RJ and J

19.1 Indicator bulb can be changed after removing lens

20 Parking lamp: bulb renewal – UK models

1 The UK models are equipped with a parking position on the main switch which allows the machine to be parked with the tail lamp and a low wattage bulb in the headlamp illuminated. To gain access to the front parking lamp bulb, detach the headlamp as described in the appropriate headlamp bulb renewal section. The parking lamp bulbholder has a bayonet fitting in the back of the reflector. Once this has been released the bulb can be removed. This too has a bayonet fitting.

UK XJ750 model only

The parking lamp takes the form of a separate auxiliary lamp. Refer to Section 14 for details.

21 Circuit testing: general procedures

1 The following Sections describe tests on specific circuits relating to the various lighting, signalling and control systems, using flow charts or diagrams to denote the test sequence or connections. It is assumed that a multimeter will be available for these tests, though in many instances an alternative arrangement using a small dry battery in conjunction with a bulb and a pair of probe leads will suffice. The accompanying diagram shows the appropriate connections for both methods to make a continuity check. It should be noted that when checking for continuity there should be no power to that section of the circuit. This is best assured by disconnecting the machine's battery. Failure to do so may damage the multimeter or the test lamp arrangement.

2 Other tests require that battery voltage is checked on various terminals. This is easily checked using the multimeter set on a dc volts scale, usually the 0 – 20 volts position. Remember that the machine has a negative (–) earth electrical system, and thus the meter's negative, black, probe lead should always be connected to a **sound** earth point on the frame, and the red positive probe to the connection to be tested. In the absence of a meter, use a spare low wattage 12 volt bulb and bulb holder. Connect to it two leads for use as probes. Using this method polarity is of no importance. When the connection being tested has battery voltage, the lamp will light up.

Continuity test procedure
 Using multimeter
1 Check circuit is not live (disconnect battery)
2 Set meter A on resistance (ohms) scale
3 Connect probe leads B to terminals or leads under test
4 Continuity is indicated by zero resistance (zero ohms)
5 An open circuit is indicated by an infinite resistance reading

 Using dry battery and bulb arrangement
1 Connect dry battery C and bulb D as shown
2 Check that circuit is not live (disconnect battery)
3 Connect probe leads E to terminals or leads under test
4 Bulb will light if continuity exists, but will remain off with open circuit

Voltage test procedure
 Using multimeter
1 Set meter F to 0-20 volts dc scale
2 Connect negative (–) probe G to earth
3 Connect positive (+) probe H to terminal or lead to be tested
4 Check that system is switched on
5 Battery voltage should be indicated on meter

 Using low wattage 12v bulb and probe leads
1 Assemble test lamp as shown below
2 Connect one lead to earth and the other to terminal or lead to be tested
3 Check that system is switched on
4 Bulb will illuminate to indicate battery voltage

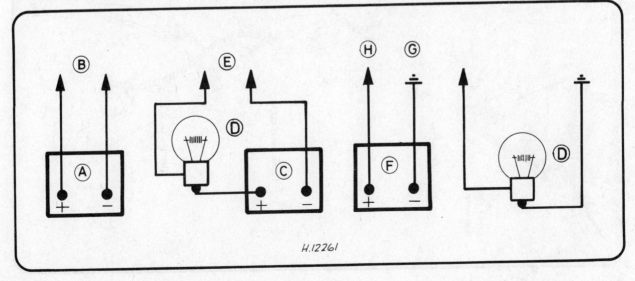

H.12261

232

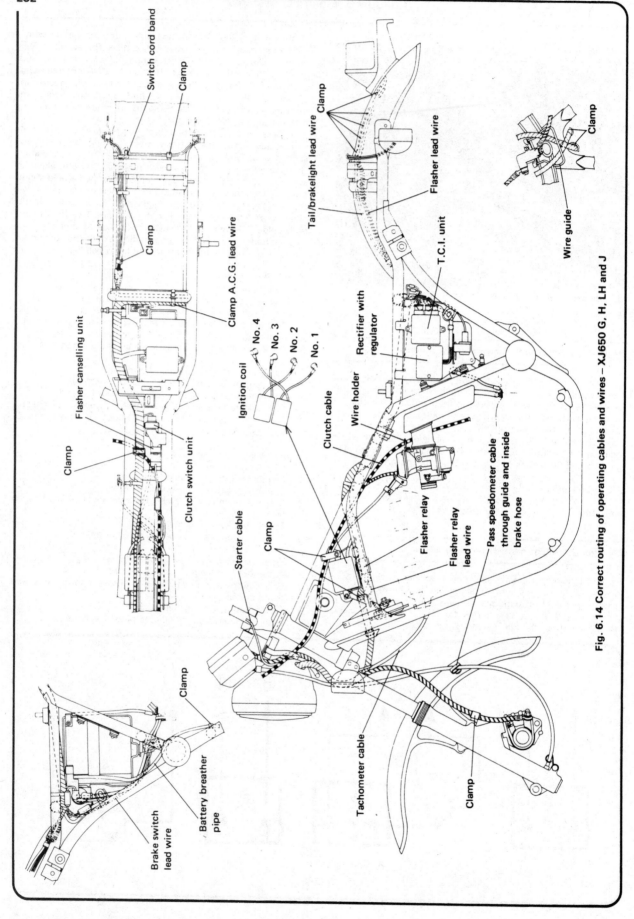

Fig. 6.14 Correct routing of operating cables and wires – XJ650 G, H, LH and J

Switch cord band

Clamp

Clamp

Clamp A.C.G. lead wire

Flasher cancelling unit

Clamp

Clutch switch unit

Tail/brakelight lead wire

Clamp

Flasher lead wire

T.C.I. unit

Wire guide

Clamp

Ignition coil

No. 4

No. 3

No. 2

No. 1

Rectifier with regulator

Clutch cable

Wire holder

Starter cable

Clamp

Flasher relay

Flasher relay lead wire

Pass speedometer cable through guide and inside brake hose

Tachometer cable

Clamp

Clamp

Battery breather pipe

Brake switch lead wire

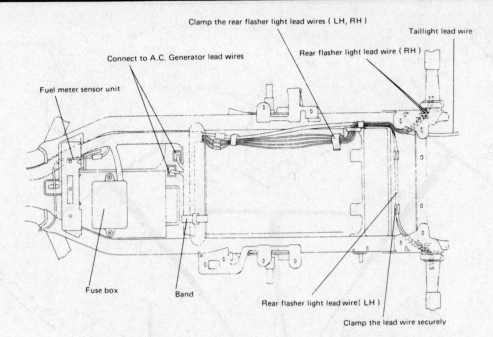

Clamp the rear flasher light lead wires (LH, RH)

Connect to A.C. Generator lead wires

Taillight lead wire

Rear flasher light lead wire (RH)

Fuel meter sensor unit

Fuse box

Band

Rear flasher light lead wire (LH)

Clamp the lead wire securely

Fig. 6.15 Frame top tubes component wiring – XJ750 RH and RJ

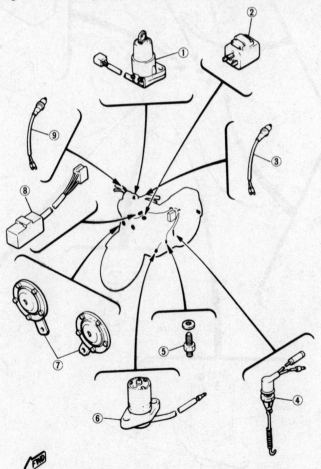

Fig. 6.16 Location of electrical components – XJ650(UK)

1	Ignition switch	4	Rear stoplamp switch	6	Oil level switch	8	Flasher cancelling unit
2	Flasher relay	5	Neutral switch	7	Horns	9	Front stoplamp switch
3	Clutch switch						

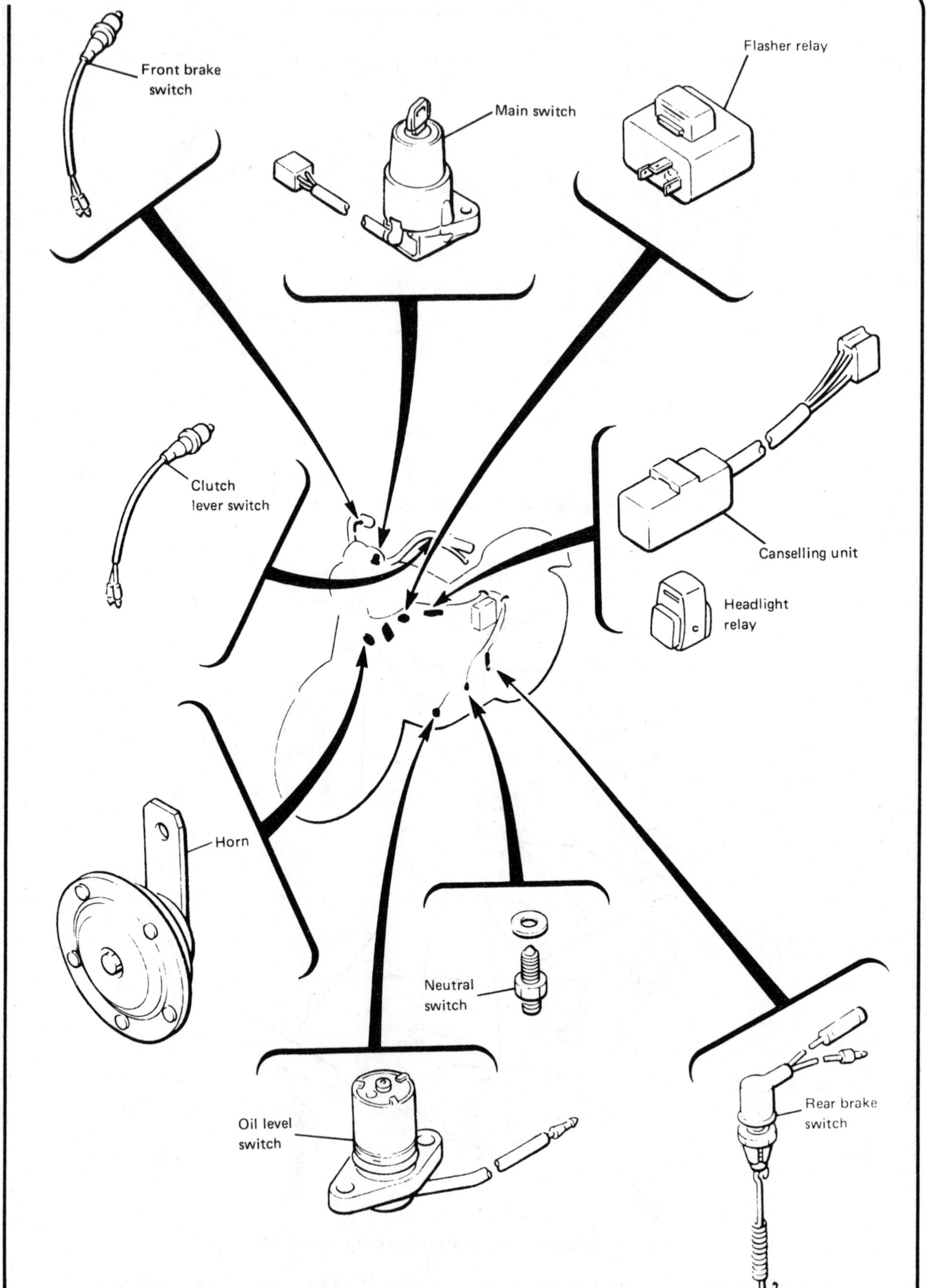

Front brake
switch

Main switch

Flasher relay

Clutch
lever switch

Canselling unit

Headlight
relay

Horn

Neutral
switch

Oil level
switch

Rear brake
switch

Fig. 6.17 Location of electrical components – XJ650 G, H, LH and J

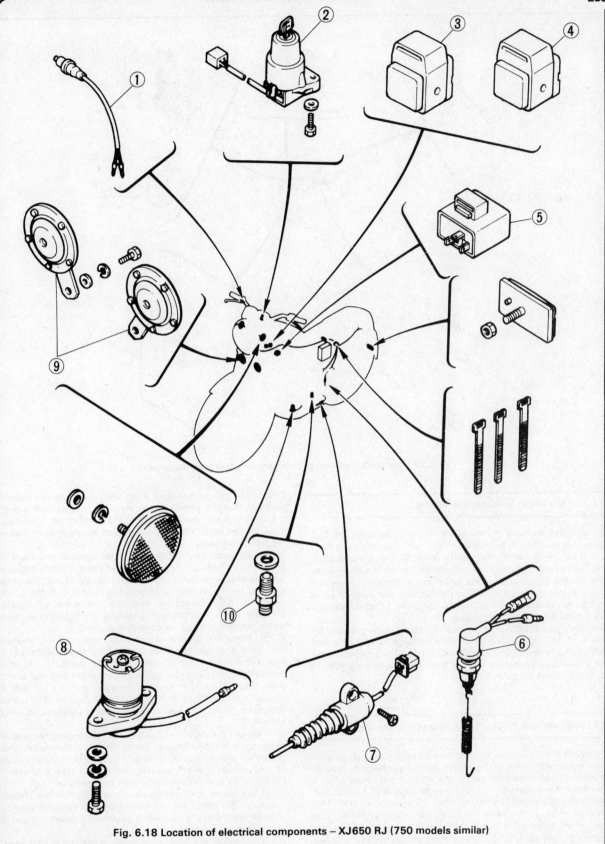

Fig. 6.18 Location of electrical components – XJ650 RJ (750 models similar)

1	Front brake switch	4	Side-stand relay	7	Side-stand switch	9	Horns
2	Main switch	5	Flasher relay	8	Oil level switch	10	Neutral switch
3	Headlamp relay	6	Rear brake switch				

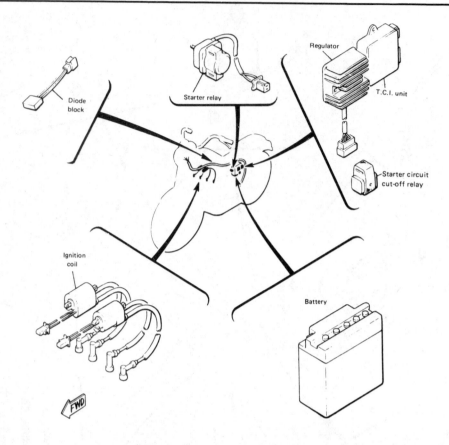

Fig. 6.19 Location of ignition components – general arrangement

22 Flasher relay: location and testing

1 The flasher relay is located on the frame beneath the fuel tank. It is retained by a moulded rubber anti-vibration mounting.
2 If the flasher unit is functioning correctly, a series of audible clicks will be heard when the indicator lamps are in action. If the unit malfunctions and all the bulbs are in working order, the usual sympton is one initial flash before the unit goes dead; it will be necessary to replace the unit complete if the fault cannot be attributed to any other cause.
3 The indicator circuit can be tested using a multimeter set on the 0-20 volts dc scale or with a low wattage 12 volt bulb and test leads, as described in Section 21. Follow the accompanying flow chart in conjunction with the appropriate wiring diagram at the end of this Chapter.

23 Self-cancelling circuit: description and testing

1 With the exception of machines sold in Germany all models are equipped with an ingenious electromechanical system which will automatically cancel the flashing indicators after a predetermined distance and/or time has elapsed. In practice, the indicators should switch off after 10 seconds or after 150 metres (164 yards) have been covered. Both systems must switch off before the indicators stop, thus at low speeds the sytem is controlled by distance, whilst at high speeds, elapsed time is the controlling factor.
2 A speedometer sensor measures the distance covered from the moment that the switch is operated. After the 150 metres have been covered, this part of the system will reset to off. The

flasher cancelling unit starts a ten second countdown from the moment that the switch is operated. As soon as both sides of the system are at the off position, the flashers are cancelled. If required, the system may be overridden manually by depressing the switch inwards.
3 In the event of malfunction the self-cancelling system can be tested as follows. The self-cancelling unit is located beneath the dualseat, next to the air intake trunking. Trace the output leads to the 6-pin connector and disconnect it. If the ignition switch is now turned on and the indicators will operate normally, albeit with manual cancelling, the flasher relay, bulbs, wiring and switch can be considered sound.
4 To check the speedometer sensor, connect a multimeter to the white/green and the black leads of the wiring harness at the 6-pin connector. Set the meter to the ohms x 100 scale, Release the speedometer cable at the wheel end and use the projecting cable end to turn the speedometer. If all is well, the needle will alternate between zero resistance and infinite resistance. If not, the sender or the wiring connections will be at fault.
5 Connect the meter probes between the yellow/red lead and earth, again on the harness side of the 6-pin connector. Check the switch and associated wiring by turning the indicator switch on and off. In the off position, infinite resistance should be shown, with zero resistance in both on positions.
6 If the above tests reveal no obvious fault in the indicator circuits or the indicators will work only as a manually operated arrangement, the self cancelling system is inoperative and the unit will require renewal.
7 If the indicators operate normally when the switch lever is first operated, but stop as soon as it is released and allowed to return to the centre position, the unit can again be considered faulty. As a temporary expedient, leave the unit disconnected and operate the indicators manually.

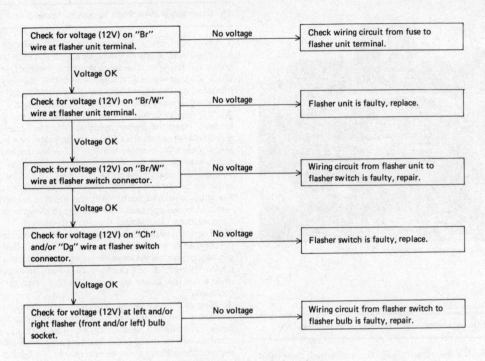

| Check for voltage (12V) on "Br" wire at flasher unit terminal. | No voltage | Check wiring circuit from fuse to flasher unit terminal. |

Voltage OK

| Check for voltage (12V) on "Br/W" wire at flasher unit terminal. | No voltage | Flasher unit is faulty, replace. |

Voltage OK

| Check for voltage (12V) on "Br/W" wire at flasher switch connector. | No voltage | Wiring circuit from flasher unit to flasher switch is faulty, repair. |

Voltage OK

| Check for voltage (12V) on "Ch" and/or "Dg" wire at flasher switch connector. | No voltage | Flasher switch is faulty, replace. |

Voltage OK

| Check for voltage (12V) at left and/or right flasher (front and/or left) bulb socket. | No voltage | Wiring circuit from flasher switch to flasher bulb is faulty, repair. |

Fig. 6.20 Flasher relay testing chart

24 Handlebar switches: maintenance and testing

1   Generally speaking, the switches give little trouble, but if necessary they can be dismantled by separating the halves which form a split clamp around the handlebars. Note that the machine cannot be started until the ignition cut-out on the right-hand end of the handlebars is turned to the central 'Run' position.

2   Always disconnect the battery before removing any of the switches, to prevent the possibility of a short circuit. Most troubles are caused by dirty contacts, but in the event of the breakage of some internal part, it will be necessary to renew the complete switch.

3   Because the internal components of each switch are very small, and therefore difficult to dismantle and reassemble, it is suggested a special electrical contact cleaner be used to clean corroded contacts. This can be sprayed into each switch, without the need for dismantling.

4   To test the operation of the switch contacts reference should be made to the appropriate wiring diagram. Each switch is shown in diagrammatic form, the contacts being shown as circles with connecting bars indicating which contacts are connected in any one switch position. Trace the wiring back to its block connector and separate it. Connect a multimeter set on the resistance scale to each pair of terminals in turn and note the readings when the switch is operated. Continuity (zero resistance) should be shown when the contacts are closed and isolation (infinite resistance) when they are open.

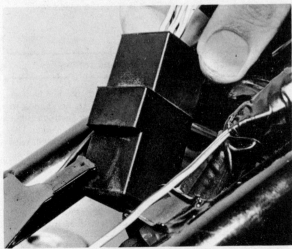

23.3 Self-cancelling unit is mounted beneath tank

24.1a LH handlebar switch assembly (UK 650)

24.1b RH handlebar switch assembly (UK 650)

25 Main switch: maintenance, testing and renewal

1   The main switch is incorporated in the instrument panel and serves to control the ignition and lighting functions. It is key operated in the interests of security and also incorporates a steering lock mechanism. Access to the switch requires a considerable amount of dismantling work and maintenance is confined to keeping it clean and occasional lubrication of the mechanism with WD40 or a similar aerosol maintenance fluid.
2   The switch functions can be checked with the switch assembly in situ, following the same procedure as described for the handlebar switches (Section 25, paragraph 4). If the switch proves to be defective or the steering lock malfunctions, renewal will be necessary. The switch can be released by removing the retaining bolts once the instrument panel has been detached to provide access. Refer to Sections 43 or 47 for further details.

25.2 Access to main switch is gained after removing the instrument panel

26 Stop lamp switches: location, testing and adjustment

1   The front brake lever and the rear brake pedal are each equipped with a switch which operates the stop lamp when the control is actuated. The front switch is incorporated in the lever stock and is non-adjustable, whilst the rear brake switch is mounted on the frame and can be adjusted to compensate for brake adjustment and changes of pedal height.
2   If either switch is suspected of failure it can be checked as follows. Trace the switch leads back to their connectors. Using a multimeter set on the 0-20 volts dc scale, or a bulb and probe leads, check for battery voltage between the brown lead and earth, noting that the ignition switch should be on. If battery voltage is indicated the supply can be taken as being in working order as far as the switch. Check for battery voltage between the grey wire and earth with the ignition switch on and the relevant brake applied. If no reading is shown the switch must be considered faulty and should be renewed.
3   The front brake switch is incorporated in the lever stock and is retained by a spring loaded pin. It can be pulled out once the pin has been pushed inwards using an electrical screwdriver or similar tool. It is not possible to adjust or repair a defective switch in the event of a malfunction, and renewal will be required.
4   The rear brake switch is mounted on the right-hand side of the frame and is operated via a short tension spring. The switch body is threaded and may be adjusted up or down to compensate for brake or brake pedal adjustment. Check that the switch operates as soon as the rear brake begins to operate.

27 Neutral switch: location, function and testing

1   The neutral switch is located on the underside of the crankcase, immediately to the left of the sump. Apart from indicating when the gearbox is in neutral, the switch forms part of the starter interlock system described in Sections 39 and 40. It follows that the switch must function properly if the engine is to be started.
2   If the switch is suspected of being faulty it can be checked using a multimeter set on the resistance scale or a battery and bulb arrangement as described in Section 21. Disconnect the neutral switch lead and connect one probe lead to the switch terminal and the other to earth. Check that when the gearbox is in neutral, continuity is indicated. If this is not the case it will be necessary to renew the switch, noting that the engine/transmission oil should be drained before it is unscrewed.

28 Clutch switch: location, function and testing

A small plunger type switch is fitted to the clutch lever stock as part of the starter interlock system. Its purpose is to prevent accidental starting of the engine when anything other than neutral is selected, unless the clutch lever is pulled in. The switch is similar to the front brake switch and can be dealt with in the same way. See Section 26 for details.

29 Side stand switch: location, function and testing – XJ650 J and RJ, XJ750(UK), XJ750 RH, J and RJ

1   A side stand switch is fitted to the above models, forming part of the starter interlock system. If the gearbox is in

anything other than neutral, the clutch must be pulled in **and** the side stand retracted before the machine can be ridden. In addition, the engine will be cut off if any gear is selected when the stand is down.

2 The switch unit is attached to the frame close to the side stand pivot. To test the operation of the switch, trace its wiring back to the connector behind the right-hand side panel. Disconnect the wiring connector and test for continuity be-

tween the two switch leads using a multimeter set on the resistance scale. When the stand is down the switch should be off, and a reading of infinite resistance should be shown. With the stand retracted, zero resistance should be indicated. If the switch proves defective it will be necessary to fit a replacement unit. As an emergency measure, a defective switch can be bypassed by connecting the blue/yellow lead to the switch directly to earth.

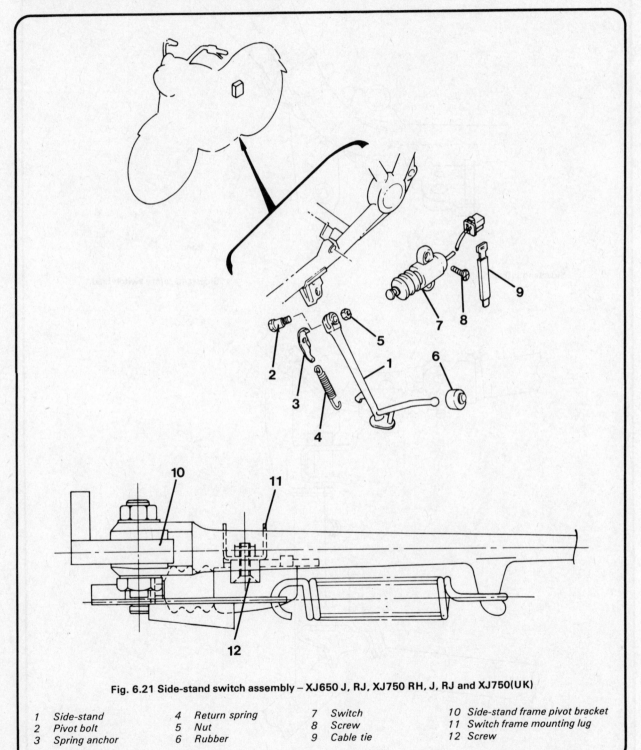

Fig. 6.21 Side-stand switch assembly – XJ650 J, RJ, XJ750 RH, J, RJ and XJ750(UK)

1	Side-stand	4	Return spring	7	Switch	10	Side-stand frame pivot bracket
2	Pivot bolt	5	Nut	8	Screw	11	Switch frame mounting lug
3	Spring anchor	6	Rubber	9	Cable tie	12	Screw

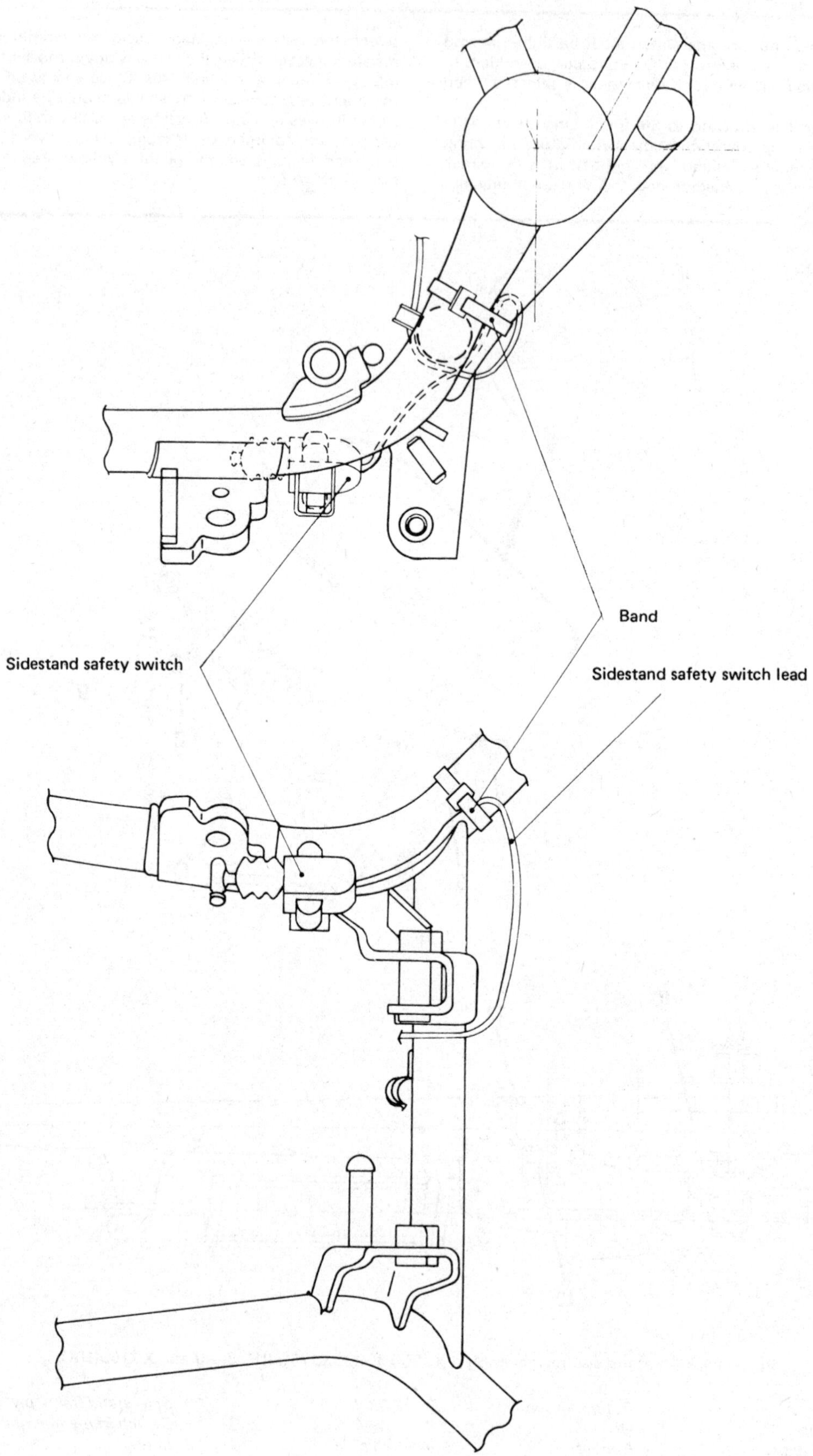

Band

Sidestand safety switch

Sidestand safety switch lead

Fig. 6.22 Side-stand switch cable routing

30 Side stand relay: location, function and testing – XJ650 J and RJ, XJ750(UK), XJ750 RH, J and RJ

The above models are equipped with a side stand relay which operates in conjunction with the side stand switch as part of the starter interlock system. When the ignition is switched on, the TCI (ignition) unit is kept earthed unless the side stand relay is on. This immobilises the engine unless the circuit is bypassed by selecting neutral.

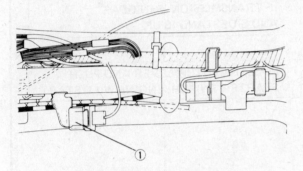

Fig. 6.23 Side-stand relay location – XJ650 J, RJ, XJ750 RH, J, RJ and XJ750(UK)

31.2a Oil level switch bolts to underside of sump

31 Engine oil level switch: general information

1   A float-type switch is incorporated in the engine sump to warn the rider when the engine oil level has reached a dangerously low level. Check that the switch is functioning properly each time the engine oil is drained, and if necessary remove and renew the switch when the sump is empty. It is worth noting that problems with the lamp or LCD panel coming on repeatedly can be reduced or eliminated by keeping the oil level at the highest mark on the side casing window.
2   The switch is secured to the sump by two bolts which pass through the flange in the switch body. After removal of the bolts, the switch can be eased from position and the single electrical lead disconnected. Note that the sump must be drained before the switch is removed, for obvious reasons.

31.2b Switch is sealed unit. Check O-ring condition

32 Front brake fluid level switch: general description – XJ750(UK), XJ750 RH, RJ and J

1   The front brake master cylinder incorporates in its fluid reservoir a small level switch. It is connected to the CMS unit and operates the warning lamp and the appropriate LCD panel if the fluid level falls to a dangerously low level. In the unlikely event of the switch failing it can be removed after the fluid reservoir has been drained. The switch is retained by a circlip fitted on the underside of the reservoir and is sealed by an O-ring. When fitting a new switch check that a new O-ring is fitted and that it is seated correctly before securing the switch to the reservoir. Switch on the ignition and check that the low level warning operates, then top up the reservoir and reset the CMS unit to make sure that the switch is off. Do not forget that it will be necessary to bleed the hydraulic system as described in Chapter 5. For further information refer to Chapter 5, Section 14.

33 Starter interlock system: general description – XJ650 J and RJ, XJ750(UK), XJ750 RH, J and RJ

1   The above models employ a sophisticated starter interlock system designed to prevent the starting or running of the engine unless certain conditions are met. The flow chart shown in Fig. 6.24. indicates the correct sequences which will permit starting. Note also that if the side stand is left down when any gear is engaged the ignition circuit will be cut off, stopping the engine. This prevents any risk of accidents due to the machine being ridden away with the stand left down.
2   The starter circuit is shown in Figs. 6.25, 6.26 and 6.27. When the ignition is switched on and the engine stop, or 'kill', switch is turned to the run position, power is fed to one terminal of the starting circuit cut-off relay. This relay will remain off, and thus no power can reach the starter relay, or solenoid, unless neutral is selected, or the clutch lever is pulled in **and** the side stand retracted. Either one of these combinations will earth the windings of the starting circuit cut-off relay causing the contacts to close and allowing power to be fed via the red/white

wire to the starter relay. If the starter switch is then pressed the engine can be started. The wiring details vary from model to model, as can be seen in the accompanying diagrams, but the basic principle remains similar throughout.

3 A second relay is connected between the side stand relay and the TCI unit. When the side stand is down, the side stand relay contacts remain closed, and the TCI unit is earthed. In practice, the engine can be started but not ridden until the stand has been retracted, since any attempt to engage a gear will kill the ignition. Fig. 6.28 shows the side stand relay circuit. The side stand relay is not fitted to XJ750(UK) models, where the switching is done direct, by the side stand switch.

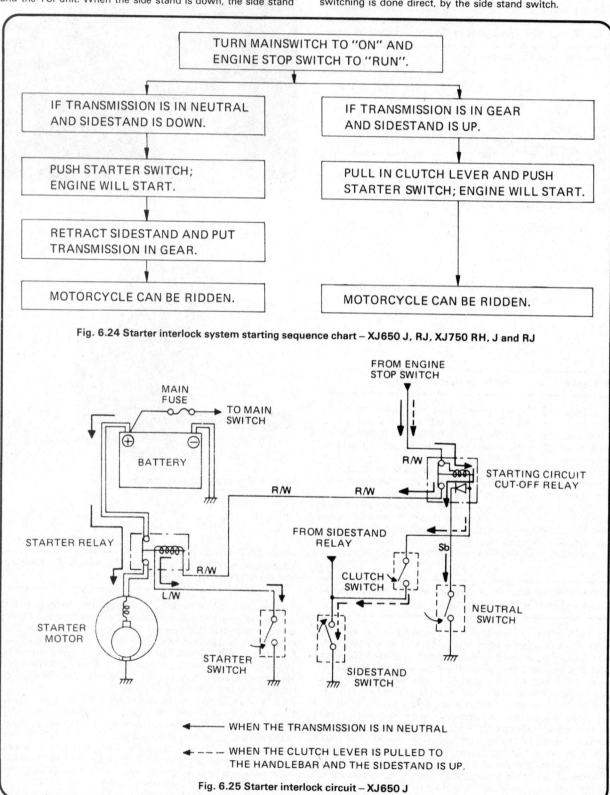

Fig. 6.24 Starter interlock system starting sequence chart – XJ650 J, RJ, XJ750 RH, J and RJ

Fig. 6.25 Starter interlock circuit – XJ650 J

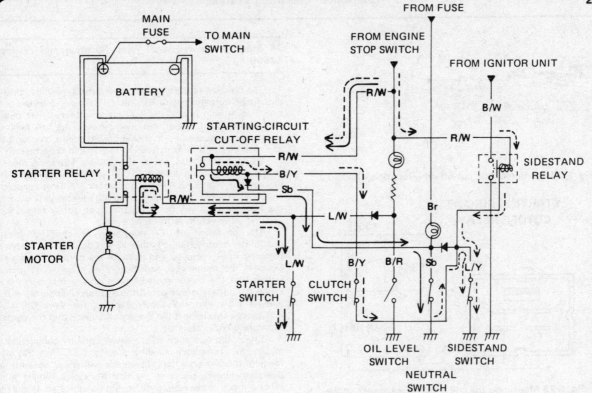

Fig. 6.26 Starter interlock circuit – XJ650 RJ

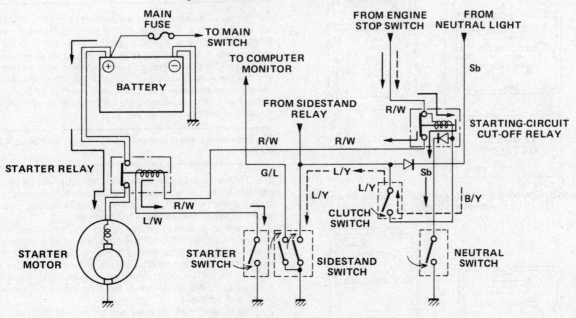

Fig. 6.27 Starter interlock circuit – XJ750 RH, J and RJ (XJ750(UK) similar)

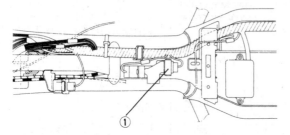

Fig. 6.28 Starting circuit cut-off relay location – XJ750 J

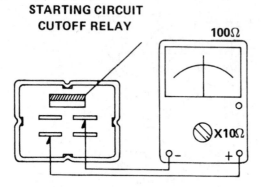

Fig. 6.29 Measuring the coil winding resistance of the starting circuit cut-off relay – XJ650 J, RJ, XJ750(UK) and XJ750 J

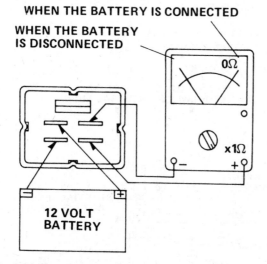

Fig. 6.30 Checking the operation of the starting circuit cut-off relay – XJ650 J

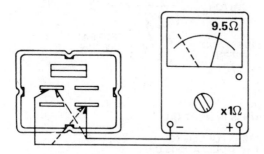

Fig. 6.31 Checking the operation of the starting circuit cut-off relay diode

34 Starting circuit cut-off relay: location and testing – XJ650 J

1   The function of the starting circuit cut-off relay is to prevent the engine from starting unless neutral has been selected or the clutch lever held in and the side stand retracted. The relay is held in a resilient rubber mounting beneath the fuel tank. If a faulty relay is suspected it should be removed from the frame and disconnected from the wiring to allow it to be tested.
2   Using a multimeter set on the Ohms x 10 range, measure the relay coil winding resistance. Connect the meter probe leads as shown in Fig. 6.30. A reading of about 100 Ohms should be indicated. If a reading of zero or infinite resistance is obtained, the coil windings will be shorted or open circuit respectively, and the relay will have to be renewed.
3   The operation of the relay can be checked using a multimeter in conjunction with a 12 volt battery. The machine's battery can be removed and used for the purposes of the test. Connect the multimeter and battery as shown in Fig. 6.31, noting that the meter should be set on the resistance scale. When the battery is connected, the relay should operate with an audible click and the meter should read zero Ohms (no resistance). Disconnect the battery and check that the contacts separate (infinite resistance).
4   Check the operation of the relay diode by connecting the multimeter probe leads as shown in Fig. 6.32. Note the meter reading, then reverse the probe leads and check again. In one direction, infinite resistance should be shown, whilst in the other a much lower resistance should be indicated. In the case of the Yamaha tester this will be about 9.5 Ohms, but may vary somewhat when other types of multimeter are used. The precise figure is not important, but make sure that the diode produces very high and very low resistances depending upon the probe connections.

35 Starting circuit cut-off relay: location and testing – XJ750(UK), XJ750 RH, J and RJ

1   The procedure for testing the starting circuit cut-off relay on the above models is essentially the same as that described in Section 34 for the XJ650 J. Note, however, that the test connections differ, and should be connected as shown in Figs. 6.33 and 6.34 below. The relay winding resistance connections are the same as those shown in Fig. 6.29. When checking the operation of the relay and diode, refer to Fig. 6.33 and 6.34 respectively. The readings obtained should be the same as those described in Section 34.

36 Side stand relay: location and testing – XJ650 J, RJ and XJ750 (UK)

1   The side stand relay is mounted beneath the fuel tank in a resilient rubber mounting. It is positioned just forward of the tank holding bolt and can be unplugged for testing once the tank has been removed from the machine.
2   The test procedure is similar to that described for the starting circuit cut-off relay. Set a multimeter to the Ohms x 10 scale and connect it as shown in Fig. 6.36. The meter should indicate the relay coil winding resistance which should be about 100 Ohms. If an open (infinite resistance) or short (zero resistance) circuit is indicated, the unit should be considered faulty and renewed.
3   The operation of the relay and its internal contacts can be checked by using a multimeter set on the Ohms x 1 scale together with the machine's battery. The two are connected as shown in Fig. 6.37. When the battery is connected the contacts should open with an audible click, and a reading of infinite resistance should be shown on the meter. When the battery is disconnected the contacts should close showing zero resistance.

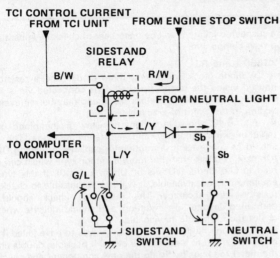

Fig. 6.32 Side-stand relay circuit – XJ750 RH, RJ and J

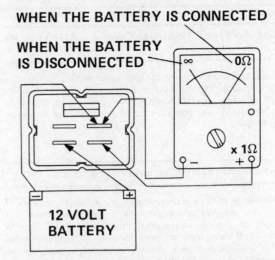

Fig. 6.33 Checking the operation of the starting circuit cut-off relay – XJ750(UK) and XJ750 RH, J and RJ

STARTING-CIRCUIT CUT-OFF RELAY

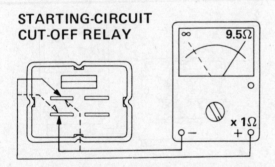

Fig. 6.34 Checking the operation of the starting circuit cut-off relay diode – XJ750(UK) and XJ750 RH, J and RJ

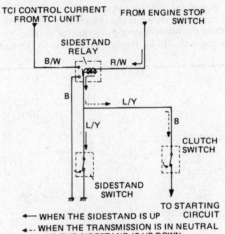

Fig. 6.35 Side-stand relay circuit – XJ650 J and RJ

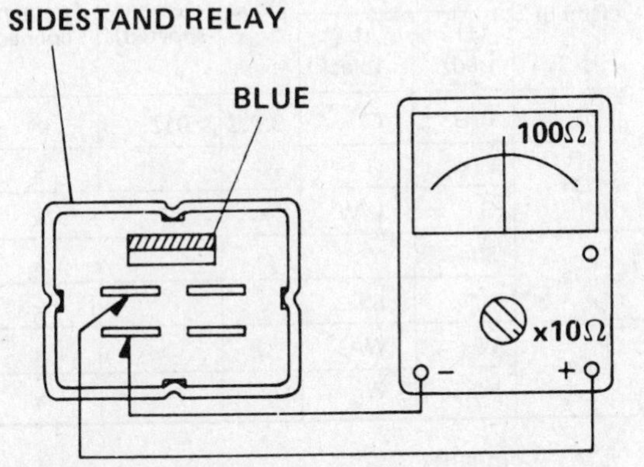

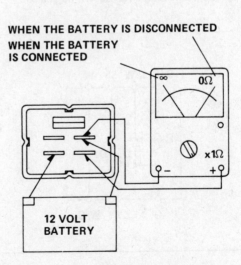

Fig. 6.36 Measuring the coil winding resistance of the side-stand relay – XJ650 J and RJ

Fig. 6.37 Checking the operation of the side-stand relay – XJ650 J and RJ

37 Side stand relay: location and testing – XJ750 RH, RJ and J

The procedure for testing the side stand relay is identical to that described for the XJ650 J in Section 36. Note, however, that the winding resistance test connections are as shown in Fig. 6.38 and the relay operation test connections are as shown in Fig. 6.39.

38 Diode block: description, location and testing

1   The design of the electrical system is such that a number of the circuits are interconnected in a manner that makes it necessary to prevent one earthing through another. This is accomplished using one or more diodes, depending on the model. The diode acts as a one way valve, allowing current to pass easily in one direction, but presenting a very high resistance in the other. In most examples, a resistor is incorporated to act as a load on one circuit.

XJ650 G, H and LH, XJ750 RH and RJ
2   These models employ a diode block using two diodes and a resistor. (See the relevant wiring diagram for details). Yamaha do not provide test data for this component, so in the event of a suspected failure it will be necessary to check by substituting a new component.

XJ650 J and RJ
3   A diode block comprising three diodes and a resistor is housed inside the headlamp shell. It is connected via a short length of cable to a connector block and may be removed for testing in the event of a suspected fault.
4   Fig. 6.40 shows a schematic view of the diode block together with continuity and resistance readings. The test should be made using a multimeter capable of reading Ohms (Ohms x 1) rather than the more common type which can only read in kilo Ohms (kOhms or Ohms x 1000). If this type of equipment is not available, it will not be possible to check the resistor with accuracy. The resistance check should be performed despite this limitation, and should indicate whether or not the resistor is serviceable.
5   If any one diode or the resistor proves to have failed it will be necessary to fit a new diode block. If possible, double check the meter readings by fitting the new component and checking that this cures the problem.

XJ750 J and all UK models
6   No test data is provided for these models, but given that a single diode is used it should not prove difficult to test. Trace and remove the diode block, following the appropriate wiring diagram. Using a multimeter set on the Ohms (resistance) scale, check that the diode shows zero resistance in one direction and infinite resistance when the probe leads are reversed.

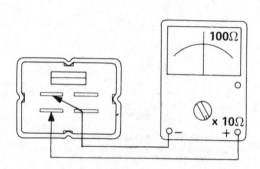

Fig. 6.38 Measuring the coil winding resistance of the side-stand relay – XJ750 RH, RJ and J

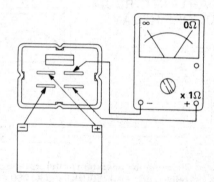

Fig. 6.39 Checking the operation of the side-stand relay – XJ750 RH, RJ and J

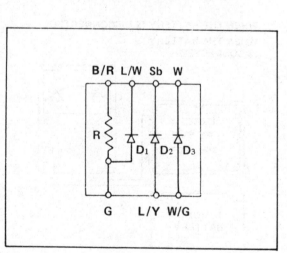

Checking element	Pocket tester connecting point		Good	Replace (element shorted)	Replace (element opened)
	(+) (red)	(-) (black)			
R	B/R	G	8.2Ω	0Ω	∞
D1	L/W	G	○	○	x
	G	L/W	x	○	x
D2	Sb	Y	○	○	x
	Y	Sb	x	○	x
D3	W	W/G	○	○	x
	W/G	W	x	○	x

○ : Continuity
x : Disconitnuity (∞)

Fig. 6.40 Testing the diode block – XJ650 J and RJ

38.6 Diode block is housed inside headlamp shell (UK 650)

39 Starter interlock system: general description – XJ650(UK), XJ650 G, H and LH

The above models employ a simplified starter interlock system in which the engine may only be started if the gearbox is in neutral or, if in gear, the clutch lever is pulled in. When the ignition and engine kill switches are turned on, power is supplied to one contact of the starting circuit cut-off relay. This component remains off until its windings are earthed through the neutral switch or the clutch switch. Once either switch closes, the relay circuit is completed and the relay contacts close. Power is fed through the relay to the starter relay (solenoid) and the engine can be started by pressing the starter switch.

40 Starter interlock system: fault diagnosis – XJ650(UK), XJ650 G, H and LH

1   In the event of a suspected fault in the starter interlock system it will be necessary to identify the faulty component by a process of elimination. The manufacturer does not give individual test procedures for each component. If, when the starter switch is pressed, the motor does not operate, trace through the system checking for battery voltage. Use either a multimeter set on the 0-20 volts dc range or a low wattage 12 volt bulb to make the tests.
2   Start by locating the starting circuit cut-off relay. On the US models this will be found behind the left-hand side panel, mounted just beneath the TCI unit. In the case of the XJ650(UK) model it is mounted just forward of the fuel tank mounting and is the smaller of the two components sharing a common mounting bracket. Note that the fuel tank will have to be removed to gain access to the relay. Positive identification can be made by ensuring that the wires entering the relay correspond with those shown in the relevant wiring diagram.
3   Switch the ignition on and off with the gearbox in neutral and listen to the relay to see whether it can be heard operating. Next, repeat the above process, this time with 1st or 2nd gear selected. Switch on the ignition and operate the clutch lever. If the clutch or neutral switches are suspect, try connecting a wire between the black/yellow terminal and a good earth point. With the ignition switched on the relay contacts should close and power should be fed to the red/white lead to the starter relay. This can be checked using a multimeter. If the starting circuit

cut-out relay still fails to work, check for power on the other red/white lead from the kill switch. If battery voltage is present but the relay fails to operate despite a sound earth connection, the relay should be considered faulty and renewed.
4   As a temporary expedient the relay can be bypassed by connecting the two red/white leads. This will allow the machine to be used while a replacement relay is being obtained. Similarly, a defective neutral or clutch switch may be bypassed by a temporary earth lead connected to the black/yellow terminal. For further information on the clutch and neutral switches, refer to Sections 28 and 27 respectively.

40.2 Engine cut-off relay is mounted beneath tank (UK 650)

41 Computer monitor system: general description – XJ750(UK), XJ750 RH, RJ and J

The above models are equipped with a microprocessor-based systems' checking and warning arrangement described by Yamaha as a computer monitor system (CMS). The instrument panel incorporates an LCD (liquid crystal display) panel showing the six areas monitored plus a four segment fuel level display. When the ignition is turned on the entire panel comes on with it and stays in this mode until the engine is started. The CMS unit then commences a sequential check of the monitored functions, starting at the top of the display and working down.

As each function is checked, the corresponding LCD flashes on and off. If any fault is noted, the relevant LCD will stay on and a warning lamp will flash on and off, indicating that the problem should be corrected immediately.

A warning control switch is provided and can be used to override the system where necessary. If pressed once, the warning lamp will stop flashing but will stay lit. If pressed a second time, the warning lamp will go out completely, whilst a third press will return it to the original warning mode with the lamp flashing. If the warning lamp is turned off to avoid distraction while riding, it should be noted that it will come on again if another fault arises. The check sequence may be initiated manually by pressing the check button on the instrument panel.

The symbols on the LCD are as follows:
STND This display warns that the side stand is down, and will go out when the stand is retracted.
BRK This indicates that the hydraulic fluid level in the front brake reservoir is low. It will go out when the fluid is topped up.
OIL This display comes on to warn that engine oil level is low. The oil level should be increased to its normal position to extinguish the display.

BATT This display comes on to warn that battery electrolyte level is low. Distilled water should be added to restore the electrolyte to the correct level and to turn off the LCD.
HEAD This LCD will come on to warn of a blown headlamp bulb.
TAIL This indicates that the tail or brake filament has blown.
FUEL This display warns that the fuel level is low. The fuel gauge display is located at the bottom of the LCD panel and is divided into four segments. Each of these approximates $\frac{1}{4}$ tank divisions.

It should be noted that the LCD or microprocessor may be damaged by the following:
1   Use of bulbs of the incorrect wattage.
2   Connecting accessories to the CMS or related circuits, ie tail lamp or headlamp circuits.

3   Ingress of water or steam from the underside of the instrument panel.
4   Shocks or pressure on the face of the LCD.
5   Magnets or magnetised objects placed on or near the LCD.

---

42 Computer monitor system: fault diagnosis

The CMS circuitry is outside the scope of normal workshop test equipment, and cannot be repaired in the event of an internal fault. In the event of a malfunction, follow the table shown below to identify the nature of the problem. In the event of renewal of the microprocessor or LCD panel, refer to Section 43 of this Chapter.

1. After the main switch is turned on.

PROBLEM	CAUSE	SOLUTION
a. Warning light doesn't come on.	Bulb is burned out.	Replace bulb.
	Low battery charge.	Recharge battery.
	Faulty coupler connection.	Clean coupler contacts.
	Broken wire.	Replace wiring.
	CMS control unit failed.	Replace CMS control unit.
b. Liquid crystal display (LCD) flashes on and off.	CMS control unit failed.	Replace CMS control unit.
c. LCD does not function.	LCD connectors incorrectly installed.	Reinstall connectors.
	Broken wire.	Replace wiring.
	Faulty contact between LCD panel and control unit.	Clean contacts.
	LCD panel failed.	Replace LCD panel.
	CMS control unit failed.	Replace CMS control unit.
d. LCD only partially displays.	LCD panel failed.	Replace LCD panel.

2. After the engine is started.

PROBLEM	CAUSE	SOLUTION
a. LCD does not cycle.	Faulty coupler connection.	Clean coupler contacts.
	Broken wire.	Replace wiring.
	CMS control unit failed.	Replace CMS control unit.

3. After the check switch is pushed.

PROBLEM	CAUSE	SOLUTION
a. LCD does not cycle.	Check switch failed.	Replace check switch.
	Faulty coupler connection.	Clean coupler contacts.
	Broken wire.	Replace wiring.
	CMS control unit failed.	Replace CMS control unit.

4. After the warning control switch is pushed.

PROBLEM	CAUSE	SOLUTION
a. Warning light continues to flash.	Warning control switch failed.	Replace warning control switch.
	Faulty coupler connection.	Clean coupler contacts.
	Broken wire.	Replace wiring.
	CMS control unit failed.	Replace CMS control unit.

Fig. 6.41 Computer monitor system fault diagnosis chart

43 Computer monitor system: removing and refitting the microprocessor and LCD panel

1   Access to the microprocessor (CMS control unit) and the LCD panel may be gained by detaching the instrument console. Start by removing the headlamp unit and placing it to one side. Release the headlamp shell holding bolts to allow the instrument console holding nuts to be reached. Separate the wiring connectors from the instrument console wiring and push the connectors through from inside the headlamp shell. Disconnect the speedometer drive cable by unscrewing the knurled ring which retains it.

2   Release the instrument console by removing the two nuts which secure it to the top yoke. The assembly can now be lifted clear and placed on a workbench for further dismantling. Note that the mounting studs pass through rubber bushes. These should be collected if they are displaced during removal.

3   Remove the bottom cover from the instrument console, having released its securing screws. Release the three nuts which retain the mounting bracket to the underside of the console assembly and lift the bracket away. Note that from this stage onwards, dismantling should be undertaken in clean conditions. It is suggested that the panel is placed on a prepared area of the bench which has been covered with a clean dust sheet or paper.

4   Slacken and remove the four cross-head screws which retain the microprocessor/LCD assembly to the console. The assembly can be lifted clear. Working from the top of the assembly, remove the four screws which secure the LCD holder.

Taking great care, turn the holder over and remove the five screws which pass through the display plate. Separate the display components and remove the display plate and LCD reflector. Remove the LCD wiring connector taking great care not to strain or damage it by pulling on the wiring. Particular care must be taken not to touch the gold plated contacts with the fingers.

5   If the LCD is to be reused, it should be cleaned, noting that extreme care must be taken. Use only an aerosol lens cleaner or a camera blower/brush to clean the LCD. These items can be obtained from photographic shops. A soft, lint-free, cloth can be used to finish off; an impregnated lens cloth such as Calotherm being ideal for this purpose. It is important not to leave dust or lint on the contacts. Note that only very light pressure should be applied to the LCD since static electricity can cause damage.

6   When reassembling the microprocessor/LCD assembly, reverse the dismantling sequence, noting the following points. Check that the face of the display aligns as shown in Fig. 6.44. This ensures that the display seals properly. Fit the LCD reflector with its shiny side towards the display. When offering up the wiring connector make absolutely certain that it seats properly on its index points. Fit the connector indexing screw first, followed by the remaining four screws. Tighten these just enough to secure the LCD unit; overtightening may crack or damage the display.

7   Once the display has been fitted to the microprocessor unit, connect up the wiring and check that the system functions properly before the instrument console is refitted. If the display fails to operate correctly, check that the connector is indexed accurately.

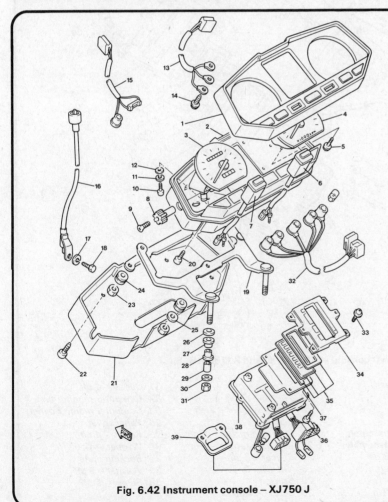

1	Instrument cover
2	Instrument housing
3	Speedometer
4	Tachometer
5	Screw and washer – 2 off
6	Control switch
7	Control switch
8	Reset knob
9	Screw
10	Screw – 2 off
11	Spring washer – 2 off
12	Washer – 2 off
13	Wiring harness
14	Screw and washer – 3 off
15	Wiring harness
16	Speedometer drive cable
17	Spring washer
18	Bolt
19	Mounting bracket
20	Screw – 5 off
21	Instrument lower cover
22	Screw – 4 off
23	Nut – 3 off
24	Grommet – 3 off
25	Washer
26	Washer – 2 off
27	Damping rubber – 2 off
28	Damping rubber – 2 off
29	Spacer – 2 off
30	Washer – 2 off
31	Nut – 2 off
32	Bulb holder assembly
33	Screw – 4 off
34	Panel cover
35	Computer monitor panel
36	Screw – 5 off
37	Backing plate
38	Computer monitor unit
39	Bracket

Fig. 6.42 Instrument console – XJ750 J

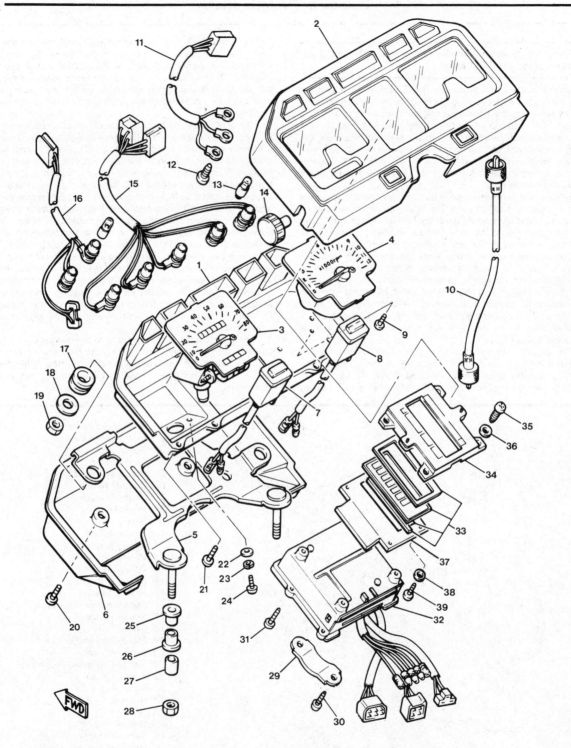

Fig. 6.43 Instrument console – XJ750(UK) and XJ750 RH, RJ

1	Instrument housing	11	Wiring harness	21	Screw – 4 off	31	Screw – 4 off
2	Instrument cover	12	Screw – 3 off	22	Washer – 2 off	32	Computer monitor unit
3	Speedometer	13	Bulb – 6 off	23	Spring washer – 2 off	33	Computer monitor panel
4	Tachometer	14	Reset knob	24	Screw – 2 off	34	Panel cover
5	Mounting bracket	15	Bulb holder assembly	25	Damping rubber – 2 off	35	Screw – 4 off
6	Instrument lower cover	16	Bulb holder assembly	26	Damping rubber – 2 off	36	Washer – 4 off
7	Control switch	17	Grommet – 3 off	27	Spacer – 2 off	37	Backing plate
8	Control switch	18	Washer – 3 off	28	Nut – 2 off	38	Washer – 5 off
9	Screw and washer – 2 off	19	Nut – 3 off	29	Bracket	39	Screw – 5 off
10	Speedometer drive cable	20	Screw – 6 off	30	Screw – 2 off		

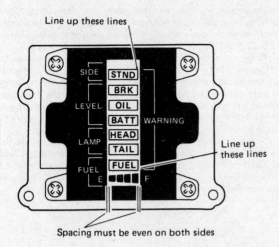

Line up these lines

SIDE

LEVEL

LAMP

FUEL

STND
BRK
OIL
BATT
HEAD
TAIL
FUEL

WARNING

Line up these lines

E   F

Spacing must be even on both sides

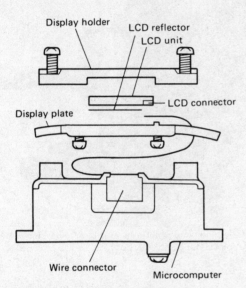

Display holder

LCD reflector
LCD unit

LCD connector

Display plate

Wire connector

Microcomputer

Fig. 6.44 Micro processor/LCD unit display panel alignment marks

44 Battery condition sensor: general description and maintenance – XJ750(UK), XJ750 RH, RJ and J

1   The battery on the above models incorporates a sensor to monitor the level of electrolyte in the battery. Once the level falls below the tip of the sensor a warning is displayed via the CMS unit and LCD panel, and the CMS warning lamp is switched on. The sensor consists of a lead terminal which is fitted in place of the fourth cell cap from the negative (–) terminal. It is important that the sensor is fitted to this cell, since a specific voltage must be picked up by the CMS unit.

2   The sensor should be removed and cleaned at 3000 mile (5000 km) intervals. Disconnect the sensor lead and unscrew the sensor from the battery. Wash the sensor in copious quantities of water to remove any residual electrolyte. Remember that sulphuric acid is highly corrosive and beware of accidental splashes on skin or clothing.

3   Dry the sensor off, then remove surface corrosion using fine abrasive paper or a wire brush to restore a clean, bright surface. Once cleaned, the sensor may be refitted and the sensor lead reconnected.

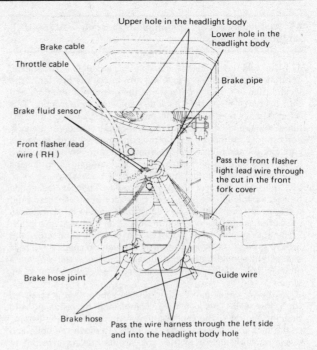

Upper hole in the headlight body

Lower hole in the headlight body

Brake cable

Throttle cable

Brake pipe

Brake fluid sensor

Front flasher lead wire (RH)

Pass the front flasher light lead wire through the cut in the front fork cover

Brake hose joint

Guide wire

Brake hose

Pass the wire harness through the left side and into the headlight body hole

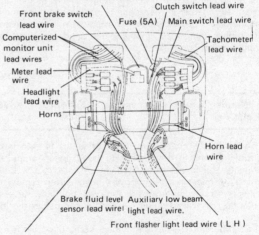

Clamp the lead wires coming from the upper left hole with the left clamp and the lead wires coming from the upper right hole with the right clamp. Keep the space between left and right leads couplers for the headlight coupler.

Front brake switch lead wire

Clutch switch lead wire

Fuse (5A)

Main switch lead wire

Computerized monitor unit lead wires

Tachometer lead wire

Meter lead wire

Headlight lead wire

Horns

Horn lead wire

Brake fluid level sensor lead wire

Auxiliary low beam light lead wire.

Front flasher light lead wire (L H)

Front flasher light lead wire (RH)

Fig. 6.45 Steering head area wiring and cable routing arrangement – XJ750 RH

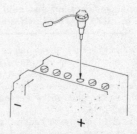

Fig. 6.46 Position of electrolyte level sensor in battery – XJ750(UK) and XJ750 RH, RJ and J

45 Fuel level sender: operation and testing – XJ750(UK), XJ750 RH, RJ and J

1   The fuel level sender consists of a variable resistance, controlled by a float on a pivoting arm inside the fuel tank. The level of fuel in the tank determines the position of the float and this in turn varies the resistance of the sender unit. This information is read by the CMS unit and is displayed on the LCD panel. As already mentioned, the fuel gauge section of the LCD consists of four blocks, each representing about $\frac{1}{4}$ of the tank's capacity. In addition, when the fuel level drops to a prescribed level, the **FUEL** panel comes on and the warning lamp flashes as a reminder that refuelling will soon be necessary.
2   The manufacturer does not provide specific resistance figures for the sender unit, so in the event of a suspected fault a degree of cunning will be required. It should be possible to check the sender using a multimeter set on the resistance scale. Trace the sender leads back to the two-pin connector, and attach a meter probe lead to each. Though no specific figure is available it should be noted that zero resistance will indicate a short circuit, whilst infinite resistance will denote an open circuit.
3   A better check can be made if the sender unit is removed from the tank. This can be accomplished by removing and draining the tank. The sender unit is mounted on the underside of the tank and is secured by four bolts. Remove the bolts and ease the sender clear of the tank surface, taking care not to damage the seal. Carefully manoeuvre the unit clear of the tank noting that the float arm is easily damaged if forced. Repeat the resistance check described above, this time moving the float arm up and down. If the sender is in good working order the resistance should rise and fall progressively as the arm is moved. No change in resistance or erratic operation indicates that the unit is defective and that renewal will be necessary. If the problem cannot be traced to the sender unit, check the wiring connections back to the CMS unit.

46 Warning lamps: bulb renewal – XJ750(UK), XJ750 J, RJ and RH

1   A bank of warning and indicator lamps is built into the instrument console, together with two similar lamps which illuminate the instruments at night. Access to the bulbs is gained by releasing the bottom cover after removing its retaining screws, noting that this will normally necessitate headlamp removal to obtain working space below the instruments.
2   Each bulb holder has a moulded rubber base and is a push fit in the instrument panel. Once the bulb holder has been pulled clear the bayonet-fitting bulb can be released by pushing it inwards and twisting it anti-clockwise.

47 Instrument panel: dismantling and bulb renewal – All models except XJ750(UK), XJ750 J, RH and RJ

1   Access to the warning lamp bulbs is gained after removing the plastic cover which is fitted between the speedometer and tachometer heads. The cover is retained by two screws and can be lifted away once these have been removed. The bulbs are bayonet fitting and can be released by depressing them and twisting them anti-clockwise.
2   The speedometer and tachometer heads each incorporate a bulb and bulb holder for internal illumination. To gain access for bulb renewal, first detach the relevant drive cable from the instrument head. Release the two domed nuts which secure the

instrument in its cup, and lift away the plain and rubber washers. Ease the instrument out of its cup, feeding the lighting wire through the base of the cup. Once clear of the cup, remove the bulb holder by pulling it out of the instrument base.
3   **Note:** do not remove the two domed nuts which secure the instrument cup to the mounting bracket. This is not necessary to change the bulbs. If the cups are to be removed, note that they are secured by bolts passed through from the inside and located by special washers. The instrument head should be removed first, allowing the bolt heads to be held down while the nuts are removed.
4   When reassembling the instrument heads, check first that the cup mounting nuts are secure. Refit the instrument head noting the anti-vibration seal between the instrument bezel and the cup. Fit the retaining nuts together with the rubber and plain washers. Reconnect the drive cable.

47.1a Warning lamp panel is secured by screws – instrument panel removed for clarity (UK 650)

47.1b Warning lamp bulbs are simple bayonet fitting

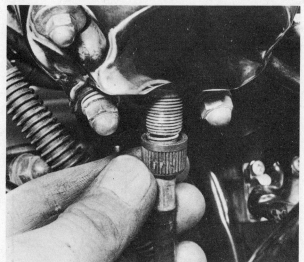
47.2a Free the instrument drive cable ...

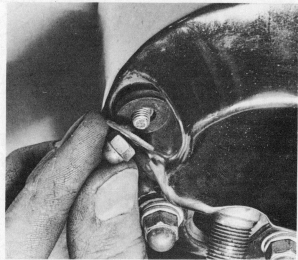
47.2b ... and release domed nuts to free instrument

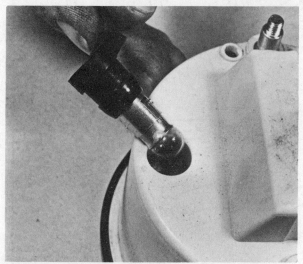
47.2c Bulb holder can be pulled clear of instrument case

47.3a **Do not** unscrew instrument cup nuts ...

47.3b ... unless bolt heads can be held inside cup

47.4 Note anti-vibration seal around edge of cup

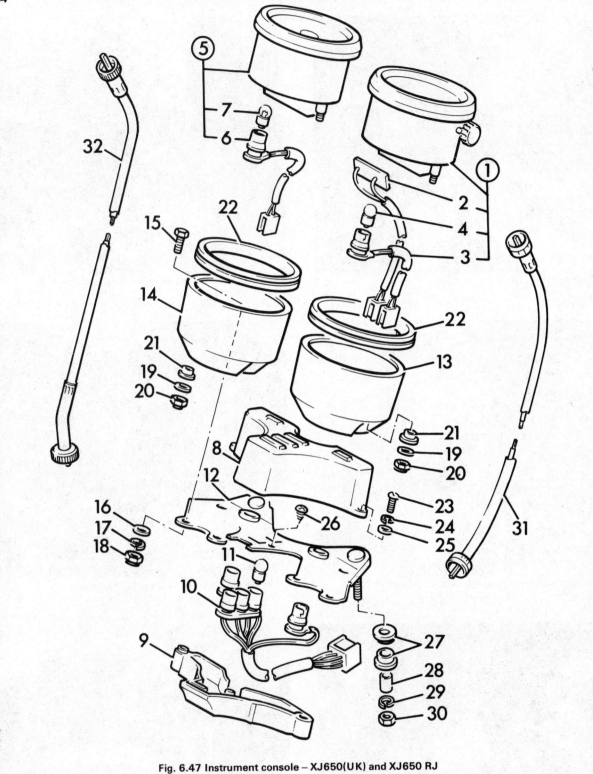

Fig. 6.47 Instrument console – XJ650(UK) and XJ650 RJ

1	Speedometer assembly	9	Instrument lower cover	17	Spring washer – 4 off	25	Washer – 2 off
2	Sender unit	10	Bulb holder assembly	18	Nut – 4 off	26	Screw – 2 off
3	Bulb holder	11	Bulb – 5 off	19	Washer – 4 off	27	Damping rubber – 4 off
4	Bulb	12	Mounting bracket	20	Nut – 4 off	28	Spacer – 2 off
5	Tachometer assembly	13	Speedometer housing	21	Damping rubber – 4 off	29	Spring washer – 2 off
6	Bulb holder	14	Tachometer housing	22	Damping rubber – 2 off	30	Nut – 2 off
7	Bulb	15	Bolt – 4 off	23	Screw – 2 off	31	Speedometer drive cable
8	Instrument panel	16	Washer – 4 off	24	Spring washer – 2 off	32	Tachometer drive cable

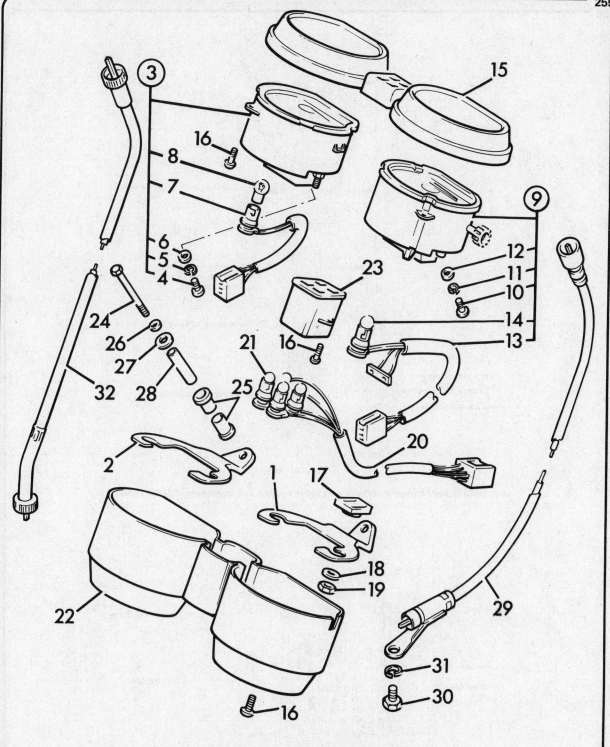

Fig. 6.48 Instrument console – XJ650 G, H, LH and J

1 Speedometer mounting bracket	9 Speedometer assembly	17 Damping rubber	25 Damping rubbers
2 Tachometer mounting bracket	10 Screw	18 Washer – 4 off	26 Spring washer
3 Tachometer assembly	11 Spring washer	19 Nut – 4 off	27 Washer
4 Screw	12 Washer	20 Wiring harness	28 Spacer
5 Spring washer	13 Bulb holder	21 Bulb – 4 off	29 Speedometer drive cable
6 Washer	14 Bulb	22 Instrument lower cover	30 Bolt
7 Bulb holder	15 Instrument cover	23 Warning lamp panel	31 Spring washer
8 Bulb	16 Screw	24 Bolt	32 Tachometer drive cable

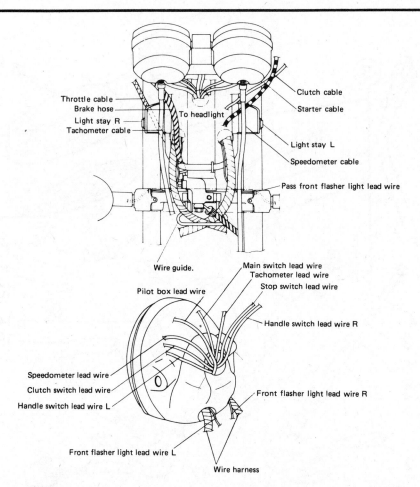

Fig. 6.49 Steering head area wire and cable routing arrangement – XJ650 G and J

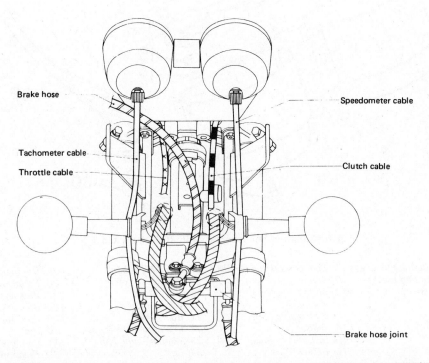

Fig. 6.50 Steering head area wire and cable routing arrangement – XJ650 RJ and XJ650(UK)

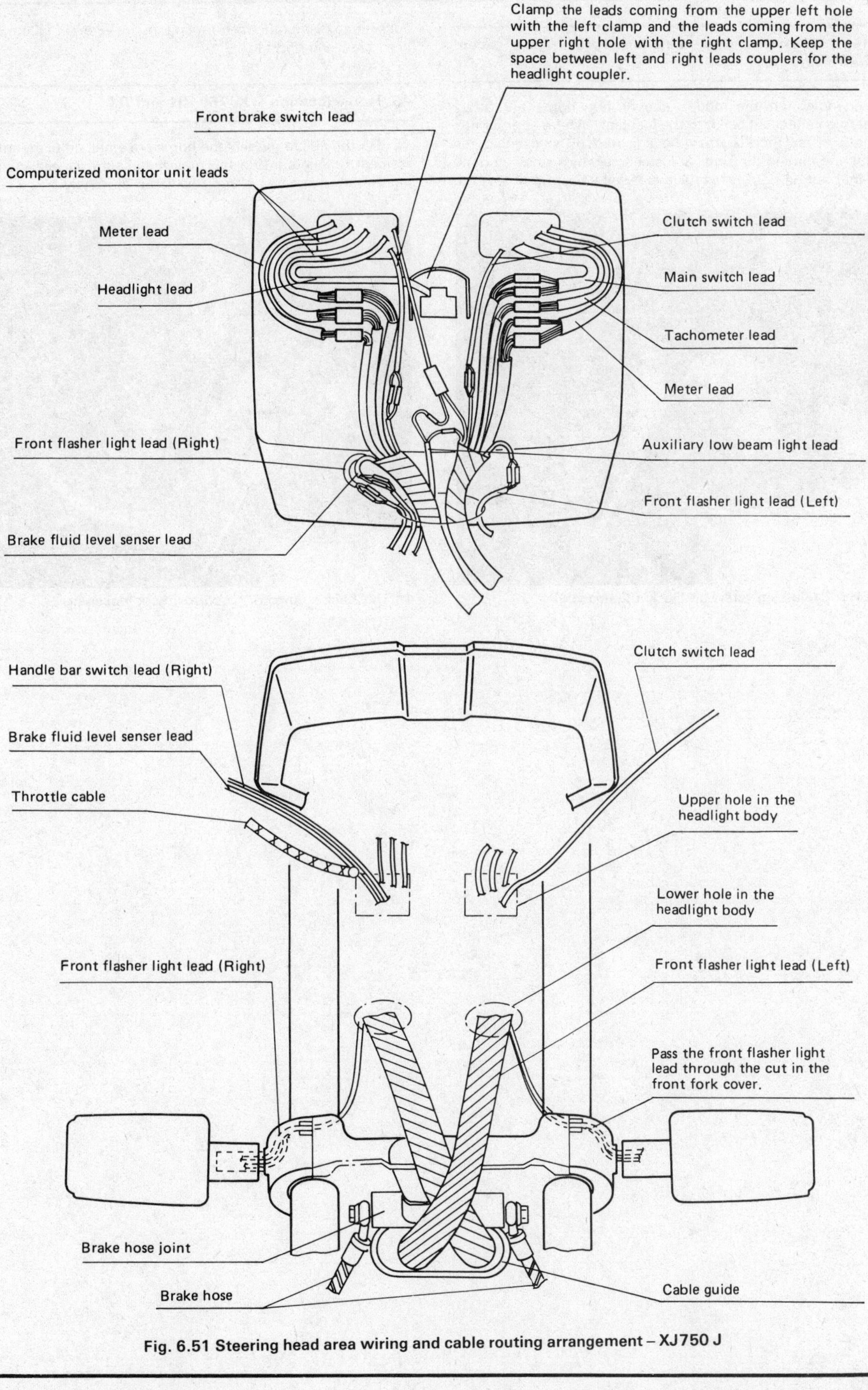

Clamp the leads coming from the upper left hole with the left clamp and the leads coming from the upper right hole with the right clamp. Keep the space between left and right leads couplers for the headlight coupler.

Front brake switch lead

Computerized monitor unit leads

Meter lead

Headlight lead

Clutch switch lead

Main switch lead

Tachometer lead

Meter lead

Front flasher light lead (Right)

Auxiliary low beam light lead

Front flasher light lead (Left)

Brake fluid level senser lead

Handle bar switch lead (Right)

Clutch switch lead

Brake fluid level senser lead

Throttle cable

Upper hole in the headlight body

Lower hole in the headlight body

Front flasher light lead (Right)

Front flasher light lead (Left)

Pass the front flasher light lead through the cut in the front fork cover.

Brake hose joint

Brake hose

Cable guide

Fig. 6.51 Steering head area wiring and cable routing arrangement – XJ750 J

48 Horn: location and examination – All models except XJ750 RH and RJ

1   Depending on the model, one or two horns are fitted, mounted on a bracket beneath the fuel tank. Where appropriate, the note of the horn(s) may be adjusted by slackening the locknut and turning the small adjuster screw by a small amount. Note that not all horn types have external adjustment. It is not

possible to dismantle the horn unit. In the event of failure, a new horn should be fitted.

49 Horn: location – XJ750 RH and RJ

1   On the above models the horns are small units mounted on resilient brackets inside the main headlamp nacelle. Access is gained after removal of the headlamp assembly.

48.1a Horn adjusting screw and locknut (arrowed)

48.1b Check connector for corrosion or looseness

Chapter 7 The 1983 US models

Contents

Specifications

Note: *information is given only where it differs from that listed in the Specifications Section of Chapters 1 to 6 under the heading of the equivalent 1982 model, eg for information relevant to XJ750 K or MK models, refer to that given for the XJ750 J*

Model dimensions and weights
Overall length – XJ750 MK .... 2195 mm (86.4 in)

Specifications relating to Chapter 5

Tyre pressures – XJ750 K, MK:

	Front	Rear
Up to 198 lb (90 kg)	26 psi (1.8 kg/cm^2)	28 psi (2.0 kg/cm^2)
198 – 511 lb (90 – 232 kg)	28 psi (2.0 kg/cm^2)	33 psi (2.3 kg/cm^2)
High speed riding	33 psi (2.3 kg/cm^2)	36 psi (2.5 kg/cm^2)

Tyre pressures – XJ750 RK:

Without fairing and luggage .... As XJ750 RJ
With fairing and luggage .... Not available, check with amended owner's manual or an authorized Yamaha dealer for full information

Torque wrench settings – XJ750 K, MK

	kgf m	lbf ft
Caliper mounting bolts	3.5	25
Pad retaining bolts	2.25	16

1 Introduction

The first six Chapters of this manual cover models sold in the UK from 1980 to 1984, and in the US from 1980 to 1982. This Chapter describes the models sold in the US in 1983, where information is available and where specifications or working procedures significantly differ from those that apply to the earlier models. When working on a later model refer first to this Chapter to note any relevant changes before referring to the relevant part of Chapters 1 to 6.

Note: Although XJ650 and XJ750 models remained available in the UK up to 1984, these are the same models as those described in Chapters 1 to 6. Refer, therefore to Chapters 1 to 6 for information on any UK model.

XJ650 K Maxim
This model remained substantially the same as the XJ650 J. Apart from the minor modifications to the engine/gearbox unit, all working procedures are the same as those described for the XJ650 J. The engine/frame numbers start at 5N8-050101.

XJ750 K Maxim
Although based on the XJ750 J this model is no longer fitted with the computer monitor system described in Chapter 6; instead it is fitted with conventional instruments, a round headlamp and turn signals and a different tail lamp assembly, all of which are similar to those described for the earlier 650 Maxim models. Also, it is now fitted with conventional one-piece handlebars and simpler front forks which no longer have

the adjustable damping facility, in addition to the other modifications listed in this Chapter. Its engine/frame numbers start at 22R-000101.

XJ750 MK Midnight Maxim

This model was introduced in 1983 but is merely a restyled version of the XJ750 K. Apart from the few minor modifications listed in this Chapter it is mechanically identical to the standard Maxim but can be distinguished easily by its striking black and gold finish. Its engine/frame numbers start at 22R-100101.

XJ750 RK Seca

Apart from the minor modifications listed in this Chapter, this model is mechanically identical to the XJ750 RJ. Its engine/frame numbers start at 5G2-150101.

2 Engine/gearbox unit: modifications

1   On all later models the sump (oil pan) is now located by two dowel pins.
2   With reference to Fig. 1.19, there are now four change pins instead of three (in addition to the single, longer, pin) fitted between the selector drum and cam. On the XJ750 RK model only, the drum bearing is now of the ball journal type.
3   The XJ750 MK model crankshaft end covers have spacers and rubber rings behind them, each being retained by a locking plate and a single screw.

3 Carburettor: modifications

Note that on some later US XJ750 models the main and pilot air jets (items 15 and 16, Fig. 2.2) are no longer shown. No information is available to determine whether this is because the jets are no longer available as separate items, or because they have been modified and can no longer be removed from the carburettor body.

4 Air filter: general

With reference to Fig. 2.10, note that the air filter fitted to the XJ750 K, MK and RK models is now of the type fitted to the XJ750 (UK) model.

5 Front forks: general – XJ750 K and MK

1   The forks fitted to these models no longer have the adjustable damping facility of the XJ750 J and are of the simpler type shown in Fig. 4.9 (XJ650 J).
2   Refer to Chapter 4, Sections 12 and 13 for information on dismantling and reassembly. Note that the only significant difference in working procedure is that the fork top plug is now threaded and is removed by unscrewing, using a spanner applied to the hexagon in the plug top surface. On refitting, renew the sealing O-ring if damaged or worn and tighten the plug to a torque setting of 2.3 kgf m (6.5 lbf ft).
3   As far as can be determined, the fork spring free length, air pressures and oil capacity remain as specified in Chapter 4 for

the XJ750 J model. It should be noted however than an oil capacity of 278 cc (9.4 US fl oz) is specified in the owner's manual for each model; check with the importer or with an authorized Yamaha dealer if in doubt.

6 Steering damper: general – XJ750 RK

1   This model, in standard form, is fitted with a steering damper assembly which is retained by a single Allen bolt underneath the steering head.
2   All models fitted with the fairing and luggage will have been converted from the standard, round, steering damper to a piston-type damper which is bolted to the frame front downtubes and clamped to the left-hand fork stanchion.
3   No further information is available on either type of damper.

7 Fairing and luggage: general – XJ750 RK

1   All XJ750 RK models will have been fitted with a frame-mounted sports/touring fairing, a tail trunk (top box) and solidly-mounted saddlebags (panniers); the saddlebags have soft liners to assist packing and unpacking.
2   No further information is available on the fairing or luggage components and their mountings.
3   Refer to the owner's manual or to an authorized Yamaha dealer for information on tyre pressures and on the maximum permissible weights that can be carried in the tail trunk or saddlebags. Do not carry heavy items in the tail trunk or the machine's stability may be impaired, and try to distribute the load evenly between the two saddlebags, keeping heavy items as low and as far forward as possible.

8 Brake hoses: general – XJ750 models

1   The XJ750 K and MK models retain the twin disc front brake of the XJ750 J model (Fig. 5.17) but are fitted with a different type of distributor union, similar to that shown in Fig. 5.15 for the 650 Maxim models. Both lower brake hoses are attached to the underside of the union by a single long union bolt.
2   The XJ750 RK model is fitted with a front brake hose system substantially the same as that shown in Fig. 5.18, but the distributor union is now a two-piece component linked by a second metal hydraulic pipe.

9 Instruments, lighting and electrical system: general – XJ750 K and MK

1   These models, no longer fitted with the XJ750 J's sophisticated computer monitor system, have a much simpler electrical system which uses more conventional instruments and lights. While detailed information is not available, the components now fitted are similar in design and construction to those described in Chapter 6 for the 650 Maxim models.
2   Careful examination of the relevant wiring diagram will show the components and their connections in any particular circuit and, when used in conjunction with the relevant Sections of Chapter 6, should permit fault finding to be carried out with relative ease.

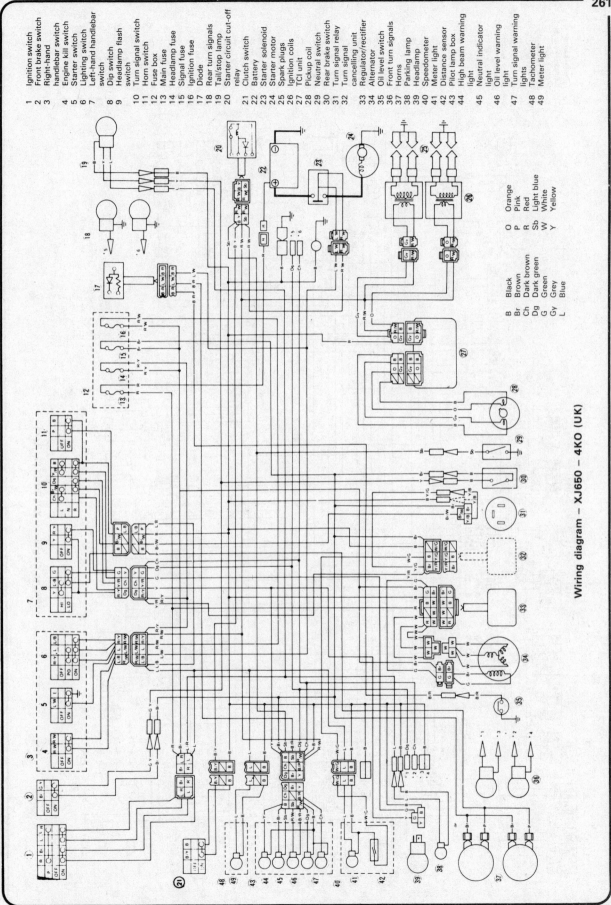

1 Ignition switch
2 Front brake switch
3 Right-hand handlebar switch
4 Engine kill switch
5 Starter switch
6 Lighting switch
7 Left-hand handlebar switch
8 Dip switch
9 Headlamp flash switch
10 Turn signal switch
11 Horn switch
12 Fuse box
13 Main fuse
14 Headlamp fuse
15 Signal fuse
16 Ignition fuse
17 Diode
18 Rear turn signals
19 Tail/stop lamp
20 Starter circuit cut-off relay
21 Clutch switch
22 Battery
23 Starter solenoid
24 Starter motor
25 Spark plugs
26 Ignition coils
27 TCI unit
28 Pickup coil
29 Neutral switch
30 Rear brake switch
31 Turn signal relay
32 Turn signal cancelling unit
33 Regulator/rectifier
34 Alternator
35 Oil level switch
36 Front turn signals
37 Horns
38 Parking lamp
39 Headlamp
40 Speedometer
41 Meter light
42 Distance sensor
43 Pilot lamp box
44 High beam warning light
45 Neutral indicator light
46 Oil level warning light
47 Turn signal warning lights
48 Tachometer
49 Meter light

B   Black
Br Brown
Ch Dark brown
Dg Dark green
G   Green
Gy Grey
L   Blue

O   Orange
P   Pink
R   Red
Sb Light blue
W   White
Y   Yellow

Wiring diagram – XJ650 – 4KO (UK)

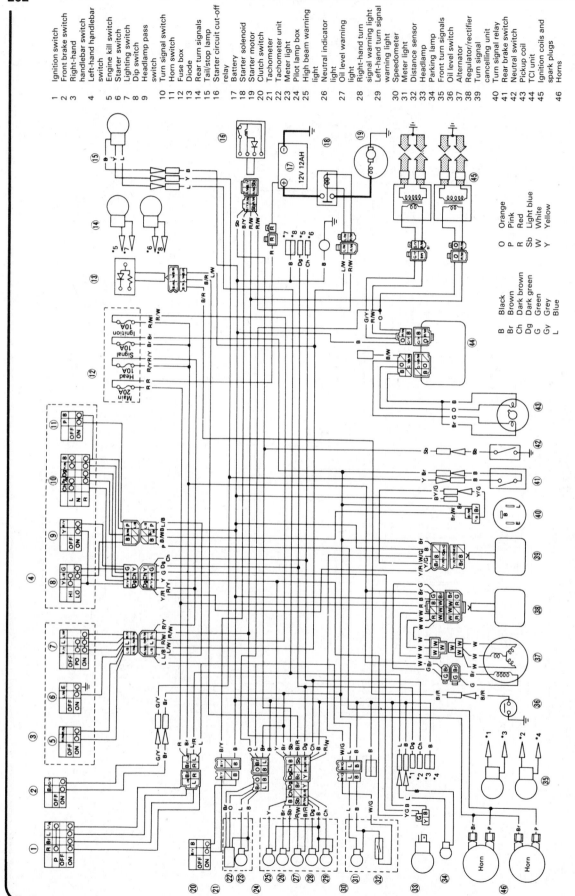

1 Ignition switch
2 Front brake switch
3 Right-hand handlebar switch
4 Left-hand handlebar switch
5 Engine kill switch
6 Starter switch
7 Lighting switch
8 Dip switch
9 Headlamp pass switch
10 Turn signal switch
11 Horn switch
12 Fuse box
13 Diode
14 Rear turn signals
15 Tail/stop lamp
16 Starter circuit cut-off relay
17 Battery
18 Starter solenoid
19 Starter motor
20 Clutch switch
21 Tachometer
22 Tachometer unit
23 Meter light
24 Pilot lamp box
25 High beam warning light
26 Neutral indicator light
27 Oil level warning light
28 Right-hand turn signal warning light
29 Left-hand turn signal warning light
30 Speedometer
31 Meter light
32 Distance sensor
33 Headlamp
34 Parking lamp
35 Front turn signals
36 Oil level switch
37 Alternator
38 Regulator/rectifier
39 Turn signal cancelling unit
40 Turn signal relay
41 Rear brake switch
42 Neutral switch
43 Pickup coil
44 TCI unit
45 Ignition coils and spark plugs
46 Horns

O   Orange
P   Pink
R   Red
Sb Light blue
W   White
Y   Yellow

B   Black
Br Brown
Ch Dark brown
Dg Dark green
G   Green
Gy Grey
L   Blue

Wiring diagram – XJ650 –11N (UK)

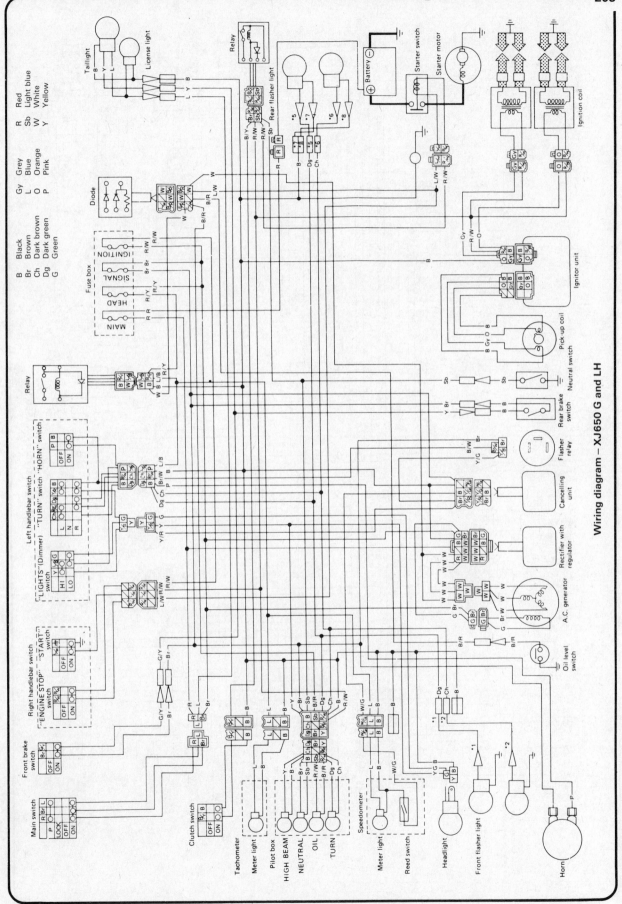

Wiring diagram – XJ650 G and LH

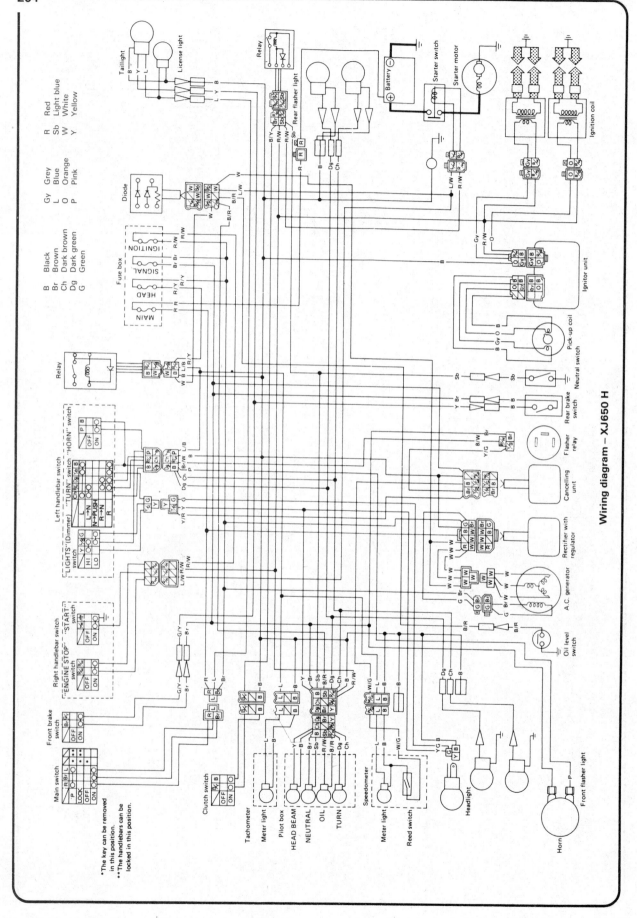

Wiring diagram – XJ650 H

R   Red
Sb   Light blue
W   White
Y   Yellow

Gy   Grey
L   Blue
O   Orange
P   Pink

B   Black
Br   Brown
Ch   Dark brown
Dg   Dark green
G   Green

*The key can be removed in this position.
**The handlebars can be locked in this position.

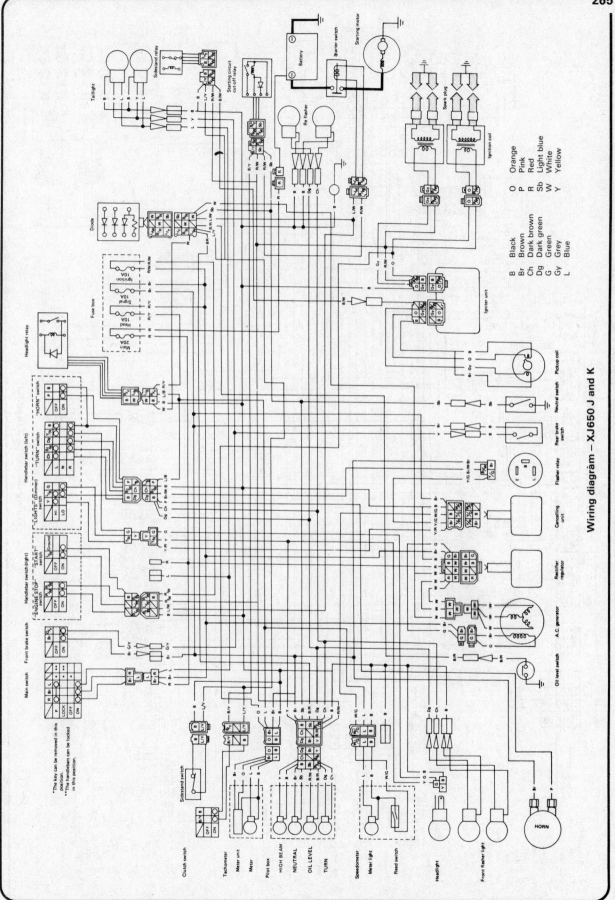

Wiring diagram – XJ650 J and K

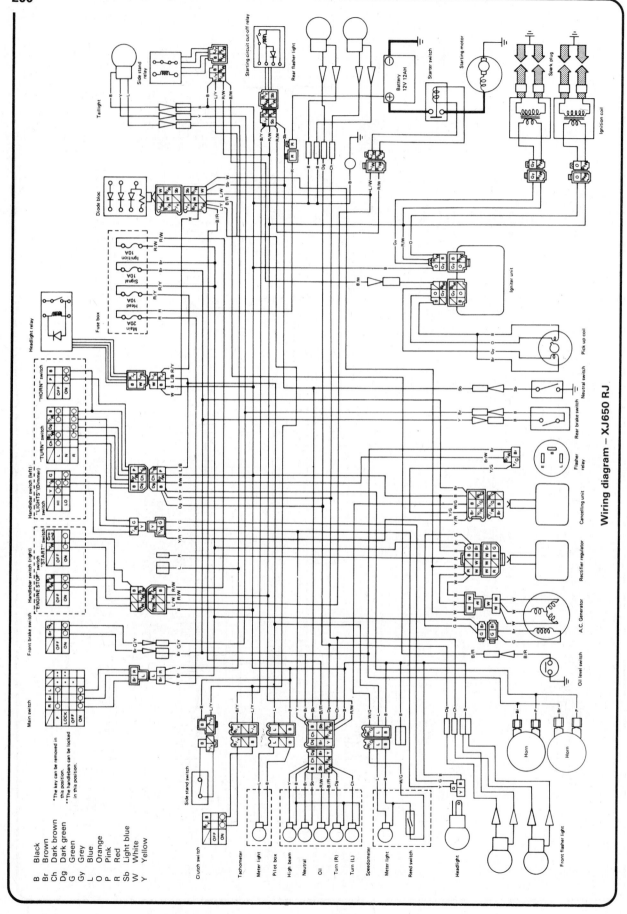

Wiring diagram – XJ650 RJ

B	Black
Br	Brown
Ch	Dark brown
Dg	Dark green
G	Green
Gy	Grey
L	Blue
O	Orange
P	Pink
R	Red
Sb	Light blue
W	White
Y	Yellow

*The key can be removed in this position.
**The handlebars can be locked in this position.

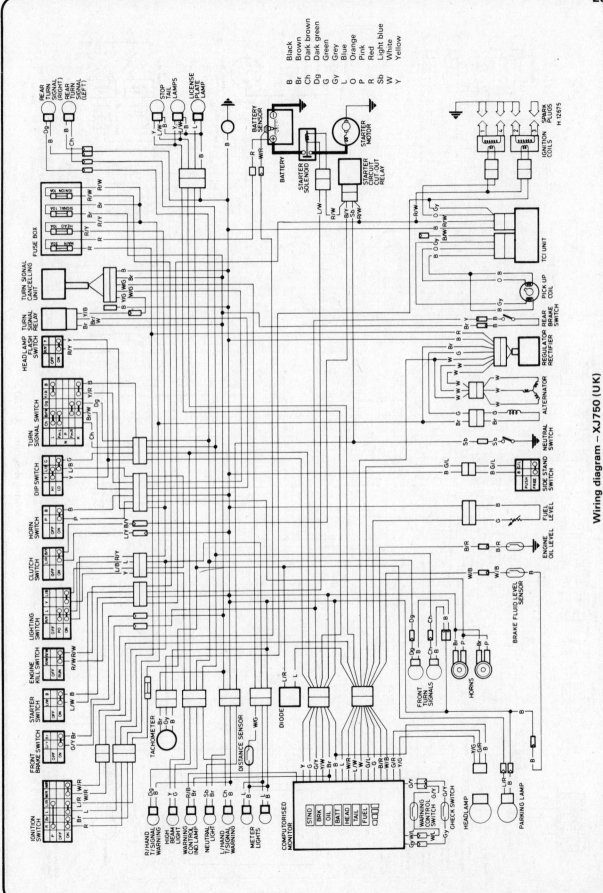

Wiring diagram – XJ750 (UK)

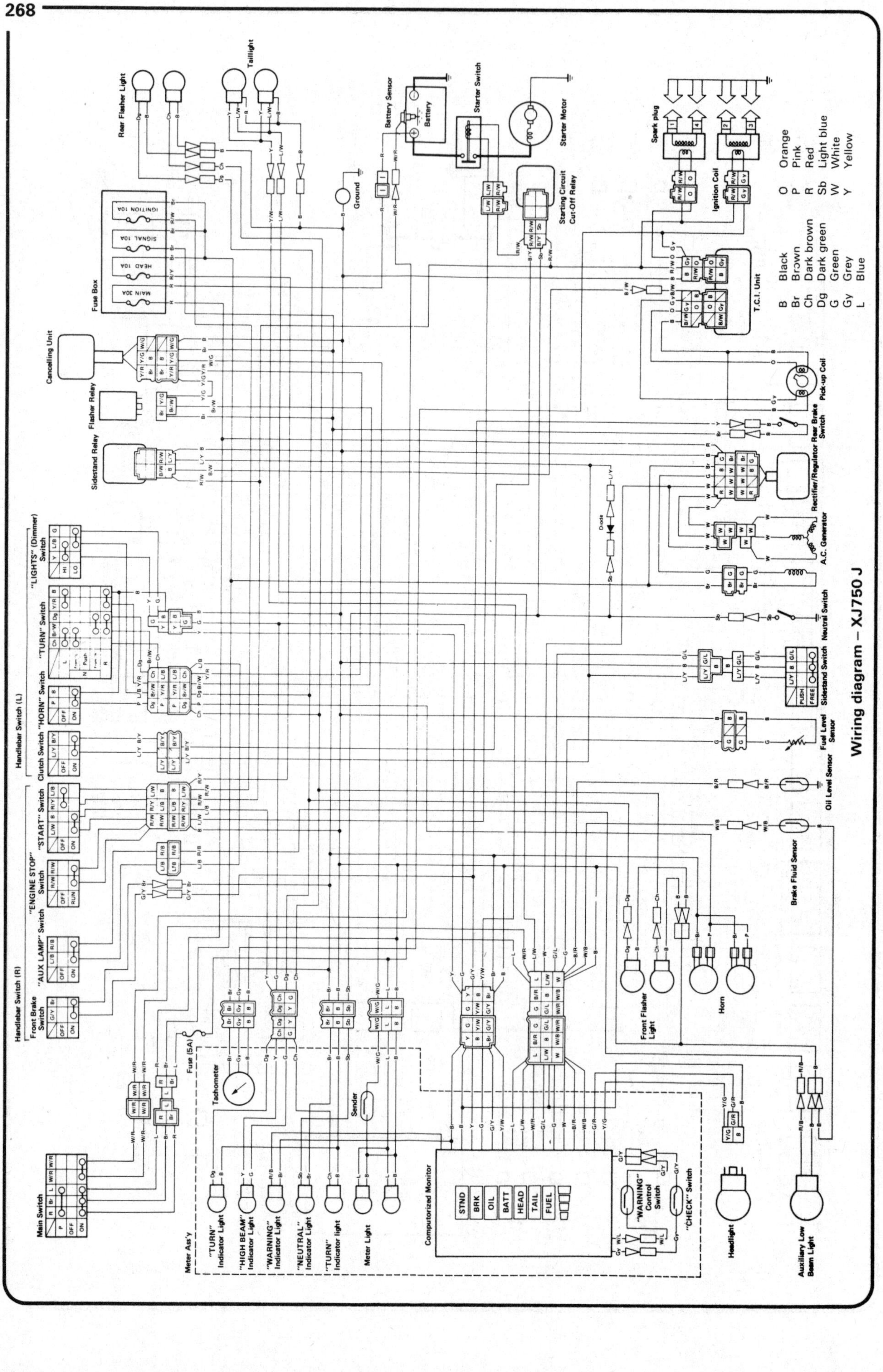

Rear Flasher Light

Taillight

Battery Sensor

Battery

Starter Switch

Starter Motor

Spark plug

Ignition Coil

Ground

Starting Circuit Cut-Off Relay

O	Orange
P	Pink
R	Red
Sb	Light blue
W	White
Y	Yellow

B	Black
Br	Brown
Ch	Dark brown
Dg	Dark green
G	Green
Gy	Grey
L	Blue

IGNITION 10A
SIGNAL 10A
HEAD 10A
MAIN 30A

Fuse Box

T.C.I. Unit

Cancelling Unit

Flasher Relay

Sidestand Relay

Pick-up Coil

Rear Brake Switch

Rectifier/Regulator

A.C. Generator

Neutral Switch

Sidestand Switch

Diode

Fuel Level Sensor

Oil Level Sensor

Brake Fluid Sensor

"LIGHTS" (Dimmer) Switch

"TURN" Switch

"HORN" Switch

Clutch Switch

Handlebar Switch (L)

"START" Switch

"ENGINE STOP" Switch

Handlebar Switch (R)

"AUX LAMP" Switch

Front Brake Switch

Front Flasher Light

Horn

Main Switch

ON OFF P

Meter Ass'y

Fuse (5A)

Tachometer

Sender

Computerized Monitor

STND
OIL
BRK
BATT
HEAD
TAIL
FUEL

"WARNING" Control Switch

"CHECK" Switch

Headlight

Auxiliary Low Beam Light

"TURN" Indicator Light

"HIGH BEAM" Indicator Light

"WARNING" Indicator Light

"NEUTRAL" Indicator Light

"TURN" Indicator Light

Meter Light

Wiring diagram – XJ750 J

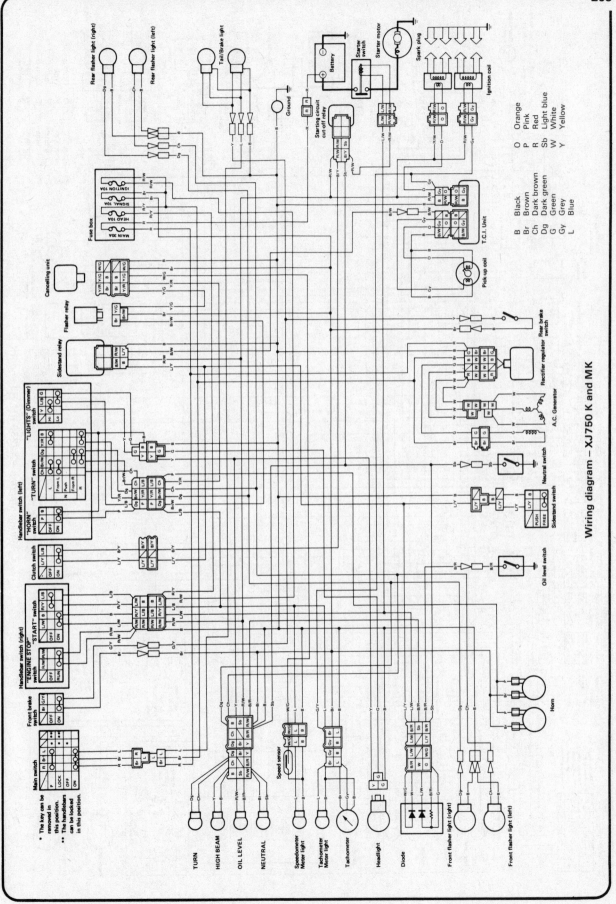

Wiring diagram – XJ750 K and MK

Wiring diagram — XJ750 RH, RJ and RK

Conversion factors

Length (distance)

Inches (in)	X	25.4	= Millimetres (mm)	X 0.0394	= Inches (in)
Feet (ft)	X	0.305	= Metres (m)	X 3.281	= Feet (ft)
Miles	X	1.609	= Kilometres (km)	X 0.621	= Miles

Volume (capacity)

Cubic inches (cu in; in³)	X	16.387	= Cubic centimetres (cc; cm³)	X 0.061	= Cubic inches (cu in; in³)
Imperial pints (Imp pt)	X	0.568	= Litres (l)	X 1.76	= Imperial pints (Imp pt)
Imperial quarts (Imp qt)	X	1.137	= Litres (l)	X 0.88	= Imperial quarts (Imp qt)
Imperial quarts (Imp qt)	X	1.201	= US quarts (US qt)	X 0.833	= Imperial quarts (Imp qt)
US quarts (US qt)	X	0.946	= Litres (l)	X 1.057	= US quarts (US qt)
Imperial gallons (Imp gal)	X	4.546	= Litres (l)	X 0.22	= Imperial gallons (Imp gal)
Imperial gallons (Imp gal)	X	1.201	= US gallons (US gal)	X 0.833	= Imperial gallons (Imp gal)
US gallons (US gal)	X	3.785	= Litres (l)	X 0.264	= US gallons (US gal)

Mass (weight)

Ounces (oz)	X	28.35	= Grams (g)	X 0.035	= Ounces (oz)
Pounds (lb)	X	0.454	= Kilograms (kg)	X 2.205	= Pounds (lb)

Force

Ounces-force (ozf; oz)	X	0.278	= Newtons (N)	X 3.6	= Ounces-force (ozf; oz)
Pounds-force (lbf; lb)	X	4.448	= Newtons (N)	X 0.225	= Pounds-force (lbf; lb)
Newtons (N)	X	0.1	= Kilograms-force (kgf; kg)	X 9.81	= Newtons (N)

Pressure

Pounds-force per square inch (psi; lbf/in²; lb/in²)	X	0.070	= Kilograms-force per square centimetre (kgf/cm²; kg/cm²)	X 14.223	= Pounds-force per square inch (psi; lbf/in²; lb/in²)
Pounds-force per square inch (psi; lbf/in²; lb/in²)	X	0.068	= Atmospheres (atm)	X 14.696	= Pounds-force per square inch (psi; lbf/in²; lb/in²)
Pounds-force per square inch (psi; lbf/in²; lb/in²)	X	0.069	= Bars	X 14.5	= Pounds-force per square inch (psi; lbf/in²; lb/in²)
Pounds-force per square inch (psi; lbf/in²; lb/in²)	X	6.895	= Kilopascals (kPa)	X 0.145	= Pounds-force per square inch (psi; lbf/in²; lb/in²)
Kilopascals (kPa)	X	0.01	= Kilograms-force per square centimetre (kgf/cm²; kg/cm²)	X 98.1	= Kilopascals (kPa)
Millibar (mbar)	X	100	= Pascals (Pa)	X 0.01	= Millibar (mbar)
Millibar (mbar)	X	0.0145	= Pounds-force per square inch (psi; lbf/in²; lb/in²)	X 68.947	= Millibar (mbar)
Millibar (mbar)	X	0.75	= Millimetres of mercury (mmHg)	X 1.333	= Millibar (mbar)
Millibar (mbar)	X	0.401	= Inches of water (inH₂O)	X 2.491	= Millibar (mbar)
Millimetres of mercury (mmHg)	X	0.535	= Inches of water (inH₂O)	X 1.868	= Millimetres of mercury (mmHg)
Inches of water (inH₂O)	X	0.036	= Pounds-force per square inch (psi; lbf/in²; lb/in²)	X 27.68	= Inches of water (inH₂O)

Torque (moment of force)

Pounds-force inches (lbf in; lb in)	X	1.152	= Kilograms-force centimetre (kgf cm; kg cm)	X 0.868	= Pounds-force inches (lbf in; lb in)
Pounds-force inches (lbf in; lb in)	X	0.113	= Newton metres (Nm)	X 8.85	= Pounds-force inches (lbf in; lb in)
Pounds-force inches (lbf in; lb in)	X	0.083	= Pounds-force feet (lbf ft; lb ft)	X 12	= Pounds-force inches (lbf in; lb in)
Pounds-force feet (lbf ft; lb ft)	X	0.138	= Kilograms-force metres (kgf m; kg m)	X 7.233	= Pounds-force feet (lbf ft; lb ft)
Pounds-force feet (lbf ft; lb ft)	X	1.356	= Newton metres (Nm)	X 0.738	= Pounds-force feet (lbf ft; lb ft)
Newton metres (Nm)	X	0.102	= Kilograms-force metres (kgf m; kg m)	X 9.804	= Newton metres (Nm)

Power

Horsepower (hp)	X	745.7	= Watts (W)	X 0.0013	= Horsepower (hp)

Velocity (speed)

Miles per hour (miles/hr; mph)	X	1.609	= Kilometres per hour (km/hr; kph)	X 0.621	= Miles per hour (miles/hr; mph)

Fuel consumption*

Miles per gallon, Imperial (mpg)	X	0.354	= Kilometres per litre (km/l)	X 2.825	= Miles per gallon, Imperial (mpg)
Miles per gallon, US (mpg)	X	0.425	= Kilometres per litre (km/l)	X 2.352	= Miles per gallon, US (mpg)

Temperature

Degrees Fahrenheit = (°C x 1.8) + 32   Degrees Celsius (Degrees Centigrade; °C) = (°F - 32) x 0.56

*It is common practice to convert from miles per gallon (mpg) to litres/100 kilometres (l/100km),
where mpg (Imperial) x l/100 km = 282 and mpg (US) x l/100 km = 235

English/American terminology

Because this book has been written in England, British English component names, phrases and spellings have been used throughout. American English usage is quite often different and whereas normally no confusion should occur, a list of equivalent terminology is given below.

English	American	English	American
Air filter	Air cleaner	Number plate	License plate
Alignment (headlamp)	Aim	Output or layshaft	Countershaft
Allen screw/key	Socket screw/wrench	Panniers	Side cases
Anticlockwise	Counterclockwise	Paraffin	Kerosene
Bottom/top gear	Low/high gear	Petrol	Gasoline
Bottom/top yoke	Bottom/top triple clamp	Petrol/fuel tank	Gas tank
Bush	Bushing	Pinking	Pinging
Carburettor	Carburetor	Rear suspension unit	Rear shock absorber
Catch	Latch	Rocker cover	Valve cover
Circlip	Snap ring	Selector	Shifter
Clutch drum	Clutch housing	Self-locking pliers	Vise-grips
Dip switch	Dimmer switch	Side or parking lamp	Parking or auxiliary light
Disulphide	Disulfide	Side or prop stand	Kick stand
Dynamo	DC generator	Silencer	Muffler
Earth	Ground	Spanner	Wrench
End float	End play	Split pin	Cotter pin
Engineer's blue	Machinist's dye	Stanchion	Tube
Exhaust pipe	Header	Sulphuric	Sulfuric
Fault diagnosis	Trouble shooting	Sump	Oil pan
Float chamber	Float bowl	Swinging arm	Swingarm
Footrest	Footpeg	Tab washer	Lock washer
Fuel/petrol tap	Petcock	Top box	Trunk
Gaiter	Boot	Torch	Flashlight
Gearbox	Transmission	Two/four stroke	Two/four cycle
Gearchange	Shift	Tyre	Tire
Gudgeon pin	Wrist/piston pin	Valve collar	Valve retainer
Indicator	Turn signal	Valve collets	Valve cotters
Inlet	Intake	Vice	Vise
Input shaft or mainshaft	Mainshaft	Wheel spindle	Axle
Kickstart	Kickstarter	White spirit	Stoddard solvent
Lower leg	Slider	Windscreen	Windshield
Mudguard	Fender		

Index